Texts in Applied Mathematics 53

Texts in Applied Mathematics

1. *Sirovich*: Introduction to Applied Mathematics.
2. *Wiggins*: Introduction to Applied Nonlinear Dynamical Systems and Chaos.
3. *Hale/Koçak*: Dynamics and Bifurcations.
4. *Chorin/Marsden*: A Mathematical Introduction to Fluid Mechanics, 3rd ed.
5. *Hubbard/West*: Differential Equations: A Dynamical Systems Approach: Ordinary Differential Equations.
6. *Sontag*: Mathematical Control Theory: Deterministic Finite Dimensional Systems, 2nd ed.
7. *Perko*: Differential Equations and Dynamical Systems, 3rd ed.
8. *Seaborn*: Hypergeometric Functions and Their Applications.
9. *Pipkin*: A Course on Integral Equations.
10. *Hoppensteadt/Peskin*: Modeling and Simulation in Medicine and the Life Sciences, 2nd ed.
11. *Braun*: Differential Equations and Their Applications, 4th ed.
12. *Stoer/Bulirsch*: Introduction to Numerical Analysis, 3rd ed.
13. *Renardy/Rogers*: An Introduction to Partial Differential Equations.
14. *Banks*: Growth and Diffusion Phenomena: Mathematical Frameworks and Applications.
15. *Brenner/Scott*: The Mathematical Theory of Finite Element Methods, 2nd ed.
16. *Van de Velde*: Concurrent Scientific Computing.
17. *Marsden/Ratiu*: Introduction to Mechanics and Symmetry, 2nd ed.
18. *Hubbard/West*: Differential Equations: A Dynamical Systems Approach: Higher-Dimensional Systems.
19. *Kaplan/Glass*: Understanding Nonlinear Dynamics.
20. *Holmes*: Introduction to Perturbation Methods.
21. *Curtain/Zwart*: An Introduction to Infinite-Dimensional Linear Systems Theory.
22. *Thomas*: Numerical Partial Differential Equations: Finitc Difference Methods.
23. *Taylor*: Partial Differential Equations: Basic Theory.
24. *Merkin*: Introduction to the Theory of Stability of Motion.
25. *Naber*: Topology, Geometry, and Gauge Fields: Foundations.
26. *Polderman/Willems*: Introduction to Mathematical Systems Theory: A Behavioral Approach.
27. *Reddy*: Introductory Functional Analysis with Applications to Boundary-Value Problems and Finite Elements.
28. *Gustafson/Wilcox*: Analytical and Computational Methods of Advanced Engineering Mathematics.
29. *Tveito/Winther*: Introduction to Partial Differential Equations: A Computational Approach.
30. *Gasquet/Witomski*: Fourier Analysis and Applications: Filtering, Numerical Computation, Wavelets.

(continued after index)

Grigorios A. Pavliotis • Andrew M. Stuart

Multiscale Methods

Averaging and Homogenization

Springer

Grigorios A. Pavliotis
Department of Mathematics
Imperial College London
London SW7 2AZ
United Kingdom
g.pavliotis@imperial.ac.uk

Andrew M. Stuart
Mathematics Institute
University of Warwick
Coventry CV4 7AL
United Kingdom
a.m.stuart@warwick.ac.uk

Series Editors

J.E. Marsden
Control and Dynamical Systems, 107–81
California Institute of Technology
Pasadena, CA 91125
USA
marsden@cds.caltech.edu

L. Sirovich
Division of Applied Mathematics
Brown University
Providence, RI 02912
USA
lawrence.sirovich@mssm.edu

S.S. Antman
Department of Mathematics
and
Institute for Physical Science
and Technology
University of Maryland
College Park, MD 20742-4015
USA
ssa@math.umd.edu

ISBN: 978-1-4419-2532-9 e-ISBN: 978-0-387-73829-1
DOI: 10.1007/978-0-387-73829-1

Mathematics Subject Classification (2000): 34, 35, 60

Printed on acid-free paper

9 8 7 6 5 4 3 2 1

springer.com

Series Preface

Mathematics is playing an ever more important role in the physical and biological sciences, provoking a blurring of boundaries between scientific disciplines and a resurgence of interest in the modern as well as the classical techniques of applied mathematics. This renewal of interest, both in research and teaching, has led to the establishment of the series Texts in Applied Mathematics (TAM).

The development of new courses is a natural consequence of a high level of excitement on the research frontier as newer techniques, such as numerical and symbolic computer systems, dynamical systems, and chaos, mix with and reinforce the traditional methods of applied mathematics. Thus, the purpose of this textbook series is to meet the current and future needs of these advances and to encourage the teaching of new couses.

TAM will publish textbooks suitable for use in advanced undergraduate and beginning graduate courses, and will complement the Applied Mathematical Sciences (AMS) series, which will focus on advanced textbooks and research-level monographs.

Pasadena, California — J.E. Marsden
New York, New York — L. Sirovich
College Park, Maryland — S.S. Antman

To my parents $A\rho\gamma\upsilon\rho\eta$ and $\Sigma o\upsilon\lambda\tau\alpha\nu\alpha$ and to my brother $\Gamma\iota\omega\rho\gamma o$.
Carry Home. $\Gamma\rho\eta\gamma o\rho\eta\varsigma$.

For my children Natalie, Sebastian, and Isobel.
AMS.

Preface

The aim of these notes is to describe, in a unified fashion, a set of methods for the simplification of a wide variety of problems that all share the common feature of possessing multiple scales.[1] The mathematical methods we study are often referred to as the methods of **averaging** and of **homogenization**. The methods apply to partial differential equations (PDEs), stochastic differential equations (SDEs), ordinary differential equations (ODEs), and Markov chains. The unifying principle underlying the collection of techniques described here is the approximation of **singularly perturbed** linear equations. The unity of the subject is most clearly visible in the application of perturbation expansions to the approximation of these singular perturbation problems. A significant portion of the notes is devoted to such perturbation expansions. In this context we use the term **Result** to describe the conclusions of a formal perturbation argument. This enables us to derive important approximation results without the burden of rigorous proof, which can sometimes obfuscate the main ideas. However, we will also study a variety of tools from analysis and probability, used to place the approximations derived on a rigorous footing. The resulting theorems are proved using a range of methods, tailored to different settings. There is less unity to this part of the subject. As a consequence, considerable background is required to absorb the entire rigorous side of the subject, and we devote a significant part of the book to this background material.

The first part of the notes is devoted to the **Background**; the second to the **Perturbation Expansions**, which provide the unity of the subject matter; and the third to the **Theory** justifying these perturbative techniques. We do not necessarily recommend that the reader covers the material in this order. A natural way to get an overview of the subject is to read through Part II of the book on perturbation

---

[1] In this book we will apply the general methodology to problems with two widely separated characteristic scales. The extension to systems with many separated scales is fairly straightforward and will be discussed in a number of the Discussion and Bibliography sections, which conclude each chapter. In all cases, the important assumption will be that of scale separation.

expansions, referring to the background material as needed. The theory can then be studied, after the form of the approximations is understood, on a case-by-case basis.

Part I (Background) contains the elements of the theory of analysis, probability, and stochastic processes, as required for the material in these notes, together with basic introductory material on ODEs, Markov chains, SDEs, and PDEs. Part II (Perturbation Expansions) illustrates the use of ideas from averaging and homogenization to study ODEs, Markov chains, SDEs, and PDEs of elliptic, parabolic, and transport type; invariant manifolds are also discussed and are viewed as a special case of averaging. Part III (Theory) contains illustrations of the rigorous methods that may be employed to establish the validity of the perturbation expansions derived in Part II. The chapters in Part III relate to those in Part II in a one-to-one fashion. It is possible to pick particular themes from this book and cover subsets of chapters devoted only to those themes. The reader interested primarily in SDEs should cover Chapters 6, 10, 11, 17, and 18. Markov chains are covered in Chapters 5, 9, and 16. The subject of homogenization for elliptic PDEs is covered in Chapters 12 and 19. Homogenization and averaging for parabolic and transport equations are covered in Chapters 13, 14, 20, and 21.

The subject matter in this set of notes has, for the most part, been known for several decades. However, the particular presentation of the material here is, we believe, particularly suited to the pedagogical goal of communicating the subject to the wide range of mathematicians, scientists, and engineers who are currently engaged in the use of these tools to tackle the enormous range of applications that require them. In particular we have chosen a setting that demonstrates quite clearly the wide applicability of the techniques to PDEs, SDEs, ODEs, and Markov chains, as well as highlighting the unity of the approach. Such a wide-ranging setting is not undertaken, we believe, in existing books, or is done so less explicitly than in this text. We have chosen to use the phrasing **Multiscale Methods** in the title of the book because the material presented here forms the backbone of a significant portion of the amorphous field that now goes by that name. However, we recognize that there are vast parts of the field we do not cover. In particular, scale separation is a fundamental requirement in all of the perturbation techniques presented in this book. Many applications, however, possess a continuum of scales, with no clear separation. Furthermore, many of the problems arising in multiscale analysis are concerned with the interfacing of different mathematical models appropriate at different scales (such as quantum, molecular, and continuum); the tools presented in these notes do not directly address problems arising in such applications, as our starting point is a single mathematical model in which scale separation is present.

These notes are meant to be an introduction, aimed primarily toward graduate students. Part I of the book (where we lay the theoretical foundations) and Part III (where we state and prove theorems concerning simplified versions of the models studied in Part II) are necessarily terse; otherwise it would be impossible to present the wide range of applications of the ideas and illustrate their unity. Extensions and generalizations of the results presented in these notes, as well as references to the literature, are given in the Discussion and Bibliography section at the end of each chapter. With the exception of Chapter 1, all chapters are supplemented with exercises.

We hope that the format of the book will make it appropriate for use both as a textbook and for self-study.

Acknowledgments

We are especially grateful to Konstantinos Zygalakis who read and commented on much of the manuscript, typed parts of it, and helped to create some of the figures. Special thanks also are due to Martin Hairer, Valeriy Slastikov, Endre Süli, and Roger Tribe, all of whom read portions of the book in great detail, resulting in many constructive suggestions for improvement. We are also grateful to Niklas Branstrom, Mikhail Cherdanstev, Dror Givon, Weinan E, David Epstein, Liam Jones, Raz Kupferman, Brenda Quinn, Florian Theil, and David White for many helpful comments that substantially improved these lecture notes. The book grew out of courses on multiscale methods and on homogenization that we taught at Imperial College London and at Warwick University during the last four years; we thank all the students who attended these courses. Some of this material, in turn, grew out of the review article [125], and we are grateful to Dror Givon and Raz Kupferman for their collaboration in this area.

Parts of the book were used as a basis for two short courses on multiscale methods that we taught in April 2007 at the Mathematical Sciences Research Institute (MSRI), Berkeley, CA, and the Mathematics Research Centre (MRC), Warwick University. We thank all the participants at these courses, together with the MSRI, MRC, London Mathematical Society, and the UK Engineering and Physical Sciences Research Council (EPSRC) for administrative and financial support for these short courses.

GP has received financial support from the EPSRC, and AMS has received financial support from the EPSRC and from the U.S. Office of Naval Research during the writing of this book. This funded research has helped shape much of the material presented in the book, and the authors are grateful for the financial support.

GA Pavliotis and AM Stuart, August 2007

Contents

1

Introduction

1.1 Overview

This chapter introduces the type of problems that are studied in the book, as well as the methods we will use to study them. Section 1.2 contains four motivating examples, illustrating a range of interrelated problems in partial differential equations (PDEs), deterministic dynamics, and stochastic dynamics. In Section 1.3 we discuss the methods of averaging and homogenization, as applied to singularly perturbed linear differential equations. This highlights the unity of the material in the book, as presented in Part II, through perturbation expansions. The chapter closes, in Section 1.4, with bibliographical remarks.

The notation we employ, including in this chapter, is explained in Section 2.2. We also make reference in this chapter to concepts from the theory of ODEs, PDEs, and SDEs that will be fully elaborated later in the text. However, we believe it is useful to introduce the subject matter of the book here, without the burden of setting up all the mathematical machinery in detail.

1.2 Motivating Examples

In this section we describe four examples that illustrate the range of problems we study in the notes. The first illustrates homogenization in the context of a linear second-order uniformly elliptic PDE. The second illustrates related ideas in the context of a time-dependent PDE of parabolic type, the advection–diffusion equation. Through the connection between hyperbolic (transport) PDEs and ordinary differential equations (ODEs), via the method of characteristics, and the connection between parabolic PDEs and stochastic differential equations (SDEs), via the Itô formula, we show that the methods of averaging and homogenization developed for the study of linear PDEs can also be applied to study dimension reduction for ODEs and SDEs; this is illustrated in the third example. We finish with a fourth example concerning variable elimination for dynamical systems.

1.2.1 Example I: Steady Heat Conduction in a Composite Material

To introduce ideas related to homogenization we consider the problem of steady heat conduction in a composite material whose properties vary rapidly compared to the macroscopic scale. If $\Omega \subset \mathbb{R}^d$ denotes the domain occupied by the material, then the size of the domain defines a macroscopic length scale L (for example, we can define L through $\mathrm{vol}(\Omega) = L^d$). On the other hand, the characteristic length scale of the heterogeneities defines a microscopic length scale ε of the problem. We assume that $L = O(1)$ with respect to $\varepsilon \ll 1$. The phenomenon of steady heat conduction can be described by the following elliptic boundary value problem for the temperature field $u^\varepsilon(x)$:

$$-\nabla \cdot (A^\varepsilon \nabla u^\varepsilon) = f \quad \text{for } x \in \Omega, \tag{1.2.1a}$$

$$u^\varepsilon = 0 \quad \text{for } x \in \partial\Omega. \tag{1.2.1b}$$

Here A^ε is assumed uniformly positive-definite in $\varepsilon > 0$.

To make a concrete problem, amenable to analysis, we assume that $A^\varepsilon = A(x/\varepsilon)$ and that the matrix $A(y)$, the *thermal conductivity tensor*, is periodic with period 1 in all d directions and positive-definite. The purpose of homogenization theory is to study the limit of u^ε as $\varepsilon \to 0$. In particular it is desirable to identify the equation satisfied by u^ε in this limit. From a physical point of view, the limit $\varepsilon \to 0$ corresponds to the case where the heterogeneities become vanishingly small. Thus our aim is to replace the original, highly *heterogeneous* material, characterized by the rapidly oscillating coefficients $A(x/\varepsilon)$, by an effective, *homogeneous* material characterized by constant coefficients $\overline{A}$. Hence the name **homogenization**.

In Chapters 12 and 19 we will show that, under appropriate assumptions on $A(y)$, $f(x)$, and Ω, the homogenized equation is

$$-\nabla \cdot \left(\overline{A} \nabla u\right) = f \quad \text{for } x \in \Omega, \tag{1.2.2a}$$

$$u = 0 \quad \text{for } x \in \partial\Omega. \tag{1.2.2b}$$

The constant **homogenized conductivity tensor** $\overline{A}$ is given by the formula:

$$\overline{A} = \int_{\mathbb{T}^d} A(y)\left(I + (\nabla \chi(y))^T\right) dy. \tag{1.2.3}$$

The **(first-order) corrector** $\chi(y)$ is a vector field solving the periodic **cell problem**

$$-\nabla_y \cdot \left(\nabla_y \chi A^T\right) = \nabla_y \cdot A^T, \quad y \in \mathbb{T}^d. \tag{1.2.4}$$

Here $\mathbb{T}^d$ denotes the d-dimensional torus and the PDE (1.2.4) is equipped with periodic boundary conditions.

The calculation of the effective coefficients $\overline{A}$ involves the solution of a partial differential equation posed on the unit torus, together with computation of the integral (1.2.3). Hence, finding the homogenized solution u requires the solution of two elliptic PDEs: the periodic cell problem (1.2.4), which allows construction of $\overline{A}$ given by (1.2.3), and the Dirichlet problem (1.2.2). The important point is that these elliptic equations *do not depend on the small scale* ε. In some cases the two elliptic PDEs can be solved explicitly. Even when this is not the case, they are amenable to rigorous analysis or direct numerical solution; because they do not involve rapidly varying coefficients, this is far less computationally expensive than the direct numerical solution of (1.2.1).

In addition to deriving the homogenized equations by use of perturbation expansions, we will prove that the solution $u^\varepsilon(x)$ of (1.2.1) converges to the solution $u(x)$ of the homogenized Equation (1.2.2) as $\varepsilon \to 0$, in an appropriate sense. The homogenized equation will be derived using perturbation expansions in Chapter 12. The rigorous homogenization theorem for second-order elliptic PDEs with rapidly oscillating coefficients will be proved in Chapter 19.

1.2.2 Example II: Homogenization for Advection–Diffusion Equations

We now show how the ideas of homogenization can be used to study evolution PDEs of parabolic type. Consider a chemical immersed in an incompressible fluid, for example, a pollutant in the atmosphere, or a dye[1] (such as ink) in water. Under the assumption that the pollutant/dye does not affect the fluid velocity field $v(x,t)$ and that it is both transported by the fluid and subject to molecular bombardment, its concentration field $T(x,t)$ satisfies the advection–diffusion equation

$$\frac{\partial T}{\partial t} + v \cdot \nabla T = D\Delta T \quad \text{for } (x,t) \in \mathbb{R}^d \times (0,T), \tag{1.2.5a}$$

$$T = T_0 \quad \text{for } (x,t) \in \mathbb{R}^d \times \{0\}. \tag{1.2.5b}$$

Here $D > 0$ is the molecular diffusion coefficient and the Laplacian term captures the spreading of the pollutant/dye in the case where the fluid is not moving – diffusion. The first derivative term captures the advection of the pollutant/dye caused by the fluid motion. It is the interaction of advection and diffusion that we wish to study.

We assume that the fluid velocity is smooth, steady, and periodic with period 1 in all directions and that it is incompressible (or divergence-free). Thus $v(x,t) = -b(x)$ with $\nabla \cdot b(x) = 0$.[2]

Note that T constant is a solution of the equation, if T_0 is constant. Imagine now that $T_0(x) = g(\varepsilon x)$ so that initially the concentration is slowly varying in space. It is then reasonable to expect that the concentration will only vary significantly on large length and time scales. Furthermore, if b averages to zero over the unit cube, then homogenization techniques enable us to show that the rescaled concentration field $T(x/\varepsilon, t/\varepsilon^2)$ – scaled so as to bring out order-one variations in concentration on large length and time scales – converges, as ε tends to 0, to the solution $\overline{T}$ of the heat equation

$$\frac{\partial \overline{T}}{\partial t} = \mathcal{K} : \nabla\nabla\overline{T} \quad \text{for } (x,t) \in \mathbb{R}^d \times (0,T),$$
$$\overline{T} = g \quad \text{for } (x,t) \in \mathbb{R}^d \times \{0\}.$$

---

[1] Indeed many experiments designed to visualize fluid motion use this methodology.

[2] This notation is chosen to be consistent with that used in Chapters 13 and 20.

The effective behavior is hence purely diffusive. Here $\mathcal{K}$ denotes the **effective diffusion tensor**

$$\mathcal{K} = DI + \int_{\mathbb{T}^d} b(y) \otimes \chi(y)\, dy.$$

The (vector) corrector field $\chi(y)$ solves the **cell problem**

$$-D\Delta_y \chi(y) - b(y) \cdot \nabla_y \chi(y) = b(y), \quad y \in \mathbb{T}^d.$$

Again the field $\chi(y)$ is periodic.

The comments made in Example I apply equally well here: finding the homogenized field requires solving two PDEs (one elliptic with periodic boundary conditions, the other parabolic) that are independent of ε, and hence amenable to analysis, exact solution, or numerical solution. Furthermore, the ideas leading to the approximate problem can be made rigorous, and error estimates found. The homogenized equation will be derived in Chapter 13, using perturbation expansions. The rigorous theory of homogenization for parabolic PDEs with rapidly oscillating coefficients is the subject of Chapter 20.

1.2.3 Example III: Averaging, Homogenization, and Dynamics

Consider Equation (1.2.5) in the case where the velocity field $v(x,t) = -b(x)$ is steady and periodic but not necessarily mean zero. Under the rescaling $x \to x/\varepsilon$ and $t \to t/\varepsilon^a$, the equation becomes

$$\frac{1}{\varepsilon^{2-a}} \frac{\partial T}{\partial t} - \frac{1}{\varepsilon} b^\varepsilon \cdot \nabla T = D \Delta T \quad \text{for } (x,t) \in \mathbb{R}^d \times (0,T), \tag{1.2.6a}$$

$$T = f \quad \text{for } (x,t) \in \mathbb{R}^d \times \{0\}. \tag{1.2.6b}$$

Here $b^\varepsilon = b\,(x/\varepsilon)$. In the case $D = 0$ it is natural to choose $a = 1$, then this equation can be solved by the method of characteristics; the characteristics are obtained by solving the ODE

$$\frac{dx^\varepsilon}{dt} = b\Big(\frac{x^\varepsilon}{\varepsilon}\Big).$$

Since $b(x)$ is periodic and x/ε varies rapidly on the scale of the period, it is natural to try and average the equation to eliminate these rapid oscillations. Thus we see that eliminating fast scales in a time-dependent transport PDE is intimately related to averaging for ODEs. See Chapters 14 and 21 for further development of these connections. It is intuitively reasonable that the cases when b averages to zero and does not average to zero (in an appropriate sense) will require the study of (1.2.6) at different time scales (choice of a) to observe interesting dynamical behavior; this issue is discussed in Chapter 14.

In the case where $D > 0$ and the vector field b has mean zero, it is natural to choose $a = 2$. Equation (1.2.6) is the backward Kolmogorov equation for x solving the SDE[3]

$$\frac{dx^\varepsilon}{dt} = \frac{1}{\varepsilon} b\Big(\frac{x^\varepsilon}{\varepsilon}\Big) + \sqrt{2D}\frac{dW}{dt}, \tag{1.2.7}$$

where W(t) is a standard Brownian motion on $\mathbb{R}^d$. This means that, for $x^\varepsilon(0) = x$, the solution $T^\varepsilon(x,t)$ of (1.2.6) can be written as

$$T(x,t) = \mathbb{E}(f(x^\varepsilon(t))|x^\varepsilon(0) = x),$$

where $\mathbb{E}$ denotes averaging with respect to the measure on Brownian motion (Wiener measure). Again we can try to eliminate the rapidly varying quantity x/ε. If b is periodic, divergence-free, and mean zero, then the result described in the previous example shows that $x^\varepsilon(t)$, the solution of (1.2.7), converges, in the limit as $\varepsilon \to 0$, to $X(t)$, where $X(t)$ is a Brownian motion with diffusion coefficient $\sqrt{2\mathcal{K}}$:

$$\frac{dX}{dt} = \sqrt{2\mathcal{K}}\frac{dW}{dt}.$$

Furthermore we have that $\mathcal{K} \geqslant DI$ (in the sense of matrices) so that the diffusion is enhanced, over molecular diffusion, by the presence of a divergence-free advection field. The connection between homogenization in parabolic PDEs and SDEs is discussed in Chapters 11 and 13. Rigorous homogenization theorems for SDEs are proved in Chapter 18, for parabolic PDEs in Chapter 20.

1.2.4 Example IV: Dimension Reduction in Dynamical Systems

The methods applied to derive homogenized elliptic and parabolic PDEs can also be used to average out, or homogenize, the fast scales in systems of ODEs and SDEs. Doing so leads to effective equations that do not contain the small parameter ε and are hence more amenable to numerical solution or analysis. The prototypical example is a dynamical system of the form

$$\begin{aligned}\frac{dx}{dt} &= f(x,y),\\ \frac{dy}{dt} &= \frac{1}{\varepsilon}g(x,y).\end{aligned}$$

In situations of this type, where $\varepsilon \ll 1$ so that there is a scale separation, it is often the case that y can be eliminated and an approximate equation for the evolution of x can be found. We write the approximate equation in the form

$$\frac{dX}{dt} = F(X).$$

[3] In fact, the advection–diffusion Equation (1.2.5) is here viewed as an equation for the density field of an advecting and diffusing quantity and, as such, is the forward Kolmogorov (Fokker–Planck) equation. However, since $v(x,t)$ is divergence-free, the forward and backward Kolmogorov equations differ only by the sign of the advection term, and (1.2.6a) is indeed the backward Kolmogorov equation corresponding to the SDE (1.2.7).

In the simplest situation y is eliminated through an **invariant manifold**; in more complex systems it is eliminated through **averaging**.

In some situations $F \equiv 0$, and it is then necessary to scale the equations to a longer time $t \to t/\varepsilon$ to see nontrivial effects. The starting point is then

$$\frac{dx}{dt} = \frac{1}{\varepsilon} f(x, y),$$
$$\frac{dy}{dt} = \frac{1}{\varepsilon^2} g(x, y).$$

In this situation f essentially averages to zero when y is eliminated and the x equation is scaled to see the fluctuations in f. If the behavior of y is mixing in a sufficiently strong sense, then the approximate equation takes the form

$$\frac{dX}{dt} = F(X) + A(X)\frac{dW}{dt},$$

where W is a standard unit Brownian motion. We refer to derivation of this equation as **homogenization**.

The perturbation expansions underlying these ideas are fleshed out in Chapters 8, 10, and 11. As for linear PDEs, techniques from analysis enable us to estimate errors between the original and simplified equations. Rigorous results are given in Chapters 15, 17, and 18. Averaging results for Markov chains, analogous to those for ODEs and SDEs, are described in Chapters 9 and 16.

1.3 Averaging Versus Homogenization

The unifying principle underlying the derivation of most of the effective equations in this book concerns formal perturbation expansions for linear operator equations of the form

$$\mathcal{L}^{\varepsilon} u^{\varepsilon} = f \tag{1.3.1}$$

or

$$\frac{\partial u^{\varepsilon}}{\partial t} = \mathcal{L}^{\varepsilon} u^{\varepsilon}. \tag{1.3.2}$$

In particular, we will be interested in cases where the operator $\mathcal{L}^{\varepsilon}$ has the form

$$\mathcal{L}^{\varepsilon} = \frac{1}{\varepsilon}\mathcal{L}_0 + \mathcal{L}_1 \tag{1.3.3}$$

or

$$\mathcal{L}^{\varepsilon} = \frac{1}{\varepsilon^2}\mathcal{L}_0 + \frac{1}{\varepsilon}\mathcal{L}_1 + \mathcal{L}_2. \tag{1.3.4}$$

In both cases we assume that $\mathcal{L}_0$ has a nontrivial null space and interest focuses on capturing the behavior of the solution within this subspace. We will refer to the first case (1.3.3) as **averaging**, or **first-order perturbation theory**. It can often be thought of as a form (or consequence) of the law of large numbers. The second,

(1.3.4), will be referred to as **homogenization** or **second-order perturbation theory**. It can often be thought of as a form (or consequence) of the central limit theorem.

Our main interest will be in the evolution Equation (1.3.2). This will apply to the study of averaging and homogenization for parabolic PDEs and transport PDEs, and also to averaging and homogenization for ODEs, Markov chains, and SDEs, via the Kolmogorov equations for SDEs and variants (the forward equation, for Markov chains, and the method of characteristics, for ODEs). We will also study Equation (1.3.1), which arises in the problem of homogenization for elliptic PDEs. The case where a second-order time derivative appears (wave equations) will not be covered explicitly herein, but the techniques developed do apply; references to the literature will be given. Here we highlight differences between averaging and homogenization and demonstrate the importance of the Fredholm alternative when carrying out these expansions. We apply the methods that will be analyzed in this book to a singularly perturbed system of linear ODEs. We thus introduce key ideas in this simplified context. The following two subsections illustrate the perturbation expansions that we use repeatedly in Part II.

1.3.1 Averaging for Systems of Linear Equations

Consider a system of linear ODEs

$$\frac{du^\varepsilon}{dt} = \mathcal{L}^\varepsilon u^\varepsilon, \tag{1.3.5a}$$

$$\mathcal{L}^\varepsilon = \frac{1}{\varepsilon}\mathcal{L}_0 + \mathcal{L}_1. \tag{1.3.5b}$$

Here, for simplicity, we consider the finite-dimensional case where $\mathcal{L}_i \in \mathbb{R}^{d\times d}$ for $i = 0, 1$. Assume that the null space of $\mathcal{L}_0$, $\mathcal{N}(\mathcal{L}_0)$, is one-dimensional and spanned by ϕ and that $\mathcal{N}(\mathcal{L}_0^T)$ is spanned by ψ. Assume further that $\langle \phi, \psi\rangle \neq 0$.

If we seek a solution in the form

$$u^\varepsilon = u_0 + \varepsilon u_1 + \mathcal{O}(\varepsilon^2),$$

then

$$\frac{1}{\varepsilon}\mathcal{L}_0 u_0 + \left(\mathcal{L}_0 u_1 + \mathcal{L}_1 u_0 - \frac{du_0}{dt}\right) + \mathcal{O}(\varepsilon) = 0.$$

Equating coefficients of respective powers of ε to zero gives

$$\mathcal{O}(1/\varepsilon) \qquad \mathcal{L}_0 u_0 = 0,$$

$$\mathcal{O}(1) \qquad \mathcal{L}_0 u_1 = \frac{du_0}{dt} - \mathcal{L}_1 u_0.$$

The first equation implies that $u_0 = \alpha\phi$ for some $\alpha = \alpha(t) \in \mathbb{R}$. Applying the Fredholm alternative to the second equation implies that a solution u_1 exists if and only if

$$\langle \psi, \frac{du_0}{dt} - \mathcal{L}_1 u_0 \rangle = 0,$$

where $\langle \cdot, \cdot \rangle$ denotes the inner product on $\mathbb{R}^d$. Substituting $u_0 = \alpha\phi$ we deduce that the *amplitude* α satisfies the equation

$$\frac{d\alpha}{dt} = \frac{\langle \psi, \mathcal{L}_1 \phi \rangle}{\langle \psi, \phi \rangle} \alpha. \tag{1.3.6}$$

This equation has a nontrivial solution provided that

$$\langle \psi, \mathcal{L}_1 \phi \rangle \neq 0. \tag{1.3.7}$$

Thus we have found the approximate solution

$$u^\varepsilon(t) \approx u_0(t) = \alpha(t)\phi,$$

where $\alpha(t)$ solves Equation (1.3.6). This derivation exemplifies the method of averaging as employed in Chapters 9, 10, and 14. Notice that the amplitude $\alpha(t)$ can be either exponentially growing or exponentially decaying. The behavior of $u^\varepsilon(t)$, the solution of (1.3.5a) for $\varepsilon \ll 1$, is determined by the behavior of $\alpha(t)$. This is a first example of the usefulness of the methods presented in this book; we can determine the qualititative properties of solutions to the original complicated Equation (1.3.5a) by studying the much simpler Equation (1.3.6).

1.3.2 Homogenization for Systems of Linear Equations

If (1.3.7) fails, the averaged equation becomes

$$u_0 = \text{const.}$$

In order to observe interesting, nontrivial dynamics it is necessary to scale $t \to t/\varepsilon$ and seek a solution to (1.3.5a). This leads to the system of linear equations

$$\frac{du^\varepsilon}{dt} = \mathcal{L}^\varepsilon u^\varepsilon,$$

with

$$\mathcal{L}^\varepsilon = \frac{1}{\varepsilon^2}\mathcal{L}_0 + \frac{1}{\varepsilon}\mathcal{L}_1.$$

Here $\mathcal{L}_i \in \mathbb{R}^{d \times d}$ for $i = 0, 1$. Because there is no loss of generality in doing so, we study the slightly more general system

$$\frac{du^\varepsilon}{dt} = \mathcal{L}^\varepsilon u^\varepsilon, \tag{1.3.8a}$$

$$\mathcal{L}^\varepsilon = \frac{1}{\varepsilon^2}\mathcal{L}_0 + \frac{1}{\varepsilon}\mathcal{L}_1 + \mathcal{L}_2. \tag{1.3.8b}$$

Here $\mathcal{L}_i \in \mathbb{R}^{d \times d}$ for $i = 0, 1, 2$. We make the same assumptions concerning the null spaces of $\mathcal{L}_0$ and $\mathcal{L}_0^T$ as in the previous subsection. We also assume that the **centering condition**

$$\langle \psi, \mathcal{L}_1 \phi \rangle = 0 \tag{1.3.9}$$

holds. If we now seek a solution of (1.3.8a) in the form

$$u^\varepsilon = u_0 + \varepsilon u_1 + \varepsilon^2 u_2 + \mathcal{O}(\varepsilon^3),$$

then

$$\frac{1}{\varepsilon^2}\mathcal{L}_0 u_0 + \frac{1}{\varepsilon}\Big(\mathcal{L}_0 u_1 + \mathcal{L}_1 u_0\Big) + \Big(\mathcal{L}_0 u_2 + \mathcal{L}_1 u_1 + \mathcal{L}_2 u_0 - \frac{du_0}{dt}\Big) + \mathcal{O}(\varepsilon) = 0.$$

Equating coefficients of respective powers of ε to zero gives

$$\begin{aligned}
\mathcal{O}(1/\varepsilon^2) \qquad & \mathcal{L}_0 u_0 = 0,\\
\mathcal{O}(1/\varepsilon) \qquad & \mathcal{L}_0 u_1 = -\mathcal{L}_1 u_0,\\
\mathcal{O}(1) \qquad & \mathcal{L}_0 u_2 = \frac{du_0}{dt} - \mathcal{L}_1 u_1 - \mathcal{L}_2 u_0.
\end{aligned}$$

The first equation implies that $u_0 = \alpha\phi$ for some $\alpha = \alpha(t) \in \mathbb{R}$. By (1.3.9) and the Fredholm alternative, we deduce that the second equation is solvable, so we write

$$u_1 = -\alpha\eta, \quad \mathcal{L}_0 \eta = \mathcal{L}_1 \phi.$$

The equation for η is called the **cell problem**. We choose to make the solution unique by asking that η is orthogonal to ϕ; however other normalizations may be used without affecting the formula for the approximate solution u_0.

Moving to the third equation, we find that, again by application of the Fredholm alternative,

$$\frac{d\alpha}{dt} = \frac{\langle \psi, \mathcal{L}_2 \phi - \mathcal{L}_1 \eta \rangle}{\langle \psi, \phi \rangle}\alpha. \tag{1.3.10}$$

Thus we have found the approximate solution

$$u^\varepsilon(t) \approx u_0(t) = \alpha(t)\phi.$$

Note that, because of (1.3.9), the choice of normalization for η does not affect the final formula for u_0 – adding multiples of ϕ to η does not effect the formula for u_0. This derivation exemplifies the method of homogenization as employed in Chapters 11, 12, and 13.

1.4 Discussion and Bibliography

The use of formal multiscale expansions in the study of homogenization was developed systematically and applied to many different problems, especially from the theory of PDEs, in [33]. A thorough development of these ideas for SDEs may be found in the papers [241, 240, 246]. Standard books on averaging/homogenization include [66, 153, 27, 279]. Averaging and homogenization in the context of stochastic processes and SDEs in particular is developed in [111]. Homogenization for SDEs is presented in the lecture notes [238, 186].

A prevalent idea throughout this book is the connection between parabolic and elliptic PDEs and SDEs through the Kolmogorov and Fokker–Planck equations, [117, 236]. The relation between homogenization and averaging for PDEs and limit theorems for (Markov) stochastic processes has been recognized and studied extensively since the early 1960s [110, 165]. A rather general theory of homogenization and averaging for stochastic processes was developed in the 1970s [181, 240, 241, 246]. The mathematical theory is presented in the monograph [94]. See also [261, 168].

Variable elimination for either stochastic or deterministic systems has become a very important area in applied mathematics. Applications include atmosphere/ocean science [202, 205], molecular dynamics [284, 285], and mathematical finance [106]. Various aspects of stochastic and deterministic mode reduction, including numerical algorithms for these problems, are discussed in the review articles [125, 148, 150]. Earlier work on the problem of mode reduction is discussed in the review paper [321].

The subject of multiscale methods comprises a broad and disparate field, subsuming, but by no means limited to, the topics of averaging and homogenization that comprise this book. There are many important topics related to homogenization and averaging for ODEs, SDEs, and PDEs that will not be discussed here. We outline some of these topics and provide references to the literature.

A very important aspect of multiscale methods is the development of efficient numerical algorithms for the study of problems with many characteristic lengths and time scales. An excellent overview of certain aspects of current research activity in the area of multiscale numerical methods can be found in [83]; see also [82] and [119].

Deterministic, nonperiodic homogenization will not be covered here. The interested reader may consult [66, ch. 13; 75; 311] and the references therein. We will not discuss homogenization problems for random PDEs (i.e., PDEs whose coefficients are rapidly oscillating random fields); see [247, 177]. A pathwise approach to multiscale stochastic systems is presented in [34].

One of the main applications of the theory of homogenization is to materials science. Monographs where the applications of homogenization to elasticity and materials science are discussed include [27, 7, 67, 237]. Various multiscale/homogenization approaches to the study of composite materials are presented in [229]. In that book a thorough discussion of bounds on the homogenized coefficients for composite materials and related models can also be found. Applications of the techniques developed in this book to various problems in chemical engineering are overviewed in [265, 325].

We will only touch on the theory of homogenization for nonlinear PDEs; the interested reader can consult [95] and the references therein. Variational methods are very well suited for the study of homogenization problems for nonlinear elliptic PDEs. Information about this problem and the related theory of Γ-convergence can be found in [214, 72, 48, 49]. Related concepts, such as H and G convergence, will not be developed here; see [296, 297, 311, 310] for material on this topic.

More detailed pointers to the literature will be provided in the Discussion and Bibliography sections concluding each chapter.

Part I

Background

2

Analysis

2.1 Setup

In this chapter we collect a variety of definitions and theorems (mainly without proofs) from analysis. The topics are selected because they will be needed in the sequel. Proofs and additional material can be found in the books cited in Section 2.7. The presentation is necessarily terse, and it is not recommended that this chapter is read in its entirety on a first read through the book; rather it should be referred to as needed when reading later chapters. The rigorous setting developed in this chapter is required primarily in Chapters 15–21 (Part III) where we prove a number of results concerning dimension reduction for dynamical problems, together with averaging and homogenization results for PDEs. The averaging and homogenization results themselves are derived in Chapters 8–14 (Part II) by means of perturbation expansions, and in that context much of the material in this chapter is not required. However, the notation introduced here, together with the Fredholm alternative, will be used througout the book, in Part II in particular.

The natural setting for the rigorous study of asymptotic problems for PDEs is the theory of weak convergence in Hilbert and Banach spaces; in this context we will also develop an appropriate kind of weak convergence, that of two-scale convergence, which is very useful for problems related to periodic homogenization. A complementary chapter on probability, Chapter 3, provides the appropriate analytical tools for the study of SDEs and Markov chains; in particular we develop background material on weak convergence of probability measures.

In Section 2.2 we describe the notation employed throughout much of the book; in particular the index-free notation for differential operators, such as the divergence, the gradient, and Laplacian, is described. Banach and Hilbert space theory is overviewed in Section 2.3, and Section 2.4 describes specific instances of function spaces that will be useful. Section 2.5 introduces the concept of two-scale convergence, highlighting its importance for the study of periodic highly oscillatory problems. The chapter ends with a discussion of two fundamental tools for the study of linear equations in Hilbert spaces: the Lax–Milgram theorem and the Fredholm alternative.

2.2 Notation

In this book we will encounter scalar, vector, and matrix fields. In the following all summations are over indices from the set $\{1, 2, \ldots, d\}$, d being the dimension of the space. We use $\mathbb{R}^d$ to denote the d-dimensional Euclidean space. We denote by $\mathbb{T}^d$ the d-dimensional unit torus found by taking a unit cube in $\mathbb{R}^d$ and identifying opposite faces.

We will denote by $\{e_i\}_{i=1}^d$ the standard basis in $\mathbb{R}^d$. Thus every $\xi \in \mathbb{R}^d$ can be written as $\xi = \sum_i \xi_i e_i$, where $\xi_i = \langle \xi, e_i \rangle$ and $\langle \cdot, \cdot \rangle$ denotes the *standard inner product* on $\mathbb{R}^d$. We also use $\cdot$ to denote the inner product between two vectors, so that

$$a \cdot b = \sum_i a_i b_i.$$

The norm induced by this inner product is the Euclidean norm

$$|a| = \sqrt{a \cdot a},$$

and it follows that

$$|a|^2 = \sum_i a_i^2, \quad a \in \mathbb{R}^d.$$

We will also use the usual ℓ^p norms on $\mathbb{R}^d$, $1 \leqslant p \leqslant \infty$, and denote them by $|\cdot|_p$. Note that $|\cdot| = |\cdot|_2$. In addition, for $A : \mathbb{R}^d \to \mathbb{R}^m$, we use $|\cdot|_p$ to denote the associated operator norm defined by

$$|A|_p = \sup_{x \neq 0} \frac{|Ax|_p}{|x|_p}.$$

In particular we use the notation $|A| = |A|_2$ for operator norms induced by the Euclidean vector norm.

The *inner product between matrices* is denoted by

$$A : B = \mathrm{tr}(A^T B) = \sum_{ij} a_{ij} b_{ij}.$$

The norm induced by this inner product is the *Frobenius norm*

$$|A|_F = \sqrt{\mathrm{tr}(A^T A)}. \tag{2.2.1}$$

On occasion it will be important to note that, for any symmetric matrix T and any matrix S,

$$S : T = S^T : T = \frac{1}{2}(S + S^T) : T. \tag{2.2.2}$$

The *outer product* between two vectors a and b is the matrix $a \otimes b$ defined by

$$\Big(a \otimes b\Big) c = (b \cdot c) a \quad \forall c \in \mathbb{R}^d.$$

More generally, for $A, B \in \mathbb{R}^{m\times d}$, $A \otimes B \in \mathbb{R}^{m\times m}$ satisfies

$$\Big(A \otimes B\Big)c = AB^T c \quad \forall c \in \mathbb{R}^m.$$

Let ∇ and $\nabla\cdot$ denote *gradient* and *divergence* in $\mathbb{R}^d$. The gradient lifts a scalar (resp. vector) to a vector (resp. matrix) while the divergence contracts a vector (resp. matrix) to a scalar (resp. vector). The gradient acts on scalar-valued functions $\phi(z)$, or vector-valued functions $v(z)$, via

$$(\nabla\phi)_i = \frac{\partial\phi}{\partial z_i}, \qquad (\nabla v)_{ij} = \frac{\partial v_i}{\partial z_j}.$$

The divergence acts on vector-valued functions $v(z)$, or matrix-valued functions $A(z)$, via

$$\nabla\cdot v = \sum_i \frac{\partial v_i}{\partial z_i}, \qquad (\nabla\cdot A)_i = \sum_j \frac{\partial A_{ij}}{\partial z_j}.$$

Given vector fields a, v we use the notation

$$a\cdot\nabla v := (\nabla v)a.$$

Thus we define the quantity by calculating $a\cdot\nabla v_k$ for each component of the vector v. Likewise we can extend to the notation

$$a\cdot\nabla\Theta,$$

where Θ is a tensor[1] field, by defining componentwise.

Since the gradient is defined for scalars and vectors, we readily make sense of the expression

$$\nabla\nabla\phi$$

for any scalar ϕ; it is the *Hessian* matrix with entries $\partial^2\phi/\partial x_i\partial x_j$. Similarly, we can also make sense of the expression

$$\nabla\nabla v$$

by applying $\nabla\nabla$ to each scalar component of the vector v, or indeed

$$\nabla\nabla\Theta,$$

again componentwise. We define the Laplacian of a scalar or vector field by

$$\Delta\phi = \nabla\cdot\nabla\phi; \quad \Delta v = \nabla\cdot\nabla v.$$

It follows that $\Delta\phi = I : \nabla\nabla\phi$ (see Exercise 1). Applying this definition componentwise allows for the definition of $\Delta\Theta$.

In many instances in what follows it will be necessary to distinguish between the gradient or divergence with respect to a variety of different independent variables. We use ∇_z to denote the gradient or divergence with respect to z coordinates alone, and similarly for other independent variables. We will retain the notation ∇, without suffixes, when the choice of independent variable is clear. A similar notation will be used for other differential operators, such as the Laplacian.

[1] In this text we use the word **tensor** to denote second-order tensors, or square matrices.

2.3 Banach and Hilbert Spaces

We assume that the reader is already familiar with the definitions of a norm, an inner product, and of vector, metric, normed, and inner product spaces. In the sequel X will denote a normed vector space, and $\|\cdot\|$ will denote its norm. We say that a sequence $\{x_j\}_{j=1}^{\infty} \subset X$ *converges strongly* to $x \in X$, written as

$$x_j \to x,$$

provided that

$$\lim_{j\to\infty} \|x_j - x\| = 0.$$

Furthermore, we say that $\{x_j\}_{j=1}^{\infty} \subset X$ is a *Cauchy sequence* provided that for each $\varepsilon > 0$ there exists an $N \in \mathbb{N}$ such that, for all $j, k \geqslant N$,

$$\|x_j - x_k\| < \varepsilon.$$

Every convergent sequence in a normed space X is a Cauchy sequence. The converse in not always true. If, however, every Cauchy sequence in X is convergent, then the space X is called *complete*.

2.3.1 Banach Spaces

Definition 2.1. *A* Banach space *X is a complete normed vector space.*

Definition 2.2. *Let X be a Banach space with norm $\|\cdot\|$. We say that a map $\ell : X \to \mathbb{R}$ is a* bounded linear functional *on X provided that*

i) $\ell(\alpha x + \beta y) = \alpha\ell(x) + \beta\ell(y)\ \forall x, y \in X,\ \alpha, \beta \in \mathbb{R}.$
ii) $\exists C > 0 : |\ell(x)| \leqslant C\|x\|\ \forall x \in X.$

Definition 2.3. *The collection of all bounded linear functionals on a Banach space X is called the* dual space *and is denoted by X^*.*

Theorem 2.4. *The dual space X^* of a Banach space X is, when equipped with the norm*

$$\|\ell\| = \sup_{x\neq 0} \frac{|\ell(x)|}{\|x\|},$$

a Banach space.

Definition 2.5. *Let X be a Banach space. A Banach space X is called* reflexive *if the dual of its dual is isomorphic to X:*

$$(X^*)^* = X.$$

The concept of the dual space enables us to introduce another topology on X, the so-called *weak topology*.

Definition 2.6. *A sequence* $\{x_n\}_{n=1}^{\infty}$ *is said to* converge weakly *to* $x \in X$*, written*

$$x_n \rightharpoonup x,$$

if

$$\ell(x_n) \to \ell(x) \quad \forall \ell \in X^*.$$

Every strongly convergent sequence is also weakly convergent. However, the converse is not true. The importance of weak convergence stems from the following theorem, in particular part (ii).

Theorem 2.7. *Let* X *be a Banach space.*

(i) Every weakly convergent sequence in X *is bounded.*
*(ii) (***Eberlein–Smuljan***) Assume that* X *is reflexive. Then from every bounded sequence in* X *we can extract a weakly convergent subsequence.*

In addition to the weak topology on X, the duality between X and X^* enables us to define a topology on X^*.

Definition 2.8. *Let* X *be a Banach space. A sequence* $\{\ell_n\}_{n=1}^{\infty} \subset X^*$ *is said to* converge weak–$*$ *to* $\ell \in X^*$*, written*

$$\ell_n \overset{*}{\rightharpoonup} \ell,$$

if

$$\lim_{n\to\infty} \ell_n(x) = \ell(x) \quad \forall x \in X.$$

We remark that if X is reflexive then weak–$*$ convergence coincides with weak convergence on X.

A compactness result similar to Theorem 2.7(ii), but without reflexivity, holds for bounded sequences in X^*, provided that X is *separable*. To define this concept we recall that a subset X_0 of X is called *dense* if for every $x \in X$ there exists a sequence $\{x_j\}_{j=1}^{\infty} \subset X_0$ that converges to x. In other words, X_0 is dense in X if its closure is X: $\overline{X_0} = X$.

Definition 2.9. *A Banach space* X *is called* separable *if it contains a countable dense subset.*

The compactness theorem for sequences in X^* can be stated as follows.

Theorem 2.10. *Let* X *be a separable Banach space. Then from any bounded sequence in* X^* *we can extract a weak–$*$ convergent subsequence.*

Every weakly convergent sequence in X^* is weak–$*$ convergent, but the converse is not true unless X is reflexive.

2.3.2 Hilbert Spaces

Definition 2.11. *A* Hilbert space *is a complete inner product space.*

We denote the inner product by $(\cdot,\cdot)$. Clearly, every Hilbert space is a Banach space with a norm on H given by:

$$\|x\| = (x,x)^{\frac{1}{2}}.$$

Furthermore, all elements of H satisfy the **Cauchy–Schwarz** inequality:

$$|(u,v)| \leqslant \|u\|\|v\|.$$

A very important property of a Hilbert space H is that we can identify the dual of H with itself through the Riesz representation theorem.

Theorem 2.12. (Riesz representation.) *For every $\ell \in H^*$ there exists a unique $y \in H$ such that*

$$\ell(x) = (x,y) \quad \forall x \in H.$$

We will usually denote the action of ℓ on $x \in H$ by[2] $\langle\cdot,\cdot\rangle_{H^*,H}$, referred to as the *dual pairing*

$$\langle \ell, x\rangle_{H^*,H} := \ell(x).$$

The Riesz representation theorem implies that every Hilbert space is reflexive, and consequently Theorem 2.7 applies. Furthermore, the definition of weak convergence simplifies to

$$x_n \rightharpoonup x \Leftrightarrow (x_n - x, y) \to 0 \quad \forall y \in H.$$

2.4 Function Spaces

2.4.1 Spaces of Continuous Functions

Let Ω be an open subset of $\mathbb{R}^d$. We will denote by $C(\overline{\Omega})$ the space of continuous functions $f : \overline{\Omega} \to \mathbb{R}$. This space, when equipped with the supremum norm

$$\|f\|_{C(\overline{\Omega})} = \sup_{x\in\overline{\Omega}} |f(x)|,$$

is a Banach space. Similarly, we can define the space $C^k(\overline{\Omega})$ of k-times continuously differentiable functions. We will denote by $C^\infty(\overline{\Omega})$ the space of smooth functions.

The notation $C_0^k(\overline{\Omega})$ will be used to denote the space of k-times continuously differentiable functions over $\overline{\Omega}$ with compact support. The notation $C_0^\infty(\overline{\Omega})$ extends this to smooth functions over $\overline{\Omega}$ with compact support. We will use the notation

[2] In $\mathbb{R}^d$ we use $\langle\cdot,\cdot\rangle$ to denote the inner product, and it is useful to retain this convention; in the infinite-dimensional setting we will always use $(\cdot,\cdot)$ for the inner product and $\langle\cdot,\cdot\rangle_{H^*,H}$ for the dual pairing between H and H^*.

$C_b^k(\mathbb{R}^d)$ to denote the set of k-times continuously differentiable functions from $\mathbb{R}^d$ to $\mathbb{R}$ for which all derivatives up to order k are bounded. The notation $C_b^\infty(\mathbb{R}^d)$ denotes the space of bounded continuous functions having bounded continuous derivatives of all orders. More generally, let (E,ρ) be a metric space. Then $C_b(E)$ will denote the space of bounded continuous functions on (E,ρ). We may extend all of the preceding to functions $f : \Omega \to X$ where $X = \mathbb{R}^d$ or $\mathbb{R}^{d\times d}$. In all these cases we will simply write $C(\overline{\Omega})$ when the meaning is clear, or $C(\overline{\Omega};X)$ and so forth, when needed. See Definition 2.22 for the general case of Banach space–valued functions f.

2.4.2 L^p Spaces

Let $1 \leqslant p \leqslant \infty$, let $f : \Omega \to \mathbb{R}$ be a Lebesgue measurable function (see Chapter 3 for a definition of measurable), and define the L^p-norm

$$\|f\|_{L^p(\Omega)} := \begin{cases} \left(\int_\Omega |f|^p\,dx\right)^{\frac{1}{p}} & \text{for } 1 \leqslant p < \infty \\ \operatorname{ess\,sup}_\Omega |f| & \text{for } p = \infty. \end{cases}$$

In the preceding definition we used the notation

$$\operatorname*{ess\,sup}_\Omega = \inf\{C, |f| \leqslant C \text{ a.e. on } \Omega\}.$$

Here *a.e.* is with respect to Lebesgue measure. Sometimes we drop explicit reference to the set Ω in the norm and simply write $\|\cdot\|_{L^p}$. For measurable functions $f : \Omega \to \mathbb{R}^d$ the norm is readily extended, replacing $|f|$ under the integral by the vector ℓ^p norm on $\mathbb{R}^d$. Likewise we may consider measurable $f : \Omega \to \mathbb{R}^{d\times d}$, using the operator ℓ^p-norm. In all these cases we write $L^p(\Omega)$ as shorthand for $L^p(\Omega;X)$ where $X = \mathbb{R}$, $\mathbb{R}^d$, $\mathbb{R}^{d\times d}$. See Definition 2.22 for the general case of Banach space–valued functions f. The remaining discussion in this subsection can also be extended with this definition.

Definition 2.13. *$L^p(\Omega)$ is the vector space of all measurable functions*[3] $f : \Omega \to \mathbb{R}$ *for which* $\|f\|_{L^p(\Omega)} < \infty$.

Theorem 2.14. (Basic properties of L^p spaces.)

i) The vector space $L^p(\Omega)$, equipped with the L^p-norm defined earlier, is a Banach space for every $p \in [1,\infty]$.
ii) $L^2(\Omega)$ is a Hilbert space equipped with the inner product

$$(u,v)_{L^2(\Omega)} = \int_\Omega u(x)v(x)\,dx \quad \forall\, u,v \in L^2(\Omega).$$

iii) $L^p(\Omega)$ is separable for $p \in [1,\infty)$ and reflexive for $p \in (1,+\infty)$. In particular, $L^1(\Omega)$ is not reflexive, and $L^\infty(\Omega)$ is neither separable nor reflexive.

[3] Strictly speaking $L^p(\Omega)$ is the set of equivalence classes of functions, with the equivalence classes defined by equality a.e. in Ω with respect to Lebsegue measure.

Let $p \in [1, \infty]$ and define $q \in [1, \infty]$ through

$$\frac{1}{p} + \frac{1}{q} = 1. \tag{2.4.1}$$

Then the Hölder inequality states that

$$\left| \int_{\Omega} u(x)v(x)\, dx \right| \leqslant \|u\|_{L^p(\Omega)} \|v\|_{L^q(\Omega)} \quad \forall u \in L^p(\Omega),\ v \in L^q(\Omega).$$

Let $p \in [1, \infty)$ and let q be defined through (2.4.1). *Then*

$$(L^p(\Omega))^* = L^q(\Omega).$$

The L^2 inner product is readily extended to $\mathbb{R}^d$ and $\mathbb{R}^{d\times d}$ valued functions, replacing pointwise multiplication under the integral by the appropriate inner product on $\mathbb{R}^d$ or $\mathbb{R}^{d\times d}$. We will use the notation $L^2(\Omega)$ in this situation.

The last part of the theorem, together with the fact that $L^p(\Omega)$ is a Banach space and Definition 2.6, implies the following equivalent definition of weak convergence in $L^p(\Omega)$, $p \in [1, +\infty)$.

Definition 2.15. *A sequence $\{u_n\}_{n=1}^{\infty} \subset L^p(\Omega)$, $p \in [1, \infty)$ is said to* converge weakly *to $u \in L^p(\Omega)$, written*

$$u_n \rightharpoonup u \quad \textit{weakly–}L^p(\Omega),$$

provided that

$$\int_{\Omega} u_n(x)v(x)\, dx \to \int_{\Omega} u(x)v(x)\, dx \quad \forall v \in L^q(\Omega),$$

where q is defined in (2.4.1).

Notice that, in the case $p \in (1, \infty)$, $L^p(\Omega)$ is reflexive and hence every bounded sequence in this space has a weakly convergent subsequence, in the sense of the preceding definition.

We will often simply use $(\cdot, \cdot)$ to denote the inner product on $L^2(\Omega)$. We will frequently use a *density argument* to simplify the verification of weak convergence. For example, to prove weak convergence of u_n to u in $L^2(\Omega)$ it suffices to prove that

$$(u_n - u, v) \to 0 \quad \forall v \in C(\overline{\Omega}).$$

This is because $C(\overline{\Omega})$ is dense in $L^2(\Omega)$.

Whereas weak convergence in L^∞ is very rarely used (because its dual is not an L^p-space), the notion of weak–* convergence in that space is very useful (because it is the dual of L^1, a separable space).

Definition 2.16. *A sequence $\{u_n\}_{n=1}^{\infty} \subset L^\infty(\Omega)$* converges weak–∗ *in L^∞, written*

$$u_n \overset{*}{\rightharpoonup} u \quad \textit{weak–* in } L^\infty(\Omega),$$

provided that

$$\int_\Omega u_n(x)\phi(x)\,dx \to \int_\Omega u(x)\phi(x)\,dx \quad \forall \phi \in L^1(\Omega).$$

Because $L^1(\Omega)$ is separable we deduce that every bounded sequence in $L^\infty(\Omega)$ has a weak–* convergent subsequence, in the sense of the previous definition.

2.4.3 Sobolev Spaces

We start with the definition of the weak derivative.

Definition 2.17. *Let $u, v \in L^2(\Omega)$. We say that v is the* weak derivative *of u with respect to x_i if*

$$\int_\Omega u \frac{\partial \phi}{\partial x_i}\,dx = -\int_\Omega v\phi\,dx \quad \forall \phi \in C_0^\infty(\Omega).$$

Higher-order weak derivatives can be defined similarly. In the context of a function $u \in L^2(\Omega)$ we will use the notation $\partial u/\partial x_i$ to denote the weak derivative with respect to x_i and the notation $\nabla u = \sum_i \partial u/\partial x_i e_i$ for the gradient.

Definition 2.18. *The* Sobolev space $H^1(\Omega)$ *consists of all square integrable functions $u : \Omega \to \mathbb{R}$ whose first-order weak derivatives exist and are square integrable:*

$$H^1(\Omega) = \left\{ u \middle| u, \nabla u \in L^2(\Omega) \right\}.$$

The space $H^1(\Omega)$ is a separable Hilbert space with inner product

$$(u, v)_{H^1(\Omega)} = (u, v)_{L^2(\Omega)} + (\nabla u, \nabla v)_{L^2(\Omega)}$$

and norm

$$\|u\|_{H^1(\Omega)} = \left(\|u\|^2_{L^2(\Omega)} + \|\nabla u\|^2_{L^2(\Omega)} \right)^{\frac{1}{2}}. \tag{2.4.2}$$

Since $H^1(\Omega)$ is a Hilbert space, it is reflexive. Hence every bounded sequence in $H^1(\Omega)$ contains a weakly convergent subsequence. In fact, more is true: a very useful property of $H^1(\Omega)$ is that its embedding into $L^2(\Omega)$ is compact. This implies the following.

Theorem 2.19. (Rellich compactness theorem). *From every bounded sequence in $H^1(\Omega)$ we can extract a subsequence that is strongly convergent in $L^2(\Omega)$.*

In many applications one is interested in elements of $H^1(\Omega)$ that vanish on the boundary $\partial\Omega$ of the domain. Functions with this property belong to the following subset of $H^1(\Omega)$.

Definition 2.20. *The* Sobolev space $H^1_0(\Omega)$ *is defined as the completion of $C_0^\infty(\Omega)$ with respect to the $H^1(\Omega)$ norm.*

A very important property of $H_0^1(\Omega)$ is the fact that we can control the L^2-norm of its elements in terms of the L^2-norm of their gradient. We now present this result. In later sections and chapters we will generalize it to consider the analogue of this result for periodic elements of $H^1(\mathbb{T}^d)$.

Theorem 2.21. (Poincaré inequality) *Let Ω be a bounded open set in $\mathbb{R}^d$. Then there is a constant C_Ω, which depends only on the size of Ω, such that, for every $u \in H_0^1(\Omega)$,*

$$\|u\|_{L^2(\Omega)} \leqslant C_\Omega \|\nabla u\|_{L^2(\Omega)}. \tag{2.4.3}$$

An immediate corollary of the first part of the theorem (see Exercise 9) is that $\|\nabla \cdot\|_{L^2(\Omega)}$ can be used as the norm in $H_0^1(\Omega)$:

$$\|u\|_{H_0^1(\Omega)} = \|\nabla u\|_{L^2(\Omega)}. \tag{2.4.4}$$

We will denote by $H^{-1}(\Omega)$ the dual space of $H_0^1(\Omega)$. Further, we will denote by $\langle \cdot, \cdot \rangle_{H^{-1},H_0^1}$ the pairing between $H^{-1}(\Omega)$ and $H_0^1(\Omega)$. In other words, the action of $f \in H^{-1}(\Omega)$ on $v \in H_0^1(\Omega)$ will be denoted by $\langle f, v\rangle_{H^{-1},H_0^1}$. Then $H^{-1}(\Omega)$ is a Banach space equipped with the norm

$$\|f\|_{H^{-1}} = \sup_{v \in H_0^1 \neq 0} \frac{|\langle f, v\rangle_{H^{-1},H_0^1}|}{\|v\|_{H_0^1}} = \sup_{v \in H_0^1,\, \|v\|_{H_0^1} \leqslant 1} \left|\langle f, v\rangle_{H^{-1},H_0^1}\right|.$$

(The representations follow from Theorem 2.4 and some elementary properties of the supremum therein.) Thus the following *Cauchy–Schwarz*–like inequality holds

$$|\langle f, v\rangle_{H^{-1},H_0^1}| \leqslant \|f\|_{H^{-1}} \|v\|_{H_0^1} \quad \forall f \in H^{-1}(\Omega),\ \forall v \in H_0^1(\Omega). \tag{2.4.5}$$

2.4.4 Banach Space–Valued Spaces

It is possible to define L^p-spaces of functions varying over spaces Ω more general than $\mathbb{R}^d$, replacing the Lebesgue integral by an integral with respect to another measure on Ω. It is also possible to work with L^p-spaces of functions taking values in an arbitrary Banach space. We illustrate these ideas.

The spaces defined here appear often in applications. They are particularly relevant for the rigorous analysis of periodic homogenization since we have to deal with functions of two arguments (one slowly varying and one rapidly varying).

Definition 2.22. *Let X be a Banach space with norm $\|\cdot\|_X$ and let Ω denote a subset of $\mathbb{R}^d$, not necessarily bounded. The space $Y := L^p(\Omega; X)$ with $p \in [1, +\infty]$ consists of all measurable functions $u : x \in \Omega \to u(x) \in X$ such that $\|u(x)\|_X \in L^p(\Omega)$.*

The space Y defined earlier has various nice properties, which we list here.

Theorem 2.23. *Let $Y = L^p(\Omega; X)$ with X and Ω as in Definition 2.22. Then*

(i) Y equipped with the norm

$$\|u\|_Y := \begin{cases} \left(\int_\Omega \|u(x)\|_X^p \, dx\right)^{\frac{1}{p}} & \text{for } 1 \leqslant p < \infty \\ \text{ess}\sup_{x\in\Omega} \|u(x)\|_X & \text{for } p = \infty \end{cases}$$

is a Banach space.

(ii) If X is reflexive and $p \in (1, +\infty)$ then Y is also reflexive.

(iii) If X is separable and $p \in [1, +\infty)$ then Y is also separable.

The following two examples show how these ideas may be generalized.

Example 2.24. Banach spaces of the form $Y = H^1(\Omega; X)$, where X is a Banach space, can be defined in a similar fashion to that employed in Theorem 2.23 to define $L^p(\Omega; X)$. The norm on such a space is

$$\|u\|_Y = \left(\int_\Omega \left(\|\nabla_x u(x)\|_X^2 + \|u(x)\|_X^2\right) dx\right)^{\frac{1}{2}}. \quad \square$$

Example 2.25. If $(\Omega, \mathcal{F}, \mu)$ is a probability space (see Chapter 3) then $L^p(\Omega; X)$ will denote a Banach space, defined similarly to Theorem 2.23(i), but found by defining the L^p integration with respect to the probability measure μ on Ω, rather than with respect to Lebesgue measure as earlier (see Chapter 3). In the case where we wish to emphasize the measure with respect to which L^p integration is to be understood, and when the space X is clear, we will also sometimes write $L^p(\Omega; \mu)$. If μ is Lebesgue measure then we will sometimes write $L^p(\Omega; \text{Leb})$. $\square$

We will also use the following Banach spaces built from the notion of continuous function.

Definition 2.26. *Let $\mathcal{I}$ denote an open or closed interval in $\mathbb{R}$ and $\Omega \subset \mathbb{R}^d$ an open or closed set. Then:*

- *$C(\mathcal{I}; X)$ will denote the Banach space of continuous functions from $\mathcal{I}$ into a Banach space X;*
- *$C^{p,k}(\overline{\Omega} \times \mathcal{I}, \mathbb{R})$ will denote the space of real-valued functions that are k and p times continuously differentiable in $t \in \mathcal{I}$ and $x \in \Omega$, respectively.*

Example 2.27. Let $T > 0$; then function spaces of the form $L^p((0, T); X)$, $C^{p,k}(\overline{\Omega} \times [0, T]; \mathbb{R})$, and $C([0, T]; X)$ appear in the analysis of evolution PDEs of parabolic and hyperbolic type, see Chapter 7. $\square$

2.4.5 Sobolev Spaces of Periodic Functions

We will use the notation $\mathbb{T}^d$ to denote the d-dimensional unit torus, sometimes calling it the *unit cell*. Functions $f : \mathbb{R}^d \to \mathbb{R}$ that satisfy

$$f(y + e_i) = f(y) \quad \forall y \in \mathbb{R}^d, i \in \{1, \ldots, d\}$$

are called 1-*periodic functions*. Thus we may also view f as a function from $\mathbb{T}^d$ into $\mathbb{R}$. In this section we study some properties of 1-periodic functions $f : \mathbb{T}^d \to \mathbb{R}$.

We will denote by $C^\infty_{per}(\mathbb{T}^d)$ the restriction to $\mathbb{T}^d$ of smooth functions $C^\infty(\mathbb{R}^d)$ that are 1-periodic. Then the space $L^p_{per}(\mathbb{T}^d)$ is defined to be the completion of $C^\infty_{per}(\mathbb{T}^d)$ with respect to the L^p-norm. A similar definition holds for $H^1_{per}(\mathbb{T}^d)$.

The Poincaré inequality does not hold in the space H^1_{per}. It does hold, however, if we remove the constants from this space. With this in mind we define

$$H = \left\{ u \in H^1_{per}(\mathbb{T}^d) \Big| \int_{\mathbb{T}^d} u\, dy = 0 \right\}. \tag{2.4.6}$$

There exists a constant $C_p > 0$ such that

$$\|u\|_{L^2(\mathbb{T}^d)} \leqslant C_p \|\nabla u\|_{L^2(\mathbb{T}^d)} \quad \forall u \in H. \tag{2.4.7}$$

Hence, we can use

$$\|u\|_H = \|\nabla u\|_{L^2(\mathbb{T}^d)} \quad u \in H, \tag{2.4.8}$$

as the norm in H. The dual of H may be shown to comprise all elements of $(H^1_{per}(\mathbb{T}^d))^*$ that are orthogonal to constants:

$$H^* = \left\{ u \in (H^1_{per}(\mathbb{T}^d))^* | \langle u, 1 \rangle_{(H^1_{per})^*, H^1_{per}} = 0 \right\}. \tag{2.4.9}$$

The space

$$L^2(\Omega; L^2(\mathbb{T}^d)) := L^2(\Omega \times \mathbb{T}^d)$$

will be used in our study of periodic homogenization. This is a Hilbert space with inner product

$$(u, v)_{L^2(\Omega \times \mathbb{T}^d)} = \int_\Omega \int_{\mathbb{T}^d} u(x, y) v(x, y)\, dy dx,$$

together with the corresponding norm

$$\|u\|^2_{L^2(\Omega \times \mathbb{T}^d)} = \int_\Omega \int_{\mathbb{T}^d} |u(x, y)|^2 dy dx.$$

On occasion we will need the set of all functions $u \in L^2(\Omega; C_{per}(\mathbb{T}^d))$. By Theorem 2.23 the norm on this space is

$$\|u\|^2_{L^2(\Omega; C_{per}(\mathbb{T}^d))} = \int_\Omega \left(\sup_{y \in \mathbb{T}^d} |u(x, y)| \right)^2 dx.$$

This is a separable Banach space that is dense in $L^2(\Omega \times \mathbb{T}^d)$. In the case where $u \in L^2(\Omega \times \mathbb{T}^d)$ is rapidly varying in the second component, a number of useful results hold. We now detail these.

Theorem 2.28. *Let* $u \in L^2(\Omega; C_{per}(\mathbb{T}^d))$, $\varepsilon > 0$, *and define* $u^\varepsilon(x) = u(x, x/\varepsilon)$. *Then*

(i) $u^\varepsilon \in L^2(\Omega)$ *and* $\|u^\varepsilon\|_{L^2(\Omega)} \leqslant \|u\|_{L^2(\Omega;C_{per}(\mathbb{T}^d))}$.
(ii) $u^\varepsilon(x)$ *converges to* $\int_{\mathbb{T}^d} u(x,y)\,dy$ *weakly in* $L^2(\Omega)$ *as* $\varepsilon \to 0$.
(iii) We have

$$\|u^\varepsilon\|_{L^2(\Omega)} \to \|u\|_{L^2(\Omega\times\mathbb{T}^d)}$$

as $\varepsilon \to 0$.

The results in Theorem 2.28 capture the notion that fast oscillations decouple from slow variation, asymptotically, provided the oscillations are sufficiently regular. This idea is central to the theory of homogenization for PDEs.

The following property of periodic functions will be used in the sequel.

Theorem 2.29. *Let* $p \in [1,\infty]$ *and* $f \in L^p_{per}(\mathbb{T}^d)$. *Set*

$$f^\varepsilon(x) = f\left(\frac{x}{\varepsilon}\right) \quad \text{a.e. on } \mathbb{R}^d.$$

Then, if $p < \infty$, *as* $\varepsilon \to 0$

$$f^\varepsilon \rightharpoonup \int_{\mathbb{T}^d} f(y)\,dy \quad \text{weakly in } L^p(\Omega),$$

for any bounded open subset Ω *of* $\mathbb{R}^d$.

We also have

$$f^\varepsilon \rightharpoonup \int_{\mathbb{T}^d} f(y)\,dy \quad \text{weak-* in } L^\infty(\mathbb{R}^d).$$

2.5 Two-Scale Convergence

A form of weak convergence that is particularly well suited for problems in periodic homogenization is **two-scale convergence**. In this section we define this concept and study some of its basic properties. Once again we consider periodic functions on $\mathbb{T}^d$. As before, Ω denotes a subset of $\mathbb{R}^d$, not necessarily bounded.[4] We start by discussing two-scale convergence for steady (time-independent) problems. We then discuss related issues for time-dependent problems.

2.5.1 Two-Scale Convergence for Steady Problems

Definition 2.30. *Let* u^ε *be a sequence in* $L^2(\Omega)$. *We say that* u^ε two-scale converges *to* $u_0(x,y) \in L^2(\Omega\times\mathbb{T}^d)$, *and write* $u^\varepsilon \overset{2}{\rightharpoonup} u_0$, *if for every test function* $\phi \in L^2(\Omega; C_{per}(\mathbb{T}^d))$ *we have*

$$\lim_{\varepsilon\to 0}\int_\Omega u^\varepsilon(x)\phi\left(x,\frac{x}{\varepsilon}\right)\,dx = \int_\Omega\int_{\mathbb{T}^d} u_0(x,y)\phi(x,y)\,dydx. \tag{2.5.1}$$

[4] Lemma 2.34 as proved applies only for bounded Ω but can be extended to the case where Ω is not bounded.

In many instances it is necessary to test the sequence u^ε against a wider class of test functions. For our purposes it will be sufficient to consider functions of the type that are described in the following lemma.

Lemma 2.31. *Let $u^\varepsilon \in L^2(\Omega)$ be a two-scale convergent sequence. Then the convergence in* (2.5.1) *holds true for all test functions $\phi(x,y)$ of the form $\psi_1(y)\psi_2(x,y)$ with $\psi_1 \in L^\infty_{per}(\mathbb{T}^d)$ and $\psi_2 \in L^2(\Omega; C_{per}(\mathbb{T}^d))$.*

Two-scale convergence implies weak convergence in $L^2(\Omega)$. In particular, we have the following lemma.

Lemma 2.32. *Let u^ε be a sequence in $L^2(\Omega)$ that two-scale converges to $u_0 \in L^2(\Omega \times \mathbb{T}^d)$. Then*

$$u^\varepsilon \rightharpoonup \overline{u}_0 \quad \text{weakly in } L^2(\Omega) \tag{2.5.2}$$

where

$$\overline{u}_0(x) = \int_{\mathbb{T}^d} u_0(x,y)\,dy.$$

Proof. Choose a test function $\phi(x) \in L^2(\Omega)$, independent of y. We use it in (2.5.1) to deduce that

$$\begin{aligned}\lim_{\varepsilon\to 0}\int_\Omega u^\varepsilon(x)\phi(x)\,dx &= \int_\Omega\int_{\mathbb{T}^d} u_0(x,y)\phi(x)\,dydx\\ &= \int_\Omega\left(\int_{\mathbb{T}^d} u_0(x,y)\,dy\right)\phi(x)\,dx\\ &= (\overline{u}_0,\phi)_{L^2(\Omega)}\,.\end{aligned}$$

This holds for every $\phi \in L^2(\Omega)$ and hence u^ε converges to $\overline{u}_0$ weakly in $L^2(\Omega)$. □

An immediate consequence of this lemma is the following.

Corollary 2.33. *Let u^ε be a sequence in $L^2(\Omega)$ that two-scale converges to $u_0 \in L^2(\Omega)$, i.e., the two-scale limit is independent of y. Then the weak L^2-limit and the two-scale limit coincide.*

Two-scale convergence is a useful tool for studying multiscale expansions of the type developed in Chapters 12, 13, and 14, where there is a periodic dependence on the "fast variable". Such rigorous analysis of these expansions is undertaken in Chapters 19, 20 and 21. The next result illustrates the role of two–scale convergence in this context.

Lemma 2.34. *Consider a function $u^\varepsilon \in C(\overline{\Omega})$ of the form*

$$u^\varepsilon(x) = u_0\left(x,\frac{x}{\varepsilon}\right) + \varepsilon u_1\left(x,\frac{x}{\varepsilon}\right),$$

where $u_j \in C(\overline{\Omega}; C_{per}(\mathbb{T}^d))$, $j = 0,1$, Ω, being a bounded open set in $\mathbb{R}^d$. Then $u^\varepsilon \overset{2}{\rightharpoonup} u_0$.

Proof. Let $\phi \in L^2(\Omega; C_{per}(\mathbb{T}^d))$ and define $f_j(x,y) = u_j(x,y)\phi(x,y)$, $j = 0, 1$. We will use the notation $f^\varepsilon(x) = f(x, x/\varepsilon)$. We clearly have

$$\int_\Omega u^\varepsilon(x)\phi\left(x, \frac{x}{\varepsilon}\right)\, dx = \int_\Omega f_0^\varepsilon(x)\, dx + \varepsilon \int_\Omega f_1^\varepsilon(x)\, dx. \tag{2.5.3}$$

Now, $f_j^\varepsilon \in L^2(\Omega; C_{per}(\mathbb{T}^d))$ for $j = 0, 1$. This implies, by Theorem 2.28, that f_0^ε converges to its average over $\mathbb{T}^d$, $\overline{f}_0(x) := \int_{\mathbb{T}^d} f_0(x,y)\, dy$, weakly in $L^2(\Omega)$. From this weak convergence it follows that

$$(\psi, f_0^\varepsilon) \to (\psi, \overline{f}_0)$$

for all $\psi \in L^2(\Omega)$. Choosing $\psi = 1$ gives

$$\begin{aligned}\int_\Omega f_0^\varepsilon(x)\, dx &\to \int_\Omega \int_{\mathbb{T}^d} f_0(x,y)\, dy dx \\ &= \int_\Omega \int_{\mathbb{T}^d} u_0(x,y)\phi(x,y)\, dy dx.\end{aligned}$$

Now consider the second integral on the right-hand side of (2.5.3). By Theorem 2.28 the sequence f_1^ε is weakly convergent in $L^2(\Omega)$. Hence it is bounded in $L^2(\Omega)$ by Theorem 2.7. Thus, again using the boundedness of Ω, together with the Cauchy–Schwarz inequality, we obtain:

$$\varepsilon \Big| \int_\Omega f_1^\varepsilon(x)\, dx \Big| \leqslant \varepsilon C \|f_1^\varepsilon\|_{L^2(\Omega)} \to 0.$$

We use the preceding two calculations in (2.5.3) to obtain:

$$\begin{aligned}\int_\Omega u^\varepsilon(x)\phi\left(x, \frac{x}{\varepsilon}\right)\, dx &= \int_\Omega \Big(f_0^\varepsilon(x) + \varepsilon f_1^\varepsilon(x) \Big) dx \\ &\to \int_\Omega \int_{\mathbb{T}^d} u_0(x,y)\phi(x,y)\, dy dx\end{aligned}$$

for all $\phi \in L^2(\Omega; C_{per}(\mathbb{T}^d))$. Hence, u^ε two-scale converges to u_0. □

We would like to find criteria that enable us to conclude that a given sequence in $L^2(\Omega)$ is two-scale convergent. The following compactness result provides us with such a criterion.

Theorem 2.35. *Let u^ε be a bounded sequence in $L^2(\Omega)$. Then there exists a subsequence, still denoted by u^ε, and function $u_0 \in L^2(\Omega \times \mathbb{T}^d)$, such that u^ε two-scale converges to u_0.*

This result concerns bounded sequences in $L^2(\Omega)$ whose two-scale limit is an element of $L^2(\Omega \times \mathbb{T}^d)$ and depends explicitly on y. It is now natural to ask whether more information on the two-scale limit can be obtained when our sequence is bounded in a stronger norm. Recall the subspace H of mean zero $H^1(\Omega)$ functions defined in (2.4.6).

Theorem 2.36. *(i) Let u^ε be a bounded sequence in $H^1(\Omega)$ and consider a subsequence with weak $H^1(\Omega)$ limit u. Then, along this subsequence, u^ε two-scale converges to its weak-$H^1(\Omega)$ limit u. In addition, there exists a function $u_1 \in L^2(\Omega; H)$ such that, possibly along a further subsequence, ∇u^ε two-scale converges to $\nabla_x u + \nabla_y u_1$.*

(ii) Let u^ε and $\varepsilon \nabla u^\varepsilon$ be bounded sequences in $L^2(\Omega)$. Then, along a subsequence, there exists a function $u_0 \in L^2(\Omega; H)$ such that u^ε and $\varepsilon\nabla u^\varepsilon$ two-scale converge to $u_0(x,y)$ and to $\nabla_y u_0$, respectively.

(iii) Let u^ε and $\varepsilon^\gamma \nabla u^\varepsilon$, $\gamma \in (0,1)$, be bounded sequences in $L^2(\Omega)$. Then the two-scale limit of u^ε is independent of y : there is a function $u_0 \in L^2(\Omega)$ such that $u^\varepsilon \overset{2}{\rightharpoonup} u_0(x)$.

The reason this theorem is so useful is as follows: if we have a function of the form

$$u^\varepsilon(x) = u(x) + \varepsilon u_1(x, x/\varepsilon) + \mathcal{O}(\varepsilon^2),$$

then, under reasonable assumptions on the $\mathcal{O}(\varepsilon^2)$ term, we expect that, for small ε, u^ε will be approximated by $u(x)$; the derivative of u^ε, however, will be approximated by $\nabla u(x) + \nabla_y u_1(x,y)\Big|_{y=x/\varepsilon}$. Examination of item (i) shows that it is set up to deal with precisely such situations.

In item (i) note that the two-scale convergence implies that the subsequence u^ε converges weakly to u in $L^2(\Omega)$, since the limit $u(x)$ depends only on x and not on y (Corollary 2.33). In fact the convergence is strong in L^2, along a further subsequence, by the Rellich compactness theorem. If this limit can be shown to be unique then convergence of the whole sequence u^ε to u occurs weakly in $H^1(\Omega)$ and strongly in $L^2(\Omega)$. In item (iii) we see that having sufficient control on the gradient of u^ε enables us to deduce that the two-scale limit of u is independent of y; this information is not enough to deduce information about the two-scale limit of the gradient, however.

2.5.2 Two-Scale Convergence for Time-Dependent Problems

When studying homogenization problems for evolutionary PDEs it is necessary to modify the concept of two-scale convergence to take into account the time dependence of the sequences of functions we consider. We present the relevant definitions and theorems. As in the previous subsection, we let Ω be a subset of $\mathbb{R}^d$, not necessarily bounded, and $\mathbb{T}^d$ denotes the unit torus. We let $\Omega_T = \Omega \times (0,T)$ and use (x,y,t) to denote a point in $\Omega \times \mathbb{T}^d \times (0,T) = \Omega_T \times \mathbb{T}^d$. Recall that H is given by (2.4.6).

Definition 2.37. *A sequence $u^\varepsilon \in L^2(\Omega_T)$* two-scale converges *to $u_0(x,y,t) \in L^2(\Omega_T \times \mathbb{T}^d)$ and we write $u^\varepsilon \overset{2}{\rightharpoonup} u_0$, if for every test function $\phi(x,y,t) \in L^2(\Omega_T; C_{per}(\mathbb{T}^d))$ we have*

$$\lim_{\varepsilon \to 0} \int_0^T \int_\Omega u^\varepsilon(x,t)\phi\left(x, \frac{x}{\varepsilon}, t\right) dx dt = \int_0^T \int_\Omega \int_{\mathbb{T}^d} u_0(x,y,t)\phi(x,y,t)\, dy dx dt. \tag{2.5.4}$$

Notice that t enters merely as a parameter in the preceding definition. As in the case of time-independent problems, a wider class of test functions than $L^2(\Omega_T; C_{per}(\mathbb{T}^d))$ can be used in (2.5.4). For example, we can use test functions of the form $\psi_1(y)\psi_2(x,y,t)$ $\psi \in L^\infty_{per}(\mathbb{T}^d)$ and $\phi \in L^2(\Omega_T; C_{per}(\mathbb{T}^d))$ in (2.5.4).

The basic compactness theorem of two-scale convergence, Theorem 2.35, is still valid.

Theorem 2.38. *Let u^ε be a bounded sequence in $L^2(\Omega_T)$. Then there exists a subsequence, still denoted by u^ε, and function $u_0 \in L^2(\Omega_T \times \mathbb{T}^d)$ such that u^ε two-scale converges to u_0. Moreover, u^ε converges weakly in $L^2(\Omega_T)$ to the average of the two-scale limit over the unit cell:*

$$u^\varepsilon \rightharpoonup \int_{\mathbb{T}^d} u_0(\cdot, y, \cdot)dy, \quad \textit{weakly in } L^2(\Omega_T).$$

In the steady case we already know that bounds on better spaces provide us with more information on the two-scale limit. The next theorem is the analogue of Theorem 2.36.

Theorem 2.39. *(i) Let u^ε be a bounded sequence in $L^2((0,T), H^1(\Omega))$ and consider a subsequence with weak $L^2((0,T), H^1(\Omega))$ limit u. Then, along this subsequence, u^ε two-scale converges to its weak-$L^2((0,T), H^1(\Omega))$ limit $u(x,t)$. In addition, there exists a function $u_1 \in L^2(\Omega_T; H)$ such that, possibly along a further subsequence, ∇u^ε two-scale converges to $\nabla_x u + \nabla_y u_1$.*

(ii) Let u^ε and $\varepsilon \nabla u^\varepsilon$ be bounded sequences in $L^2(\Omega_T)$. Then, along a subsequence, there exists a function $u_0 \in L^2(\Omega_T; H)$ such that u^ε and $\varepsilon\nabla u^\varepsilon$ two-scale converge to u_0 and to $\nabla_y u_0$, respectively.

(iii) Let u^ε and $\varepsilon^\gamma \nabla u^\varepsilon$, $\gamma \in (0,1)$, be bounded sequences in $L^2(\Omega_T)$. Then the two-scale limit of u^ε is independent of y: there is a function $u_0 \in L^2(\Omega_T)$ such that $u^\varepsilon \overset{2}{\rightharpoonup} u_0$.

The proofs of these two theorems are almost identical to the proofs of the corresponding results from the preceding subsection.

2.6 Equations in Hilbert Spaces

Many PDEs can be expressed conveniently as equations in an appropriate Hilbert space. This is a particularly useful viewpoint for the linear elliptic and parabolic PDEs that we will study in this book. It is consequently useful to develop an abstract formulation for such problems. We summarize this theory here. There are two main components: the Lax-Milgram theorem, which provides existence and uniqueness of solutions for linear equations in Hilbert spaces, and the Fredholm alternative.

2.6.1 Lax-Milgram Theory

Let H be a Hilbert space[5] with inner product $(\cdot,\cdot)$ and let $A : H \to H^*$ be a linear operator. Let $f \in H^*$ and let $\langle\cdot,\cdot\rangle_{H^*,H}$ denote the pairing between H and H^*. We are interested in studying the equation

$$Au = f. \tag{2.6.1}$$

The weak formulation of this equation is

$$(Au, v) = \langle f, v\rangle_{H^*,H} \quad \forall\, v \in H. \tag{2.6.2}$$

The linearity of A implies that the left-hand side of this equation defines a bilinear form $a : H \times H \to \mathbb{R}$ given by

$$a[u, v] = (Au, v).$$

Existence and uniqueness of solutions for equations of the form (2.6.2) can be proved by means of the following theorem.

Theorem 2.40. (Lax-Milgram). *Let H be a Hilbert space with norm $\|\cdot\|$ and inner product $(\cdot,\cdot)$. Let $\langle\cdot,\cdot\rangle_{H^*,H}$ denote the pairing between H^* and H. Let $a : H\times H \to \mathbb{R}$ be a bilinear mapping which satisfies the following properties:*

(i) (Coercivity) There exists a constant $\alpha > 0$ such that

$$a[u, u] \geqslant \alpha\|u\|^2 \quad \forall u \in H.$$

(ii) (Continuity) There exists a constant $\beta > 0$ such that

$$a[u, v] \leqslant \beta\|u\|\|v\| \quad \forall u, v \in H.$$

Now let $f : H \to \mathbb{R}$ be a bounded linear functional on H. Then there exists a unique element $u \in H$ such that

$$a[u, v] = \langle f, v\rangle_{H^*,H}$$

for all $v \in H$.

This theory is enormously powerful. In particular it is central in the theory of weak solutions for elliptic PDEs. That connection will be developed in Chapter 7, and then the setting will be used to prove homogenization theorems for elliptic PDEs in Chapter 19.

[5] Note that here we are talking about a general Hilbert space, not the specific H given by (2.4.6).

2.6.2 The Fredholm Alternative

Assume now that $A : H \to H$ is a linear operator and consider Equation (2.6.1) with $f \in H$. Let $A^* : H \to H$ denote the adjoint of A defined via the identity

$$(Au, v) = (u, A^*v) \quad \forall\, u, v \in H.$$

Now let $v \in H$ denote any element of the null space $\mathcal{N}(A^*)$ of A^* where

$$\mathcal{N}(A^*) = \{v \in H : A^*v = 0\}.$$

Equation (2.6.1) implies that

$$(f, v) = 0 \quad \forall v \in \mathcal{N}(A^*).$$

Consequently, a necessary condition for the existence of a solution for (2.6.1) is that the right-hand side of this equation is orthogonal to the null space of the adjoint operator of A. (Note that coercivity fails if $\mathcal{N}(A^*)$ is nontrivial and so the Lax-Milgram theorem does not apply).

The formal argument applies in finite dimensions and can be made rigorous in the infinite-dimensional case when A is a compact perturbation of the identity: $A = I - K$, with K compact. Thus the following definition is important.

Definition 2.41. *A bounded operator $K : H \to H$ is compact if it maps bounded sets into sets with compact closure:*

$$\overline{K(M)} \quad \textit{is compact in } H \textit{ for all bounded } M \subset H.$$

Equivalently, K is compact if and only if for every bounded sequence $\{u_n\}_{n=1}^{\infty} \in H$, the sequence $\{Ku_n\}_{n=1}^{\infty}$ has a strongly convergent subsequence in H.

We now study Equation (2.6.1) in the case where $A = I - K$.

Theorem 2.42. (Fredholm alternative) *Let H be a Hilbert space and let $K : H \to H$ be a compact operator. Then the following alternative holds.*

(i) Either the equations

$$(I - K)u = f, \tag{2.6.3a}$$

$$(I - K^*)U = F, \tag{2.6.3b}$$

have unique solutions for every f, $F \in H$; or

(ii) the homogeneous equations

$$(I - K)V_0 = 0, \quad (I - K^*)v_0 = 0$$

have the same finite number of nontrivial solutions:

$$1 \leqslant \mathit{dim}\,(\mathcal{N}(I - K)) = \mathit{dim}\,(\mathcal{N}(I - K^*)) < \infty.$$

In this case equations (2.6.3a) *and* (2.6.3b) *have a solution if and only if*

$$(f, v_0) = 0 \quad \forall v_0 \in \mathcal{N}(I - K^*)$$

and

$$(F, V_0) = 0 \quad \forall V_0 \in \mathcal{N}(I - K),$$

respectively.

This theory is particularly useful in the study of elliptic boundary value problems. It will be developed in that context in Chapter 7. It will then be used throughout Part II of the book.

2.7 Discussion and Bibliography

The index-free notation that we employ in this book is described in numerous texts on continuum mechanics; for example, see [128], chapters 1 and 2, and the references therein. The index-free notation leads to clean statements of results and derivation of formulae. However, the reader should not be afraid to dive into indices in order to verify many of the line-by-line calculations; as with any index-free presentation, one line of index-free mathematics can hide several lines of calculations in index notation.

Most of the remaining material presented in this chapter is also very standard and can be found in a variety of books on functional analysis and PDEs. For more information on Banach, Hilbert, and L^p-spaces see, for example, [53, 179, 189, 274, 337].

Since $L^1(\Omega)$ is not reflexive, a bounded sequence in $L^1(\Omega)$ does not necessarily have any weakly convergent subsequences. It is natural to ask for conditions under which we can extract a weakly convergent subsequence from a sequence in $L^1(\Omega)$. The answer to this question is given through the **Dunford–Pettis theorem**: in addition to boundedness we also need *equi-integrability*. We refer to [87] for details. The nonreflexivity of $L^1(\Omega)$ implies that this space cannot be characterized as the dual of a Banach space, and hence weak–* convergence is not a useful concept for this space. Weak–* convergence becomes, however, an extremely important concept for $(C_b(\Omega))^* =: M(\Omega)$, the space of *Radon measures* on Ω.[6] In fact, a bounded sequence in $L^1(\Omega)$ is weak–* compact in $M(\Omega)$: we can extract a weakly convergent subsequence that converges to an element of $M(\Omega)$. Probabilists refer to weak–* convergence in $M(\Omega) = (C_b(\Omega))^*$ as **weak convergence of probability measures**. This is the most useful (and natural) concept for limit theorems in probability theory and is the topic of Section 3.5 in the next chapter. The interested reader may also consult [37, 156].

Sobolev spaces of periodic functions can be defined using Fourier series. For example, the space $H^1_{per}(\mathbb{T}^d)$ can be defined as

[6] This space is larger than $L^1(\Omega)$, which is a proper subset of $M(\Omega)$.

$$H^1_{per}(\mathbb{T}^d) = \left\{ u \, : \, u = \sum_{k \in \mathbb{Z}^d} u_k e^{2\pi i k \cdot x}, \quad u_k = u_{-k}, \; \sum_{k \in \mathbb{Z}^d} |k|^2 |u_k|^2 < \infty \right\}.$$

Similarly we have

$$H = \left\{ u \in H^1_{per}(\mathbb{T}^d) \, : \, u_0 = 0 \right\}.$$

Sobolev spaces of periodic functions are discussed in [274, 315]. An exhaustive treatment of Sobolev spaces can be found in [4]. For many applications to the theory of PDEs, the material presented in [98, ch. 5] is sufficient.

The concept of two-scale convergence was introduced by Nguetseng [233, 234] and later popularized and developed further by Allaire [5, 6]. Most of the results presented in Section 2.5 are taken from [6]. Recent review articles on two-scale convergence are [197, 341].

A natural question that arises is what is the largest set of test functions that we can take in (2.5.1). This question is related to the concept of the set of *admissible test functions* , defined to be the set of $\phi \in L^2(\Omega \times \mathbb{T}^d)$ for which

$$\lim_{\varepsilon \to 0} \int_\Omega \left| \phi \left(x, \frac{x}{\varepsilon} \right) \right|^2 \, dx = \int_\Omega \int_{\mathbb{T}^d} |\phi(x, y)|^2 \, dy dx. \tag{2.7.1}$$

Notice that, by Theorem 2.28, any $\phi \in L^2(\Omega; C_{per}(\mathbb{T}^d))$ is admissible. The functions considered in Lemma 2.31 are also admissible. It is important to note, however, that the set of admissible test functions is a proper subset of $L^2(\Omega \times \mathbb{T}^d)$; there are elements of $L^2(\Omega \times \mathbb{T}^d)$ which do not satisfy (2.7.1)—see [6] for an example. A certain amount of regularity, in either the x or y variable, is required. A characterization of the set of test functions that we can use in (2.5.1) is given in [142, Theorem 2.3].

If a two-scale convergent sequence $u\,(x, x/\varepsilon)$ is also an admissible test function according to (2.7.1) and, furthermore, the two-scale limit $u_0 \in L^2(\Omega; C_{per}(\Omega))$, then u^ε is strongly two-scale convergent in the sense that

$$\lim_{\varepsilon \to 0} \left\| u^\varepsilon(\cdot) - u_0 \left(\cdot, \frac{\cdot}{\varepsilon} \right) \right\| = 0.$$

Notice however that the two-scale limit will not in general possess any further regularity. In fact, every function u_0 in $L^2(\Omega \times \mathbb{T}^d)$ is attained as a two-scale limit of some sequence in $L^2(\Omega)$ [6, lem. 1.13]. A discussion concerning interrelations among strong and weak two-scale convergence can be found in [341].

In Section 2.5.2 we considered sequences of functions that do not oscillate in time. The concept of two-scale convergence has been extended to cover the case of sequences of the form

$$u^\varepsilon(x, t) = u \left(x, \frac{x}{\varepsilon}, t, \frac{t}{\varepsilon^p} \right),$$

where $p > 0$ and $u(x, y, t, \tau)$ is periodic in both y and τ; see [142]. This extension is useful when studying homogenization problems for parabolic PDEs with rapidly oscillating coefficients in both space and time. The concept of two-scale convergence has also been extended to cover the case of functions that depend on more than two scales, i.e.,

$$u^{\varepsilon} = u\left(x, \frac{x}{\varepsilon}, \frac{x}{\varepsilon^2}, \dots\right).$$

See [8]. A concept similar to that of two-scale convergence has also been developed for nonperiodic oscillations; see [213].

The use of appropriately chosen test functions to study asymptotic problems for PDEs is a standard technique. See, for example, [95]. A very similar technique to that of two-scale convergence was introduced by Kurtz [181] in the probabilistic context. The approach is taken further in the perturbed test function method of Evans [96, 97].

Lax-Milgram theory is covered in [98]. The Fredholm alternative is absolutely central to the developments in Part II of this book. In this context it is of interest to note that it holds in normed spaces. That is, neither completeness nor the inner product structure are necessary; see, for instance, [179, sec. 8.7].

2.8 Exercises

1. Show that, for any scalar field ϕ,

$$\Delta\phi = I : \nabla\nabla\phi.$$

2. Let $\{X, \|\cdot\|\}$ be a Banach space. Show that every strongly convergent sequence is weakly convergent.
3. Let $\{X, \|\cdot\|\}$ be a finite-dimensional Banach space. Show that every weakly convergent sequence is strongly convergent.
4. Let $\Omega = (0,1) \subset \mathbb{R}$. Define

$$u(x) = \begin{cases} x & : \quad \text{for } 0 \leqslant x \leqslant \frac{1}{2}, \\ 1 - x & : \quad \text{for } \frac{1}{2} < x \leqslant 1. \end{cases}$$

Show that the weak derivative of $u(x)$ is

$$\frac{du}{dx}(x) = \begin{cases} 1 & : \quad \text{for } 0 \leqslant x \leqslant \frac{1}{2}, \\ -1 & : \quad \text{for } \frac{1}{2} < x \leqslant 1. \end{cases}$$

Is this function differentiable in the classical sense?
5. Consider the function u from the previous exercise. Show that $u \in H_0^1(\Omega)$.
6. Recall H defined in (2.4.6). Use Fourier series to prove that the Poincaré inequality holds in H:

$$\|u\|_{L^2} \leqslant \frac{1}{2\pi}\|\nabla u\|_{L^2} \quad \forall u \in H.$$

7. Let $\Omega = (0,1) \subset \mathbb{R}$ and define $u_\alpha(x) = |x|^\alpha$. For what values of α does $u_\alpha \in H^1(\Omega)$?

8. Let $L > 0$. Prove the Poincaré inequality of Theorem 2.21 for a function $f \in C^\infty(0, L) \bigcap H_0^1(0, L)$. Estimate the optimal value of the Poincaré constant C_L. Show that

$$\lim_{L \to \infty} C_L = \infty.$$

Interpret this result.

9. Prove that the norm (2.4.4) is equivalent to the norm (2.4.2).

10. Consider a function $u^\varepsilon \in L^2(\Omega)$ that admits the following two-scale expansion

$$u^\varepsilon(x) = \sum_{j=0}^{N} \varepsilon^j u_j\left(x, \frac{x}{\varepsilon}\right),$$

where $u_j \in C(\Omega; C_{per}(Y))$, $j = 0, 1, \ldots, N$, and Ω is a bounded domain in $\mathbb{R}^d$. Show that $u^\varepsilon \stackrel{2}{\rightharpoonup} u_0$ (this is a generalization of Lemma 2.34).

11. Consider the systems of linear equations (1.3.5) and (1.3.8). Assume that $\|(\mathcal{L}^\varepsilon)^{-1}\|$ is bounded independently of ε. Show that, in both cases considered,

$$\|u^\varepsilon - u_0\| \leqslant C\varepsilon.$$

3

Probability Theory and Stochastic Processes

3.1 Setup

In this chapter we present some basic definitions and results from probability theory and from the theory of stochastic processes. We define the Wiener process (Brownian motion) and develop the Itô theory of stochastic integration. We summarize the basic properties of martingales and apply these to Itô integrals. When studying dimension reduction for Markovian problems we will often work in the context of weak convergence of probability measures. We will see in later chapters that averaging and homogenization for Markov chains and SDEs are essentially limit theorems for stochastic processes and that they are intimately related to the theory of weak convergence of probability measures in metric spaces. Thus we will discuss various forms of probabilistic convergence, and of weak convergence in particular. As in Chapter 2, it is not recommended that this chapter be read in its entirety on a first read through the book; rather, it should be referred to as needed when reading later chapters.

Section 3.2 introduces ideas from probability theory, and Section 3.3 does the same for stochastic processes. Section 3.4 discusses martingales and stochastic integrals in particular. Weak convergence is developed in Section 3.5. Extensions of the theory and references to the literature are presented in Section 3.6.

3.2 Probability, Expectation, and Conditional Expectation

A collection of subsets of a set Ω is called a σ–algebra if it contains Ω and is closed under the operations of taking complements and countable unions of its elements. A sub-σ–algebra is a collection of subsets from a σ–algebra, which itself satisfies the axioms of a σ–algebra.

A *measurable space* is a pair $(\Omega, \mathcal{F})$ where Ω is a set and $\mathcal{F}$ is a σ–algebra of subsets of Ω. Let $(\Omega, \mathcal{F})$ and $(E, \mathcal{G})$ be two measurable spaces. A function $X : \Omega \mapsto E$ such that the *event*

$$\{\omega \in \Omega : X(\omega) \in A\} =: \{X \in A\}$$

belongs to $\mathcal{F}$ for arbitrary $A \in \mathcal{G}$ is called a *measurable function* or *random variable*.

Let $(\Omega, \mathcal{F})$ be a measurable space. A function $\mu : \mathcal{F} \mapsto [0,1]$ is called a *probability measure* if $\mu(\emptyset) = 1$, $\mu(\Omega) = 1$, and $\mu(\cup_{k=1}^{\infty} A_k) = \sum_{k=1}^{\infty} \mu(A_k)$ for all sequences of pairwise disjoint sets $\{A_k\}_{k=1}^{\infty} \in \mathcal{F}$. The triplet $(\Omega, \mathcal{F}, \mu)$ is called a *probability space*. Let X be a measurable function (random variable) from $(\Omega, \mathcal{F}, \mu)$ to $(E, \mathcal{G})$. If E is a Banach space then we may define *expectation* with respect to the measure μ by[1]

$$\mathbb{E}X = \int_{\Omega} X(\omega)\, d\mu(\omega).$$

More generally, let $f : E \mapsto \mathbb{R}$ be $\mathcal{G}$–measurable. Then

$$\mathbb{E}f(X) = \int_{\Omega} f(X(\omega))\, d\mu(\omega).$$

Let U be a topological space. We will use the notation $\mathcal{B}(U)$ to denote the Borel σ–algebra of U: the smallest σ–algebra containing all open sets of U. Every random variable from a probability space $(\Omega, \mathcal{F}, \mu)$ to a measurable space $(E, \mathcal{B}(E))$ induces a probability measure on E:

$$\mu_X(B) = \mathbb{P}X^{-1}(B) = \mu(\omega \in \Omega; X(\omega) \in B), \quad B \in \mathcal{B}(E).$$

The measure μ_X is called the *distribution* (or sometimes the *law*) of X.

Example 3.1. Let $\mathcal{I}$ denote a subset of the positive integers. A vector $\rho_0 = \{\rho_{0,i},\ i \in \mathcal{I}\}$ determines a measure on $\mathcal{I}$ (the distribution of an $\mathcal{I}$-valued random variable) if it has nonnegative entries and its total mass equals 1: $\sum_{i \in \mathcal{I}} \rho_{0,i} = 1$. □

We can use the distribution of a random variable to compute expectations and probabilities:

$$\mathbb{E}f(X) = \int_E f(x)\, d\mu_X(x)$$

and

$$\mathbb{P}(X \in G) = \int_G d\mu_X(x), \quad G \in \mathcal{B}(E).$$

A collection of independent and identically distributed random variables are termed *i.i.d.*

When $E = \mathbb{R}^d$ and the measure μ_X has Radon-Nikodym derivative with respect to Lebesgue measure, then we can write

$$d\mu_X(x) = \rho(x)\, dx.$$

[1] Integration with respect to measure μ is defined by means of a limiting procedure, starting from a definition appropriate for step functions.

We refer to $\rho(x)$ as the *probability density function* (pdf), or *density with respect to Lebesgue measure* for X. Furthermore, when $E = \mathbb{R}^d$, then by $L^p(\Omega; \mathbb{R}^d)$, or sometimes $L^p(\Omega; \mu)$, or even simply $L^p(\mu)$, we mean the Banach space of measurable functions on Ω with norm

$$\|X\|_{L^p} = \left(\mathbb{E}|X|^p\right)^{1/p}.$$

Example 3.2. i) Consider the random variable $X : \Omega \mapsto \mathbb{R}$ with pdf

$$\gamma_{\sigma,m}(x) := (2\pi\sigma^2)^{-\frac{1}{2}} \exp\left(-\frac{(x-m)^2}{2\sigma^2}\right).$$

Such an X is termed a **Gaussian** or **normal** random variable. The mean is

$$\mathbb{E}X = \int_{\mathbb{R}} x\gamma_{\sigma,m}(x)\,dx =: m$$

and the variance is

$$\mathbb{E}(X-m)^2 = \int_{\mathbb{R}} (x-m)^2\gamma_{\sigma,m}(x)\,dx =: \sigma^2.$$

Since the mean and variance completely specify a Gaussian random variable on $\mathbb{R}$, the Gaussian is commonly denoted by $\mathcal{N}(m, \sigma^2)$. The **standard normal** random variable is $\mathcal{N}(0, 1)$.

ii) Let $m \in \mathbb{R}^d$ and $\Sigma \in \mathbb{R}^{d\times d}$ be symmetric and positive definite. The random variable $X : \Omega \mapsto \mathbb{R}^d$ with pdf

$$\gamma_{\Sigma,m}(x) := \left((2\pi)^d \mathrm{det}\Sigma\right)^{-\frac{1}{2}} \exp\left(-\frac{1}{2}\langle \Sigma^{-1}(x-m), (x-m)\rangle\right)$$

is termed a **multivariate Gaussian** or **normal** random variable. The mean is

$$\mathbb{E}X = m \tag{3.2.1}$$

and the covariance matrix is

$$\mathbb{E}\Big((X-m)\otimes(X-m)\Big) = \Sigma. \tag{3.2.2}$$

Since the mean and covariance matrix completely specify a Gaussian random variable on $\mathbb{R}^d$, the Gaussian is commonly denoted by $\mathcal{N}(m, \Sigma)$.

iii) The case of a **degenerate Gaussian**, where Σ is only positive-semi-definite, is also sometimes required. The eigenspace of Σ corresponding to zero eigenvalues determines coordinate directions in which the probability measure is nonrandom – a **Dirac mass**. □

Example 3.3. An exponential random variable $T : \Omega \to \mathbb{R}^+$ with rate $\lambda > 0$ satisfies

$$\mathbb{P}(T > t) = e^{-\lambda t}, \quad \forall t \geqslant 0.$$

We write $T \sim \exp(\lambda)$. The related pdf is

$$f_T(t) = \begin{cases} \lambda e^{-\lambda t}, & t \geqslant 0, \\ 0, & t < 0. \end{cases} \tag{3.2.3}$$

Notice that

$$\mathbb{E}T = \int_{-\infty}^{\infty} t f_T(t) dt = \frac{1}{\lambda} \int_0^{\infty} (\lambda t) e^{-\lambda t} d(\lambda t) = \frac{1}{\lambda}.$$

If the times $\tau_n = t_{n+1} - t_n$ are i.i.d. random variables with $\tau_0 \sim \exp(\lambda)$ then, for $t_0 = 0$,

$$t_n = \sum_{k=0}^{n-1} \tau_k,$$

and it is possible to show (see Exercise 3.3) that

$$\mathbb{P}(0 \leqslant t_k \leqslant t < t_{k+1}) = \frac{e^{-\lambda t}(\lambda t)^k}{k!}. \tag{3.2.4}$$

An exponential random variable with mean $1/\lambda$, or rate λ, is denoted $\exp(\lambda)$. □

We use the notation $\sim$ to mean *distributed as*. Thus $X \sim \exp(\lambda)$ means that X is an exponential random variable with mean $1/\lambda$. Similarly $X \sim \mathcal{N}(0, \Sigma)$ is a mean zero Gaussian with covariance matrix Σ.

Assume that $\mathbb{E}|X| < \infty$ and let $\mathcal{G}$ be a sub–σ–algebra of $\mathcal{F}$. The **conditional expectation** of X with respect to $\mathcal{G}$ is defined to be the function $\mathbb{E}[X|\mathcal{G}] : \Omega \mapsto E$, which is $\mathcal{G}$–measurable and satisfies

$$\int_G \mathbb{E}[X|\mathcal{G}]\, d\mu = \int_G X\, d\mu \quad \forall G \in \mathcal{G}.$$

We can define $\mathbb{E}[f(X)|\mathcal{G}]$ and the conditional probability $\mathbb{P}[X \in F|\mathcal{G}] = \mathbb{E}[I_F(X)|\mathcal{G}]$, where I_F is the indicator function of F,[2] in a similar manner.

3.3 Stochastic Processes

Let T be an ordered set. A **stochastic process** is a collection of random variables $X = \{X_t; t \in T\}$ where, for each fixed $t \in T$, X_t is a random variable from $(\Omega, \mathcal{F})$ to $(E, \mathcal{G})$. The measurable space $\{\Omega, \mathcal{F}\}$ is called the **sample space**. The space $(E, \mathcal{G})$ is called the **state space**. In this book we will take the set T to be either $\mathbb{R}^+$ or $\mathbb{Z}^+$[3] Thus we will assume that all elements of T are nonnegative. The state space E will often be $\mathbb{R}^d$ equipped with the σ–algebra of Borel sets or, on some occasions, $\mathbb{T}^d$ or a subset $\mathcal{I}$ of the positive integers. When the ordered set T is clear

[2] This function is one on the set F and zero elsewhere.
[3] We use the convention $\mathbb{R}^+ = [0, \infty)$ and $\mathbb{Z}^+ = \{0, 1, 2, \cdots\}$.

from the context we will sometimes write $\{X_t\}$ rather than $\{X_t; t \in T\}$. Notice that X may be viewed as a function of both $t \in T$ and $\omega \in \Omega$. It is therefore sometimes convenient to write $X(t), X(t,\omega)$ or $X_t(\omega)$ instead of X_t.

The **finite dimensional distributions** of a stochastic process are the E^k-valued random variables $(X(t_1), X(t_2), \dots, X(t_k))$ for arbitrary positive integer k and arbitrary times $t_i \in T$, $i \in \{1, \dots, k\}$. A process is called **stationary** if all such collections of random variables are equal in distribution when translated in time: for any integer k and times $t_i \in T$, the distribution of $(X(t_1), X(t_2), \dots, X(t_k))$ is equal to that of $(X(s+t_1), X(s+t_2), \dots, X(s+t_k))$ for any s such that $s+t_i \in T$ for all $i \in \{1, \dots, k\}$.

Definition 3.4. *A* Gaussian process *is one for which $E = \mathbb{R}^d$ and all the finite dimensional distributions are (possibly degenerate) Gaussian.*

A Gaussian process $x(t)$ is charaterized by its mean

$$m(t) := \mathbb{E}x(t)$$

and the covariance function

$$C(t,s) = \mathbb{E}\Big(\big(x(t) - m(t)\big) \otimes \big(x(s) - m(s)\big)\Big).$$

Note that, for fixed $t, s \in T$, $m(t) \in \mathbb{R}^d$ and $C(t,s) \in \mathbb{R}^{d \times d}$.

The most important continuous-time stochastic process is **Brownian motion**. We define it now, first in one dimension and then in arbitrary finite dimensions d.

Definition 3.5. *i) A one-dimensional standard* Brownian motion $W(t) : \mathbb{R}^+ \to \mathbb{R}$ *is a real-valued stochastic process with the following properties:*

a) $W(0) = 0$;
b) $W(t)$ is continuous;
c) $W(t)$ has increments $W(t) - W(s)$ that are independent on nonoverlapping intervals. Furthermore, for every $t > s \geqslant 0$ $W(t) - W(s)$ has a Gaussian distribution with mean 0 and variance $t - s$. That is, the density of the random variable $W(t) - W(s)$ is

$$g(x; t, s) = \Big(2\pi(t-s)\Big)^{-\frac{1}{2}} \exp\left(-\frac{x^2}{2(t-s)}\right); \tag{3.3.1}$$

ii) A d-dimensional standard Brownian motion $W(t) : \mathbb{R}^+ \to \mathbb{R}^d$ is a collection of d independent one-dimensional Brownian motions:

$$W(t) = (W_1(t), \dots, W_d(t)),$$

where $W_i(t), i = 1, \dots, d$ are independent one-dimensional Brownian motions. The density of the Gaussian random vector $W(t) - W(s)$ is thus

$$g(x; t, s) = \Big(2\pi(t-s)\Big)^{-d/2} \exp\left(-\frac{\|x\|^2}{2(t-s)}\right).$$

Brownian motion is sometimes referred to as the **Wiener process**. It is a Gaussian process. Notice that, for the d-dimensional Brownian motion, and for I the $d \times d$ dimensional identity, we have (see (3.2.1) and (3.2.2))

$$\mathbb{E}W(t) = 0 \quad \forall t \geqslant 0$$

and

$$\mathbb{E}\Big((W(t) - W(s)) \otimes (W(t) - W(s))\Big) = (t-s)I. \tag{3.3.2}$$

Moreover,

$$\mathbb{E}\Big(W(t) \otimes W(s)\Big) = \min(t,s)I. \tag{3.3.3}$$

Another fundamental continuous-time process is the **Poisson process**, which we now define. Notice the connection to exponential random variables via (3.2.4).

Definition 3.6. *The Poisson process with intensity λ, denoted by $N(t)$, is an integer-valued, continuous-time, stochastic process with independent increments satisfying*

$$\mathbb{P}[(N(t) - N(s)) = k] = \frac{e^{-\lambda(t-s)}\big(\lambda(t-s)\big)^k}{k!}, \qquad t > s \geqslant 0,\ k \in \mathbb{N}.$$

Both Brownian motion and the Poisson process are *homogeneous* (or *time-homogeneous*): the law of the increments between successive times s and t depend only on $t - s$.

Let $(\Omega, \mathcal{F}, \mu)$ be a probability space, (E, ρ) a metric space, and let $T = [0, \infty)$. Let $\{X_t\}$ be a stochastic process from $(\Omega, \mathcal{F}, \mu)$ to (E, ρ) with continuous sample paths. In other words, for all $\omega \in \Omega$ we have that $X_t \in C_E := C([0,\infty); E)$. The space of continuous functions C_E is called the *path space* of the stochastic process. We can put a metric on C_E as follows:

$$\rho_E(X^1, X^2) := \sum_{n=1}^{\infty} \frac{1}{2^n} \max_{0 \leqslant t \leqslant n} \min\big(\rho(X_t^1, X_t^2), 1\big).$$

We can then define the Borel sets on C_E, using the topology induced by this metric, and $\{X_t\}$ can be thought of as a random variable on $(\Omega, \mathcal{F}, \mu)$ with state space $(C_E, \mathcal{B}(C_E))$. The probability measure $\mu X_{\cdot}^{-1}$ on $(C_E, \mathcal{B}(C_E))$ is called the *law* of $\{X_t\}$. Notice that the law of a stochastic process is a probability measure on its path space. In many instances the study of limit theorems for stochastic processes reduces to the analysis of sequences of probability measures on path space.

Example 3.7. The space of continuous functions C_E is the path space of Brownian motion (the Wiener process). The law of Brownian motion, that is, the measure that it induces on $C([0,\infty), \mathbb{R}^d)$, is known as the **Wiener measure**. □

Although Brownian motion is continuous, many stochastic processes do not have continuous sample paths and their path space is larger than the space of continuous functions C_E. We denote by D_E the space of right continuous processes $X_t : [0,\infty) \mapsto E$ with left limits.[4] The space D_E is usually called the space of

[4] That is, for all $t \in [0,\infty)$, $\lim_{s \to t+} X_s = X_t$ and $\lim_{s \to t-} X_s := X_{t-}$ exist.

càdlàg functions[5] and the stochastic processes whose sample paths are in D_E are called *càdlàg processes*. The space D_E plays a useful role in many limit theorems for stochastic processes.

Let $(\Omega, \mathcal{F})$ be a measurable space and T an ordered set. Let $X = X_t(\omega)$ be a stochastic process from the sample space $(\Omega, \mathcal{F})$ to the state space $(E, \mathcal{G})$. It is a function of two variables, $t \in T$ and $\omega \in \Omega$. For a fixed $\omega \in \Omega$ the function $t \mapsto X_t(\omega)$, $t \in T$ is the *sample path* of the process X associated with ω. Let $\mathcal{K}$ be a collection of subsets of Ω. The smallest σ–algebra on Ω that contains $\mathcal{K}$ is denoted by $\sigma(\mathcal{K})$ and is called the *σ–algebra generated* by $\mathcal{K}$. Let $X_t : \Omega \mapsto E$, $t \in T$. The smallest σ–algebra $\sigma(X_t,\ t \in T)$, such that the family of mappings $\{X_t,\ t \in T\}$ is a stochastic process with sample space $(\Omega, \sigma(X_t,\ t \in T))$ and state space $(E, \mathcal{G})$, is called the *σ–algebra generated by* $\{X_t,\ t \in T\}$.

A *filtration* on $(\Omega, \mathcal{F})$ is a nondecreasing family $\{\mathcal{F}_t, t \in T\}$ of sub–σ–algebras of $\mathcal{F}$: $\mathcal{F}_s \subseteq \mathcal{F}_t \subseteq \mathcal{F}$ for $s \leqslant t$. We set $\mathcal{F}_\infty = \sigma(\cup_{t \in T} \mathcal{F}_t)$. The *filtration generated by* X_t, where X_t is a stochastic process, is

$$\mathcal{F}_t^X := \sigma\left(X_s; s \leqslant t\right).$$

We say that a stochastic process $\{X_t; t \in T\}$ is *adapted* to the filtration $\{\mathcal{F}_t\} := \{\mathcal{F}_t, t \in T\}$ if for all $t \in T$, X_t is an $\mathcal{F}_t$-measurable random variable.

Definition 3.8. *Let $\{X_t\}$ be a stochastic process defined on a probability space $(\Omega, \mathcal{F}, \mu)$ with values in E and let $\{\mathcal{F}_t^X\}$ be the filtration generated by $\{X_t\}$. Then $\{X_t\}$ is a* Markov process *if*

$$\mathbb{P}(X_t \in \Gamma | \mathcal{F}_s^X) = \mathbb{P}(X_t \in \Gamma | X_s) \tag{3.3.4}$$

for all $t, s \in T$ with $t \geqslant s$, and $\Gamma \in \mathcal{B}(E)$.

Roughly speaking, the statistics of X_t for $t \geqslant s$ are completely determined once X_s is known; information about X_t for $t < s$ is superfluous; in other words a Markov process has no memory.

We postulate the existence of a function $P : T \times T \times E \times \mathcal{B}(E) \to \mathbb{R}^+$ satisfying the *Chapman–Kolmogorov* equation

$$\int_E P(s,t,x,dy) P(t,v,y,\Gamma) = P(s,v,x,\Gamma), \tag{3.3.5}$$

for all $x \in E$, $\Gamma \in \mathcal{B}(E)$, and $s, t, v \in T$ with $s \leqslant t \leqslant v$. A stochastic process $\{X_t\}$ is Markov with *transition function* P if P satisfies the Chapman–Kolmogorov equation and

$$\mathbb{P}\left(X_t \in \Gamma | \mathcal{F}_s^X\right) = P(s,t,X_s,\Gamma) \quad \text{a.s.},$$

for all $t, s \in T$ with $t \geqslant s$ and all $\Gamma \in \mathcal{B}(E)$.

A Markov process is homogeneous[6] if

[5] From the French: continue à droite, limites à gauche.

[6] See the discussion following Definition 3.6.

$$P(s,t,x,\Gamma) = P(0,t-s,x,\Gamma).$$

In this case we simplify the notation by setting $P(0,t,\cdot,\cdot) = P(t,\cdot,\cdot)$ and the Chapman–Kolmogorov equation becomes

$$\int_E P(s,x,dy)P(t,y,\Gamma) = P(t+s,x,\Gamma). \tag{3.3.6}$$

Example 3.9. The one-dimensional Brownian motion is a homogeneous Markov process. The transition function is the Gaussian defined in Example 3.2:

$$P(t,x,dy) = \gamma_{t,x}(y)dy. \quad \square$$

Example 3.10. The Poisson process is a homogeneous Markov process. □

Let (E,ρ) be a metric space and let $\{X_t\}$ be an E-valued homogeneous Markov process. Assume that $T = \mathbb{R}^+$. Define the one-parameter family of operators $\mathcal{T}(t)$ through

$$\mathcal{T}(t)f(x) = \int f(y)P(t,x,dy) = \mathbb{E}\left(f(X_t)|X_0 = x\right)$$

for all $f \in C_b(E)$. We assume for simplicity that $\mathcal{T}(t) : C_b(E) \to C_b(E)$. Then $\mathcal{T}(t)$ forms a semigroup of operators on $C_b(E)$. We define by $\mathcal{D}(\mathcal{L})$ the set of all $f \in C_b(E)$ such that the strong limit in $C_b(E)$

$$\mathcal{L}f = \lim_{t\to 0} \frac{\mathcal{T}(t)f - f}{t},$$

exists. The operator $\mathcal{L} : \mathcal{D}(\mathcal{L}) \to C_b(E)$ is called the **infinitesimal generator** of the operator semigroup $\mathcal{T}(t)$. Relatedly we have the following definition.

Definition 3.11. *The operator $\mathcal{L} : C_b(E) \to C_b(E)$ defined earlier is called the* generator *of the Markov process $\{X_t\}$.*

The space $C_b(E)$ is natural in a probabilistic context, but other Banach spaces often arise in applications; in particular when there is a measure μ on E, the spaces $L^p(E;\mu)$ sometimes arise. (See Chapters 4 and 5.) However, in this chapter we work in $C_b(E)$ for ease of presentation.

The generator is frequently taken as the starting point for the definition of a homogeneous Markov process. Conversely, let $\mathcal{T}(t)$ be a *contraction semigroup*[7] with $\mathcal{D}(\mathcal{T}(t)) \subset C_b(E)$, closed. Then, under mild technical hypotheses, there is an E-valued homogeneous Markov process $\{X_t\}$ associated with $\mathcal{T}(t)$ satisfying

$$\mathbb{E}\left(f(X(t))|\mathcal{F}_s^X\right) = \mathcal{T}(t-s)f(X(s))$$

for all $t, s \in T$ with $t \geqslant s$ and $f \in \mathcal{D}(\mathcal{T}(t))$.

[7] Let X be a Banach space and let $T : X \to X$ be a bounded operator. Then T is called a *contraction* provided that $\|Tf\|_X \leqslant \|f\|_X \ \forall f \in X$.

Example 3.12. Consider again the one-dimensional Brownian motion of Example (3.9). The semigroup associated to the standard Brownian motion is the heat semigroup $\mathcal{T}(t) = \exp(\frac{t}{2}\frac{d^2}{dx^2})$. The generator of this Markov process is $\frac{1}{2}\frac{d^2}{dx^2}$. □

A very important concept in the study of limit theorems for stochastic processes is that of *ergodicity*. This concept, in the context of Markov processes, provides us with information on the long-time behavior of a Markov semigroup.

Definition 3.13. *A Markov process is called* **ergodic** *if the equation*

$$\mathcal{T}(t)g = g, \quad g \in C_b(E) \quad \forall t \geqslant 0$$

has only constant solutions.

Roughly speaking ergodictiy corresponds to the case where the semigroup $\mathcal{T}(t)$ is such that $\mathcal{T}(t) - I$ has only constants in its null space or, equivalently, to the case where the generator $\mathcal{L}$ has only constants in its null space. However, the choice of function space is a subtle issue when formulating ergodicity in this way, especially for deterministic dynamics (see Chapter 4).

Under some additional compactness assumptions, an ergodic Markov process has an *invariant measure* μ on E with the property that, in the case $T = \mathbb{R}^+$,

$$\lim_{t \to +\infty} \frac{1}{t} \int_0^t g(X_s)\, ds \to \mathbb{E} g(x),$$

where $\mathbb{E}$ denotes the expectation with respect to μ. Various forms of convergence are possible, with respect to a measure on initial point X_0 and the probability measure underlying the Markov process. Furthermore if X_0 is distributed according to μ, then so is X_t for all $t > 0$. The resulting stochastic process, with X_0 distributed in this way, is stationary.

As mentioned earlier, it is sometimes useful to view a stochastic process X_t as a function of two variables $t \in T$ and $\omega \in \Omega : X(t,\omega)$. In this context it is then of interest to look at Banach space–valued spaces, as in the previous chapter (see Definition 2.22 and the discussion following it).

Example 3.14. Consider a stochastic process X taking values in the space of real valued continuous functions $E = C([0,T],\mathbb{R})$ and let $p \in [1,\infty)$. We define $L^p(\Omega, E)$ to be the Banach space equipped with norm $\|\cdot\|_{L^p(\Omega,E)}$ given by

$$\|X\|^p_{L^p(\Omega,E)} = \mathbb{E}\Big(\big(\sup_{t\in[0,T]} |X(t)|\big)^p\Big).$$

Notice that this definition is equivalent to

$$\|X\|^p_{L^p(\Omega,E)} = \mathbb{E}\Big(\sup_{t\in[0,T]} |X(t)|^p\Big),$$

and the norm is often written this way. □

3.4 Martingales and Stochastic Integrals

3.4.1 Martingales

Definition 3.15. *Let $\{\mathcal{F}_t\}$ be a filtration on the probability space $(\Omega, \mathcal{F}, \mu)$ and let $\{M_t\}$ be an $\mathbb{R}^d$-valued stochastic process adapted to $\{\mathcal{F}_t\}$ and satisfying $\mathbb{E}|M_t| < \infty$ for all $t \in T$. We say that $\{M_t\}$ is an $\mathcal{F}_t$–*martingale *if*

$$\mathbb{E}[M_t|\mathcal{F}_s] = M_s \quad \textit{for all } t, s \in T, t > s.$$

Thus

$$\mathbb{E}[M_t - M_s|\mathcal{F}_s] = 0 \quad \text{for all } t, s \in T, t > s$$

for a martingale; that is, the increments have mean zero, conditional on $\mathcal{F}_s$.

A martingale M_t is *square–integrable* if $\mathbb{E}|M_t|^2 < \infty$ for all $t \in T$.

Example 3.16. By Definition 3.5 the standard d-dimensional Brownian motion is a martingale with respect to the filtration generated by itself. □

The quadratic variation of a martingale is central in both the theory and applications of stochastic processes. In order to define it we need the following definition.

Definition 3.17. *An $\mathbb{R}^{d\times d}$-valued process $Q(t)$ with $Q(0) = 0$ and defined for $t \in [0, T]$ is* increasing *if $Q(t)$ is nonnegative for each $t \in [0, \infty)$ and if $Q(t) - Q(s)$ is nonnegative for $0 \leqslant s \leqslant t \leqslant T$.*

Definition 3.18. *An $\mathbb{R}^{d\times d}$-valued, adapted, increasing process $Q(t)$ is the* quadratic variation *of an $\mathbb{R}^d$-valued $\mathcal{F}_t$–martingale $M(t)$ if*

$$M(t) \otimes M(t) - Q(t)$$

is an $\mathcal{F}_t$–martingale. We write $Q(t) = \langle M \rangle_t$.

Theorem 3.19. *Let $M(t)$ be a continuous, $\mathbb{R}^d$-valued, square integrable $\mathcal{F}_t$–martingale. Then it has exactly one quadratic variation process.*

Example 3.20.

i) The quadratic variation of the one-dimensional standard Brownian motion is t. To see this note that $M(t) := W(t)^2 - t$ is a martingale. More generally, let $W(t)$ be the standard Brownian motion in $\mathbb{R}^d$. Then $\big(W(t) \otimes W(t) - It\big)$ is a martingale and hence $\langle W \rangle_t = It$.

ii) Let X_t be an $\mathbb{R}$-valued Markov process with generator $\mathcal{L}$ and let $\phi \in \mathcal{D}(\mathcal{L})$, the domain of definition of $\mathcal{L}$. Then

$$M_t = \phi(X_t) - \phi(X_0) - \int_0^t (\mathcal{L}\phi)(X_s)\, ds \tag{3.4.1}$$

is a martingale. Assume that $\phi^2 \in \mathcal{D}(\mathcal{L})$. Then the quadratic variation of M_t is given by the formula

$$\langle M \rangle_t = \int_0^t \big((\mathcal{L}\phi^2)(X_s) - 2\phi(X_s)(\mathcal{L}\phi)(X_s)\big)\, ds. \tag{3.4.2}$$

iii) Let $\phi(x) = x$, $X_t = W(t)$ and $d = 1$ in the notation of the two previous examples. Then $\mathcal{L}\phi(x) = 0$, $\mathcal{L}\phi^2(x) = 1$, $M_t = W(t)$, and $\langle W \rangle_t = t$ as expected. □

Every continuous martingale satisfies the **Doob martingale inequality**

$$\mathbb{P}\left[\sup_{0 \leqslant s \leqslant t} |M_s| \geqslant \lambda\right] \leqslant \frac{1}{\lambda^p} \mathbb{E}|M_t|^p, \tag{3.4.3}$$

for all $\lambda > 0$, $p \geqslant 1, t \geqslant 0$. Furthermore

$$\mathbb{E}\left(\sup_{0 \leqslant s \leqslant t} |M_s|^p\right) \leqslant \left(\frac{p}{p-1}\right)^p \mathbb{E}|M_t|^p, \tag{3.4.4}$$

for all $p > 1, t \geqslant 0$.

3.4.2 The Itô Stochastic Integral

For the rigorous analysis of stochastic differential equations it is necessary to define stochastic integrals of the form

$$I(t) = \int_0^t f(s)\, dW(s), \tag{3.4.5}$$

where $W(t)$ is a d-dimensional Brownian motion and $f(s) \in \mathbb{R}^{m \times d}$. This is not straightforward because $W(t)$ does not have paths of bounded variation. We start with the case $m = d = 1$. In order to define the stochastic integral we assume that $f(t)$ is a random process, adapted to the filtration $\mathcal{F}_t$ generated by the process $W(t)$, and such that

$$\mathbb{E}\left(\int_0^T f(s)^2\, ds\right) < \infty.$$

We now give the Itô interpretation of the stochastic integral. The Itô stochastic integral $I(t)$ is defined as the L^2–limit of the following Riemann sum approximation of (3.4.5):

$$I(t) := \lim_{K \to \infty} \sum_{k=1}^{K-1} f(t_{k-1}) \left(W(t_k) - W(t_{k-1})\right), \tag{3.4.6}$$

where $t_k = k\Delta t$ and $K\Delta t = t$. Notice that the function f is evaluated at the left end of each interval $[t_{n-1}, t_n]$ in (3.4.6). The resulting Itô stochastic integral $I(t)$ has a continuous version in t. These ideas are readily generalized to the case where $W(s)$ is a standard d-dimensional Brownian motion and suitably integrable $f(s) \in \mathbb{R}^{m \times d}$ for each s, yielding a process $I(t) \in \mathbb{R}^m$.

The resulting integral satisfies the *Itô isometry*

$$\mathbb{E}|I(t)|^2 = \int_0^t \mathbb{E}|f(s)|_F^2 ds, \tag{3.4.7}$$

where $|\cdot|_F$ denotes the Frobenius norm defined in Equation (2.2.1). Furthermore, the integral is a *martingale*:

$$\mathbb{E}I(t) = 0$$

and

$$\mathbb{E}[I(t)|\mathcal{F}_s] = I(s) \quad \forall t \geqslant s,$$

where $\mathcal{F}_s$ denotes the filtration generated by $W(s)$.

Example 3.21. Consider the Itô stochastic integral

$$I(t) = \int_0^t f(s)\, dW(s),$$

where f, W are scalar–valued. This is a martingale with quadratic variation

$$\langle I \rangle_t = \int_0^t (f(s))^2\, ds.$$

More generally, for f, W in arbitrary finite dimensions, the integral $I(t)$ is a martingale with quadratic variation

$$\langle I \rangle_t = \int_0^t (f(s) \otimes f(s))\, ds. \quad \square$$

Notice that

$$\begin{aligned} \mathbb{E}|\langle I \rangle_t| &\leqslant \int_0^t \mathbb{E}|f(s)|^2\, ds \\ &\leqslant C \int_0^t \mathbb{E}|f(s)|_F^2\, ds \\ &= C\mathbb{E}|I(t)|^2. \end{aligned}$$

Similar bounds hold for moments of the quadratic variation process, under appropriate moment bounds on $|f|_F$:

$$\mathbb{E}\Big(|\langle I \rangle_t|^{p/2}\Big) \leqslant C \tag{3.4.8}$$

for $p \geqslant 1$.

One of the most useful features of martingales is that they satisfy various path inequalities. These inequalities are particularly important when proving limit theorems. One of the most important martingale inequalities is the following.

Theorem 3.22. Burkholder-Davis-Gundy Inequality *Consider the Itô stochastic integral* (3.4.5)*, a martingale with quadratic variation process* $\langle I \rangle_t$*. For every* $p > 0$ *there are positive constants* $C_p^{\pm}$ *such that*

$$C_p^- \mathbb{E}\Big(|\langle I \rangle_t|^{p/2}\Big) \leqslant \mathbb{E}\Big(\sup_{0 \leqslant s \leqslant t} |I(s)|^p\Big) \leqslant C_p^+ \mathbb{E}\Big(|\langle I \rangle_t|^{p/2}\Big).$$

The proof of this theorem is based on the Itô formula (see Chapter 6) and the Doob Martingale inequality (3.4.4).

3.4.3 The Stratonovich Stochastic Integral

In addition to the Itô stochastic integral, the following Stratonovich integral is also sometimes useful. It is defined as the L^2–limit of a different Riemann sum approximation of (3.4.5), namely

$$I_{strat}(t) := \lim_{K\to\infty} \sum_{k=1}^{K-1} \frac{1}{2}\Big(f(t_{k-1}) + f(t_k)\Big)\left(W(t_k) - W(t_{k-1})\right), \quad (3.4.9)$$

where $t_k = k\Delta t$ and $K\Delta t = t$. Notice that the function $f(t)$ is evaluated at both endpoints of each interval $[t_{n-1}, t_n]$ in (3.4.9). The multidimensional Stratonovich integral is defined in a similar way. The resulting integral is written as

$$I_{strat}(t) = \int_0^t f(s) \circ dW(s).$$

The limit in (3.4.9) gives rise to an integral that differs from the Itô integral. Thus the situation is more complex than that arising in the standard theory of Riemann integration for functions of bounded variation; in that case the points in $[t_{k-1}, t_k]$ where the integrand is evaluated do not effect the definition of the integral, via a limiting process. In the case of integration against Brownian motion, which does not have bounded variation, the limits differ. However, when f and W are correlated through an SDE, a formula exists to convert between them; see Chapter 6. In general, it is harder to characterize the class of integrands for which the Stratonovich integral can be defined than it is for the Itô integral.

3.5 Weak Convergence of Probability Measures

A type of convergence that is very often used in probability theory is that of **weak convergence of probability measures**.

Definition 3.23. *Let (E, ρ) be a metric space with Borel σ–algebra $\mathcal{B}(E)$. Let $\{\mu_n\}_{n=1}^{\infty}$ be a sequence of probability measures on $(E, \mathcal{B}(E))$ and let μ be another measure on this space. We say that $\{\mu_n\}_{n=1}^{\infty}$* converges weakly *to μ, and write $\mu_n \Rightarrow \mu$, if*

$$\lim_{n\to\infty} \int_E f(s)\, d\mu_n(s) = \int_E f(s)\, d\mu(s),$$

for every $f \in C_b(E)$.

Definition 3.24. *Let $\{\Omega_n, \mathcal{F}_n, \mu_n\}_{n=1}^{\infty}$ be a sequence of probability spaces and let (E, ρ) be a metric space. Let $X_n : \Omega_n \mapsto E$, $n = 1, 2, \ldots$, be a sequence of random variables. Assume that $(\Omega, \mathcal{F}, \mu)$ is another probability space and let $X : \Omega \mapsto E$ be another random variable. We will say that $\{X_n\}_{n=1}^{\infty}$* converges to X in distribution, *or* weakly, *and write $X_n \Rightarrow X$, if the sequence of measures $\{\mu X_n^{-1}\}_{n=1}^{\infty}$ on $(E, \mathcal{B}(E))$ converges weakly to the measure μX^{-1} on $(E, \mathcal{B}(E))$.*

In other words, $X_n \Rightarrow X$ if and only if

$$\lim_{n \to \infty} \mathbb{E}_n f(X_n) = \mathbb{E} f(X)$$

for all $f \in C_b(E)$, where $\mathbb{E}_n$ denotes expectation under μX_n^{-1} and $\mathbb{E}$ under μX^{-1}. The following example illustrates the importance of weak convergence in the theory of probability.

Example 3.25. **(Central Limit Theorem)** Let $\{\xi_n\}_{n=1}^{\infty}$ be a sequence of i.i.d. random variables with mean zero and variance 1. Define

$$S_n := \sum_{k=1}^{n} \xi_k. \tag{3.5.1}$$

Then the sequence

$$X_n := \frac{1}{\sqrt{n}} S_n$$

converges in distribution to a standard normal random variable. □

There are various other types of convergence that are also useful in the study of limit theorems for random variables and stochastic processes. In the following, let $\{X_n\}_{n=1}^{\infty}$ be a sequence of random variables taking values in $(E, \mathcal{B}(E))$ and let X be another random variable, all on the same probability space $(\Omega, \mathcal{F}, \mathbb{P})$. Assume that E is a Banach space and let $\|\cdot\|$ denote its norm. We will say that the sequence **converges in probability** to a random variable X if, for every $\varepsilon > 0$,

$$\lim_{n \to \infty} \mathbb{P}(\|X_n - X\| > \varepsilon) \to 0.$$

We will say that $\{X_n\}_{n=1}^{\infty}$ **converges almost surely** (or with probability 1) if

$$\mathbb{P}(\lim_{n \to \infty} X_n = X) = 1.$$

Finally, we say that $\{X_n\}_{n=1}^{\infty}$ **converges in pth mean** (or in L^p) provided that

$$\lim_{n \to \infty} \mathbb{E}\|X_n - X\|^p = 0.$$

Note that this is simply strong convergence in the Banach space $L^p(\Omega; E)$, integrating with respect to μ on Ω.

Example 3.26. **(Strong Law of Large Numbers)** Let $\{\xi_n\}_{n=1}^{\infty}$ be a sequence of i.i.d. random variables with mean m. Define

$$X_n = \frac{1}{n}\sum_{k=1}^{n} \xi_k.$$

Then the sequence X_n converges to m almost surely. Assume furthermore that $\xi_n \in L^2$. Then X_n converges to 1 also in L^2. □

Remark 3.27. There are useful relations between the different notions of convergence for random variables: for example, almost sure convergence implies weak convergence; convergence in probability implies weak convergence. □

We now return to the discussion of weak convergence. In particular, we remark that it is a property preserved under continuous mappings.

Theorem 3.28. *Let $(\Omega_n, \mathcal{F}_n, \mu_n)_{n=1}^{\infty}$ be a sequence of probability spaces and let (E_i, ρ_i) be metric spaces for $i = 1, 2$. Let $X_n : \Omega_n \mapsto E_1$, $n = 1, \dots, \infty$ be a sequence of random variables. Assume that $(\Omega, \mathcal{F}, \mu)$ is another probability space and let $X : \Omega \mapsto E_1$, be another random variable. If $f : E_1 \to E_2$ is continuous then $\{f(X_n)\}_{n=1}^{\infty}$ converges to $f(X)$ in distribution if $\{X_n\}_{n=1}^{\infty}$ converges to X in distribution.*

Example 3.29. Let $E_1 = C([0,1], \mathbb{R})$ and $E_2 = \mathbb{R}$. The function $f : E_1 \to E_2$ defined by

$$f(x) := \sup_{t \in [0,1]} x(t)$$

is continuous. Hence, Theorem 3.28 applies, and we have that $X_n \Rightarrow X$ in $C([0,1], \mathbb{R})$ implies that $\sup_{t\in[0,1]} X_n(t) \Rightarrow \sup_{t\in[0,1]} X(t)$ in $\mathbb{R}$. □

In general, if $X_n \Rightarrow X$ and $Y_n \Rightarrow Y$ it does not follow that the pair (X_n, Y_n) is weakly convergent to (X, Y); correlations between (X_n, Y_n) may prevent this. However, if the limit Y is not random (the probability measure associated with the random variable Y is a point mass), then the pair does converge.

Theorem 3.30. *If $X_n \Rightarrow X$ and $Y_n \Rightarrow Y$, where Y is nonrandom, then $(X_n, Y_n) \Rightarrow (X, Y)$.*

Let $(\Omega, \mathcal{F}, \mu)$ be a probability space, $T = \mathbb{R}^+$ and (E, ρ) a metric space. Let $\{X_t^n\}_{n=1}^{\infty}$ be a family of stochastic processes and X_t another stochastic process, all with sample paths in $C_E = C([0,\infty), E)$. Recall that the sequence $\{X^n\}_{n=1}^{\infty}$ converges weakly to X, and we write $X^n \Rightarrow X$, if the sequence of probability measures $\mathbb{P}(X^n)^{-1}$ converges weakly to the probability measure $\mathbb{P}X^{-1}$ on $(C_E, \mathcal{B}(C_E))$. Sometimes we will say that $\{X_t^n\}_{n=1}^{\infty}$ converges weakly in C_E, or that $\{X_t^n\}_{n=1}^{\infty}$ converges weakly on path space.

Example 3.31. **Functional Central Limit Theorem—Invariance Principle** Consider the situation of Example 3.25 and let S_n be given by (3.5.1). Let $[t]$ denote the integer part of a real number t and define the continuous-time process

$$X_t^n := \frac{1}{\sqrt{n}} S_{[nt]} + \frac{1}{\sqrt{n}}(nt - [nt])\xi_{[nt]+1}. \tag{3.5.2}$$

This process has continuous paths. Furthermore, $\{X_t^n\}_{n=1}^{\infty}$ converges weakly in $C_{\mathbb{R}}$ to a standard one-dimensional Brownian motion. □

In Examples 3.25 and 3.31 we cover the classical central limit theorem and the functional central limit theorem (invariance principle), respectively, based on sums of independent random variables. Similar results hold even for dependent random variables, provided that the dependence between the random variables is not too strong. In many instances the martingale property, which can be viewed as a useful generalization of independence, is sufficient for a limit theorem to hold. The following theorem demonstrates this and includes the previous example as a special case.

Theorem 3.32. *Let $(\Omega, \mathcal{F}, \mu)$ be a probability space and let $\{\mathcal{F}_j,\ j \geqslant 1\}$ be a filtration. Let $\{\xi_j,\ j \geqslant 1\}$ be an ergodic adapted process such that*

$$\mathbb{E}[\xi_1]^2 = \sigma^2$$

and

$$\mathbb{E}[\xi_{k+1} | \mathcal{F}_k] = 0. \tag{3.5.3}$$

Define

$$S_n := \sum_{k=1}^{n} \xi_k. \tag{3.5.4}$$

Then

$$X^n := \frac{1}{\sqrt{n}} S_n$$

converges in distribution to a Gaussian variable with mean 0 *and variance* σ^2. *Furthermore, the process*

$$X_t^n := \frac{1}{\sqrt{n}} S_{[nt]} + \frac{1}{\sqrt{n}}(nt - [nt])\xi_{[nt]+1}$$

converges weakly in $C_{\mathbb{R}}$ to $\sigma W(t)$ where $W(t)$ is a standard Brownian motion. Finally, the process

$$X_t^n := \frac{1}{\sqrt{n}} S_{[nt]}$$

converges weakly in $D_{\mathbb{R}}$ to the same process $\sigma W(t)$.

Notice that the condition (3.5.3) implies that S_n defined in (3.5.4) is an $\mathcal{F}_j$–martingale. The preceding theorem is also valid in arbitrary dimensions $d \geqslant 1$, that is, in the case where $\{\xi_j,\, j \geqslant 1\}$ is a vector-valued sequence of stationary, ergodic random variables. In this case, the limiting process is a standard d-dimensional Brownian motion premultiplied by the square root of the covariance matrix of ξ_1.

Furthermore, a result similar to that of Theorem 3.32 can be proved for continuous-time martingales, and we now state this.

Theorem 3.33. (Martingale Central Limit Theorem) *Let $\{M(t) : \mathbb{R}^+ \mapsto \mathbb{R}^d\}$ be a continuous square integrable martingale on a probability space $(\Omega, \mathcal{F}, \mu)$ with respect to a filtration $\{\mathcal{F}_t : t \geqslant 0\}$; let $\langle M \rangle_t$ denote its quadratic variation process. Assume that:*

i) $M(0) = 0$;
ii) the process $M(t)$ has continuous sample paths and stationary increments;
iii) the scaled quadratic variation of $M(t)$ converges in $L^1(\mu)$ to some symmetric positive-definite matrix Σ:

$$\lim_{t\to\infty} \mathbb{E}\left(\left|\frac{\langle M \rangle_t}{t} - \Sigma\right|\right) = 0. \tag{3.5.5}$$

Then the process $1/\sqrt{t}M_t$ converges in distribution to an $\mathcal{N}(0, \Sigma)$ random variable. Furthermore, the rescaled martingale

$$M^{\varepsilon}(t) := \varepsilon M\left(\frac{t}{\varepsilon^2}\right)$$

converges weakly in $C_{\mathbb{R}^d}$ to $\sqrt{\Sigma}W(t)$, where $W(t)$ is a standard d-dimensional Brownian motion and $\sqrt{\Sigma}$ denotes the square root of the matrix Σ.

Roughly speaking, the Martingale central limit theorem relates the strong law of large numbers for the quadratic variation to a statement about fluctuations in the process itself. Upon combining the fact that stochastic integrals are martingales, together with an ergodicity assumption, we obtain the following corollary.

Corollary 3.34. *Let $W(t)$ denote a d-dimensional Brownian motion and let $z(t) : \mathbb{R}^+ \mapsto \mathbb{R}^d$ be a continuous, stationary, ergodic Markov process with invariant measure $\mu(dz)$, adapted to the filtration generated by $W(t)$. If $F : \mathbb{R}^d \mapsto \mathbb{R}^{m\times d}$ is a smooth bounded function, define*

$$I(t) = \int_0^t F(z(s))\, dW(s)$$

and note that, by ergodicity,

$$\lim_{t\to\infty} \frac{1}{t}\int_0^t \Big(F(z(s)) \otimes F(z(s))\Big)\, ds = \int_{\mathcal{Z}} \Big(F(z) \otimes F(z)\Big)\, \mu(dz) \quad \text{in } L^1(\mu). \tag{3.5.6}$$

Then, for every finite $T > 0$, the rescaled stochastic integral

$$I^{\varepsilon}(t) := \varepsilon I\left(\frac{t}{\varepsilon^2}\right)$$

converges in distribution to $\sqrt{\Sigma}W(t)$ where $W(t)$ is an m-dimensional standard Brownian motion and

$$\Sigma = \int_{\mathbb{R}^d} \Big(F(z) \otimes F(z)\Big)\mu(dz). \tag{3.5.7}$$

Proof. We have to check that the martingale $I(t)$ satisfies the assumptions of Theorem 3.33. We clearly have that $I(0) = 0$. Furthermore, the stationarity of the process $z(t)$ implies that $I(t)$ has independent increments. By (3.5.6) and (3.5.7)

$$\begin{aligned}
\lim_{t\to\infty} \frac{1}{t}\langle I\rangle_t &= \lim_{t\to\infty} \frac{1}{t}\int_0^t \Big(F(z(s)) \otimes F(z(s))\Big)\, ds \\
&= \int_{\mathcal{Z}} \Big(F(z) \otimes F(z)\Big)\, \mu(dz) \\
&= \Sigma.
\end{aligned}$$

Hence, Theorem 3.33 applies. □

3.6 Discussion and Bibliography

Most of the material contained in this chapter is very standard and can be found in many textbooks. Standard references on probability theory are the books [102, 103, 50, 194, 195, 36, 302]. Standard textbooks on the theory of stochastic process are the books [270, 156, 304].

A general reference on probability measures and their properties is [252]. An excellent reference on weak convergence of probability measures is [37]. Various convergence results for sequences of Markov processes can be found in [94]. The text [120] has good background on stochastic processes and weak convergence for probability measures on path space. For discussion concerning the relationships between different modes of convergence see [130]. A lot of information on limit theorems for stochastic processes can be found in the book [152]. A good introduction to the use of the theory of probability and stochastic processes in applied mathematics is the book [64].

Necessary and sufficient conditions for an operator to be the generator of a continuous semigroup are provided by the **Hille–Yosida theorem**; see, for example, [98, 337]. A systematic study of the theory of Markov processes based on the theory of semigroups was initiated by Feller in the 1950s. See [79, 196, 337]. Various limit theorems for Markov processes can be proved by using limit theorems for semigroups of operators (for example, the **Trotter–Kato theorem** [157]). Results of this type were proved, for example, in [181]. See [94].

Ergodic Markov processes with a generator that is symmetric with respect to the L^2-space weighted by the invariant measure of the process are termed *reversible*. See [303, ch. 5] for a good discussion of this topic. The symmetry of the generator of a reversible Markov process is very useful in the proof of limit theorems. See, for example, [173, 74, 127].

The Burkholder-Davis-Gundy inequality of Theorem 3.22 has a formulation for arbitary martingales not just for Itô stochastic integrals; see, for example, [156, ch. 3, Theorem 3.28]. Bounds on moments of the quadratic variation (such as (3.4.8)) and on moments of stochastic integrals may be found in [156, 210].

The proofs of the results presented in Example 3.20 can be found in [270]. The operator that appears on the right-hand side of Equation (3.4.2) is sometimes called the *Operateur carré du champ*. The proof of the martingale central limit theorem can be found in [94, ch. 7]. See also [186, 185]. A wealth of information on limit theorems for martingales can be found in [137].

In one dimension, continuous local martingales can be expressed as time-changed Brownian motions. This is sometimes referred to as the **Dambis–Dubins–Schwarz theorem** [156, ch. 3, Theorem 4.6]. Let $M = \{M_t, \mathcal{F}_t; 0 \leqslant t < \infty\}$ be an $\mathbb{R}$-valued martingale satisfying

$$\lim_{t \to \infty} \langle M \rangle_t = \infty, \ \mathbb{P}\,\text{a.s.}$$

Define, for each $0 \leqslant s < \infty$, the stopping time

$$T(s) = \inf\{t \geqslant 0; \langle M \rangle_t > s\}.$$

Then the time-changed process B_s and filtration $\mathcal{G}_s$ given by

$$B_s = M_{T(s)}, \quad \mathcal{G}_s = \mathcal{F}_{T(s)}, \quad 0 \leqslant s < \infty,$$

is a standard one-dimensional Brownian motion. In particular, $\mathbb{P}$ a.s:

$$M_t = B_{\langle M \rangle_t}.$$

This property can be used to characterize the limit of a sequence of martingales. See, for example, [134, 40]. Notice that the quantities B_t and M_t are in general highly correlated and that they are not adapted to the same filtration, because of the time change.

The martingale central limit theorem leads to a general central limit theorem for additive functionals of Markov processes: let $y(t)$ be an ergodic Markov process on $\mathcal{Y}$ with generator $\mathcal{L}$ and invariant measure $\mu(dy)$. Consider the following integral (which is an additive functional of $y(t)$):

$$x(t) = x_0 + \int_0^t f(y(s))\,ds.$$

Assume that

$$\int_{\mathcal{Y}} f(y)\mu(dy) = 0.$$

Then the rescaled process

$$x^{\varepsilon}(t) = \varepsilon x(t/\varepsilon^2)$$

converges weakly to $\sqrt{\Sigma}W(t)$, with $W(t)$ a standard Brownian motion, provided that the Poisson equation

$$-\mathcal{L}\phi = f \tag{3.6.1}$$

is well posed, in some appropriate (weak) sense. The variance of the limiting Brownian motion is given by the Dirichlet form (see Chapter 6) of the generator $\mathcal{L}$, evaluated at the solution of (3.6.1):

$$\Sigma = \int_{\mathcal{Y}} (-\mathcal{L}\phi) \otimes \phi \, \mu(dy).$$

To see this in one dimension, notice that (3.4.1) implies that the rescaled process $x^{\varepsilon}(t)$ satisfies

$$x^{\varepsilon}(t) = \varepsilon\big(x_0 - \phi(x(t/\varepsilon^2)) + \phi(x(0))\big) + \varepsilon M(t/\varepsilon^2),$$

where $M(t)$ is a martingale. The first term on the right-hand side of this equation tends to 0, provided that ϕ satisfies appropriate estimates. The second term converges to a Brownian motion with variance Σ, by the martingale central limit theorem. (An integration by parts is required with respect to the invariant measure to see this.) This theorem was proved for reversible Markov processes in [173]; see also [74]. A general approach for proving limit theorems for Markov processes based on martingale limit theorems was developed in [243]. Various extensions of this result have been proved, in particular to situations where the Markov process is not reversible, under appropriate assumptions on the generator. See [186] and the references therein.

3.7 Exercises

1. Prove the assertions in Example 3.2 concerning the means and covariances associated with Gaussian random variables.
2. Prove the assertions (3.3.2) and (3.3.3) concerning the Wiener process.
3. Prove the assertion (3.2.4) concerning the sums of exponential random variables.
4. Let $W(t) : \mathbb{R}^+ \mapsto \mathbb{R}$ be a standard Brownian motion. Calculate all moments of $W(t) - W(s)$, $t > s \geqslant 0$.
5. Carry out the same program as in the previous exercise for the d-dimensional Brownian motion.
6. Let $W(t)$ be a standard Brownian motion in one dimension. Show that

$$\mathbb{E}\left(\left|\frac{\Delta W(t)}{\Delta t}\right|\right) = \sqrt{\frac{2}{\pi \Delta t}},$$

for $\Delta W(t) = W(t+\Delta t) - W(t)$. Deduce that, with probability 1, the Brownian motion is nowhere differentiable.

7. Let $W(t)$ be a standard Brownian motion in $\mathbb{R}^d$ and let $\mathcal{F}_t$ denote the filtration generated by $W(t)$. Prove that $W(t)$ and $W(t) \otimes W(t) - It$ are both $\mathcal{F}_t$–martingales.
8. State the analogue of Theorem 3.32 in arbitrary dimensions.
9. Consider the *Brownian bridge* $BB(t) : [0, T] \mapsto \mathbb{R}$ defined by

$$BB(t) = W(t) - \frac{t}{T} W(T),$$

where $W(t)$ is a standard one-dimensional Brownian motion. Calculate the mean and covariance of the Brownian bridge.

4

Ordinary Differential Equations

4.1 Setup

In this chapter we study ordinary differential equations (ODEs) of the form

$$\frac{dz}{dt} = h(z), \quad z(0) = z_0, \tag{4.1.1}$$

where $h : \mathcal{Z} \to \mathbb{R}^d$. Typically $\mathcal{Z} = \mathbb{T}^d$, $\mathbb{R}^d$, or $\mathbb{R}^l \oplus \mathbb{T}^{d-l}$. In later chapters we will often consider the case where $z = (x^T, y^T)^T$, with $x \in \mathcal{X}$, $y \in \mathcal{Y}$. If $\mathcal{Z} = \mathbb{T}^d$ (resp. $\mathbb{R}^d$) then $\mathcal{X} = \mathbb{T}^l$ (resp. $\mathbb{R}^l$) and $\mathcal{Y} = \mathbb{T}^{d-l}$ (resp. $\mathbb{R}^{d-l}$). If $\mathcal{Z} = \mathbb{R}^l \oplus \mathbb{T}^{d-l}$ then $\mathcal{X} = \mathbb{R}^l$ and $\mathcal{Y} = \mathbb{T}^{d-l}$.

When we consider Equation (4.1.1) on the torus this is simply a convenient shorthand for the case that h is periodic in z (resp. y) and that we consider z (resp. y) as an element of the torus, by working modulo 1 in all directions. On occasion we will consider differential operators on the torus and, in this setting, the operator automatically has periodic boundary conditions.

In Section 4.2 we outline a theory of existence and uniqueness for ODEs. The idea of the generator, the Liouville equation, and the method of characteristics are introduced in Section 4.3. In Section 4.4 we discuss ergodicity for ODEs. Various extensions of the results presented in this chapter, together with bibliographical remarks, are discussed in Section 4.5.

4.2 Existence and Uniqueness

When there is a unique solution $z(t)$ to the initial value problem (4.1.1) for all initial data z_0 in $\mathcal{Z}$ and $t \in \mathbb{R}$ we write

$$z(t) = \varphi^t(z_0).$$

Thus $\varphi^t : \mathcal{Z} \to \mathcal{Z}$ is the solution operator and forms a *one-parameter group of operators*, that is,

$$\varphi^{t+s} = \varphi^t \circ \varphi^s \;\; \forall t, s \in \mathbb{R} \;\text{ and }\; \varphi^0 = \mathcal{I},$$

where $\mathcal{I} : \mathcal{Z} \to \mathcal{Z}$ denotes the identity operator. The inverse of φ^t is φ^{-t}. We may apply φ^t to sets via the definition

$$\varphi^t(A) = \bigcup_{z \in A} \varphi^t(z).$$

Once φ^t is defined it will frequently be notationally clean to write $\varphi^t(z)$ for the solution of (4.1.1) starting from initial condition z.[1]

In practice, existence and uniqueness of solutions of (4.1.1) can be verified in a wide range of different scenarios. For ease of exposition we work within the simplest setting, namely when h is Lipschitz continuous on $\mathcal{Z}$. This condition can be weakened when *a priori* bounds on the solution prevent finite time blow-up. Such bounds may be proved using Lyapunov functions; see Section 4.3.

Definition 4.1. *A function $f : \mathcal{Z} \to \mathbb{R}^d$ is* Lipschitz *on $\mathcal{Z}$ with Lipschitz constant L if*

$$|f(x) - f(y)| \leqslant L|x - y| \quad \forall x, y \in \mathcal{Z}.$$

We observe that a Lipschitz function is also linearly bounded:

$$|f(x)| \leqslant |f(0)| + L|x|.$$

Theorem 4.2. *If h is Lipschitz on $\mathcal{Z}$ then the ODE (4.1.1) has a unique solution $z(t) \in C^1(\mathbb{R}, \mathcal{Z})$ for any $z_0 \in \mathcal{Z}$. Moreover*

$$|\varphi^t(x) - \varphi^t(y)| \leqslant e^{Lt}|x - y| \quad \forall x, y \in \mathcal{Z}, t \geqslant 0. \tag{4.2.1}$$

Proof. The existence and uniqueness follow from a standard fixed-point argument (Picard iteration). To establish the bound (4.2.1) note that

$$\frac{1}{2}\frac{d}{dt}|x - y|^2 = \langle h(x) - h(y), x - y \rangle.$$

Using the Lipschitz property of h we find that

$$\langle h(x) - h(y), x - y \rangle \leqslant L|x - y|^2.$$

Hence, for $e = |x - y|^2$, we have

$$\frac{1}{2}\frac{d}{dt}|e| \leqslant L|e|.$$

Integrating this differential inequality (or using the Gronwall Lemma 4.4) gives the desired result.

[1] In this situation z should not be confused with $z(t)$.

With the assumption the f is Lipschitz only on compact subsets of $\mathcal{Z}$ it is always possible, for any $z \in \mathcal{Z}$, to define $\varphi^t(z)$ on some interval $t \in [-a, b]$ containing 0. This interval depends, in general, on z. With this in mind we make the following defintion.

Definition 4.3. *A set* $\mathcal{A}$ *is* invariant *(resp.* forward invariant, backward invariant*)) if* $\varphi^t(z)$ *is defined for* $t \in \mathbb{R}$ *(resp.* $t \in [0, \infty)$, $t \in (-\infty, 0]$*) for all* $z \in \mathcal{A}$ *and* $\varphi^t(\mathcal{A}) \equiv \mathcal{A}$ *(resp.* $\subseteq \mathcal{A}$*) for all* $t \in \mathbb{R}$ *(resp.* $t \in [0, \infty)$, $t \in (-\infty, 0]$*).*

An invaluable tool for studying evolution equations is the Gronwall lemma. This is an inequality version of the exponential solutions that may be obtained, as the previous theorem illustrates, from linear constant coefficient differential equations.

Lemma 4.4. *(***Gronwall***)*

i) (Differential form). Let $\eta(t) \in C^1([0, T]; \mathbb{R}^+)$ *satisfy the differential inequality*

$$\frac{d\eta(t)}{dt} \leqslant a\eta(t) + \psi(t), \quad \eta(0) = \eta, \tag{4.2.2}$$

where $a \in \mathbb{R}$ *and* $\psi(t) \in L^1([0, T]; \mathbb{R}^+)$*. Then*

$$\eta(t) \leqslant \exp(at)\Big(\eta + \int_0^t \exp(-as)\psi(s)\, ds\Big) \tag{4.2.3}$$

for all $t \in [0, T]$.

ii) (Integral form). Assume that $\xi(t) \in C([0, T]; \mathbb{R}^+)$ *satisfies the integral inequality*[2]

$$\xi(t) \leqslant a \int_0^t \xi(s)\, ds + b,$$

for some positive constants a, b*. Then*

$$\xi(t) \leqslant b \exp(at) \quad \text{for } t \in [0, T].$$

Proof. i). We multiply Equation (4.2.2) by $\exp(-at)$ to obtain

$$\Big(\frac{d}{dt}\eta(t)\Big) \exp(-at) \leqslant \Big(\psi(t) + a\eta(t)\Big) \exp(-at).$$

Consequently

$$\frac{d}{dt}\Big(\eta(t) \exp(-at)\Big) \leqslant \exp(-at)\psi(t).$$

Integrating this inequality from 0 to t and multiplying through by $\exp(at)$ gives (4.2.3).

ii). Define $\eta(t) = \int_0^t \xi(s)\, ds$. Then $\eta(t)$ satisfies the inequality

[2] This assumption may be weakened.

$$\frac{d\eta}{dt} \leqslant a\eta + b.$$

We apply the first part of the lemma with $\eta(0) = 0$ to deduce that

$$\eta(t) \leqslant \frac{b}{a}\Big(\exp(at) - 1\Big).$$

Consequently

$$\begin{aligned}\xi(t) &\leqslant a\eta(t) + b \\ &\leqslant b\exp(at). \quad \square\end{aligned}$$

4.3 The Generator

It is often of importance to understand how functions of $t \mapsto z(t)$ change with time. We may achieve this by using the **generator** $\mathcal{L}$:

$$\mathcal{L}v = h(z) \cdot \nabla v(z). \tag{4.3.1}$$

If $z(t)$ solves (4.1.1) and $V \in C^1(\mathcal{Z}, \mathbb{R})$ then

$$\begin{aligned}\frac{d}{dt}\Big(V(z(t))\Big) &= \nabla V(z(t)) \cdot \frac{dz}{dt}(t) \\ &= \nabla V(z(t)) \cdot h(z(t)) \\ &= \mathcal{L}V(z(t)).\end{aligned} \tag{4.3.2}$$

The generator is readily exteneded to functions taking values in $\mathbb{R}^n$. In some situations $\mathcal{L}V(z)$ can be bounded above in terms of $V(z)$, for all $z \in \mathcal{Z}$. If V is also positive then it is possible to use differential inequalities, such as the Gronwall Lemma 4.4, to obtain bounds on $V(z(t))$, and hence on $z(t)$. Then $V(z(t))$ is a *Lyapunov function.*

Example 4.5. Consider the equation

$$\frac{dz}{dt} = z - z^3. \tag{4.3.3}$$

If $V(z) = \frac{1}{2}z^2$ then

$$\mathcal{L}V(z) = z^2 - z^4 \leqslant 1 - 2V(z).$$

Thus

$$\frac{d}{dt}\Big(V(z(t)\Big) \leqslant 1 - 2V(z(t)).$$

Application of the Gronwall lemma gives

$$V(z(t)) \leqslant e^{-2t}V(z(0)) + \frac{1}{2}(1 - e^{-2t}).$$

Hence, the existence of the Lyapunov function $V(z) = \frac{1}{2}|z|^2$ leads to the *a priori* bound on the solution of (4.3.3)

$$|z(t)|^2 \leqslant 1 + e^{-2t}|z(0)|^2. \square$$

Also important is the formal L^2-adjoint operator $\mathcal{L}^*$, given by

$$\mathcal{L}^* v = -\nabla \cdot \big(hv\big). \tag{4.3.4}$$

As defined $\mathcal{L}$ and $\mathcal{L}^*$ only apply to $C^1(\mathcal{Z})$ functions. It will be useful in the sequel to extend the domain of definition to $L^\infty(\mathcal{Z})$ and $L^1(\mathcal{Z})$ functions, respectively, and we do this later.

We now show the crucial role played by $\mathcal{L}$ and $\mathcal{L}^*$ in understanding how families of solutions of (4.1.1), parameterized by the initial data, and possibly carrying a probability measure, behave. Let v be the solution of the Cauchy problem (see Chapter 7)

$$\frac{\partial v}{\partial t} = \mathcal{L}v \quad \text{for } (z,t) \in \mathcal{Z} \times (0,\infty), \tag{4.3.5a}$$

$$v(z,0) = \phi(z) \quad \text{for } z \in \mathcal{Z}. \tag{4.3.5b}$$

This equation is sometimes called the **backward equation**. We will denote the solution of (4.3.5) by $v(z,t) = (e^{\mathcal{L}t}\phi)(z)$. This is often referred to as the semigroup notation for the solution of a time-dependent linear operator equation; see Section 7.5. By a classical solution of this linear PDE we mean a function $v(z,t)$ smooth enough so that the PDE is satisfied pointwise for $(z,t) \in \mathcal{Z} \times (0,\infty)$ and is continuous in $\mathcal{Z} \times \mathbb{R}^+$; see Chapter 7.

Theorem 4.6. *Assume that the solution of* (4.1.1) *generates a one-parameter group on* $\mathcal{Z}$ *so that* $\varphi^t(\mathcal{Z}) = \mathcal{Z}$ *for all* $t \in \mathbb{R}$. *Assume also that* ϕ *is sufficiently smooth so that* (4.3.5) *has a classical solution. Then the classical solution is given by* [3]

$$v(z,t) = \phi(\varphi^t(z)) \quad \forall t \in \mathbb{R}^+,\ z \in \mathcal{Z}. \tag{4.3.6}$$

Proof. Note that (4.3.6) satisfies the initial condition $v(z,0) = \phi(\varphi^0(z)) = \phi(z)$. Using the group property of φ^t we deduce that $v(z,t)$ given by (4.3.6) satisfies $v(\varphi^{-t}(z),t) = \phi(z), \forall t \in \mathbb{R}^+, z \in \mathcal{Z}$. By differentiating with respect to t we obtain from (4.3.6) that

$$\frac{d}{dt}\Big(v(\varphi^{-t}(z),t)\Big) = 0$$

and so

[3] Note that here z is the independent spatial variable in the PDE and occurs in the formula for the solution of the PDE as the *initial condition* for the group φ^t. This use of z should not be confused with the function $z(t)$ solving (4.1.1) with initial condition z_0.

$$\frac{\partial v(\varphi^{-t}(z),t))}{\partial t} + \left\langle \nabla v(\varphi^{-t}(z),t), \frac{d}{dt}\left(\varphi^{-t}(z)\right)\right\rangle = 0.$$

But $\varphi^{-t}(z)$ is the backward time solution for (4.1.1) and hence satisfies

$$\frac{d}{dt}(\varphi^{-t}(z)) = -h(\varphi^{-t}(z)).$$

Thus

$$\frac{\partial v(\varphi^{-t}(z),t))}{\partial t} + \langle \nabla v(\varphi^{-t}(z),t), -h(\varphi^{-t}(z))\rangle = 0, \quad \forall t \in \mathbb{R}^+, z \in \mathcal{Z}.$$

This is equivalent to

$$\frac{\partial v(z,t)}{\partial t} + \langle \nabla v(z,t), -h(z)\rangle = 0, \quad \forall t \in \mathbb{R}^+, z \in \mathcal{Z},$$

showing that (4.3.6) solves the linear PDE (4.3.5).

Remark 4.7. Formula (4.3.6) represents the solution of the PDE (4.3.5) by the **method of characteristics**. Remarkably, it shows that the family of solutions of the *nonlinear* ODE (4.1.1), found by considering all $z_0 \in \mathcal{Z}$, can be represented via the solution of a *linear* PDE. Conversely, it enables us to solve the initial value problem for a linear evolution PDE of first order in both space and time (transport equation) by solving a nonlinear system of ODEs. □

We can extend the definition of the operator $e^{\mathcal{L}t}$ to act on arbitrary functions $\phi \in L^\infty(\mathcal{Z})$ by setting

$$\left(e^{\mathcal{L}t}\phi\right)(z) = \phi(\varphi^t(z)), \quad \forall t \in \mathbb{R}^+, z \in \mathcal{Z}. \tag{4.3.7}$$

The operator $e^{\mathcal{L}t}$ maps $L^\infty(\mathcal{Z})$ into itself.

The generator $\mathcal{L}$ is then defined in the following way. Let $\mathcal{D}(\mathcal{L})$ denote the set of functions ϕ for which the following limit

$$\mathcal{L}\phi = \lim_{t\to 0} \frac{e^{\mathcal{L}t}\phi - \phi}{t} \tag{4.3.8}$$

exists, strongly in $L^\infty(\mathcal{Z})$. This limit then defines $\mathcal{L}$ on $\mathcal{D}(\mathcal{L})$ and coincides with (4.3.1) on $C^1(\mathcal{Z})$; the space $C^1(\mathcal{Z})$ is not a dense set in $L^\infty(\mathcal{Z})$, however.

We now study what happens when we place a probability measure on z_0, so that $z(t)$ solving (4.1.1) is a random variable. To this end consider the adjoint of (4.3.5), namely the **Liouville equation**

$$\frac{\partial \rho}{\partial t} = \mathcal{L}^*\rho \quad \text{for } (z,t) \in \mathcal{Z} \times (0,\infty), \tag{4.3.9a}$$

$$\rho(z,0) = \rho_0(z) \quad \text{for } z \in \mathcal{Z}. \tag{4.3.9b}$$

This is also sometimes refered to as the **forward equation**. Using the semigroup notation the solution can be denoted by $\rho(z,t) = (e^{\mathcal{L}^* t}\rho_0)(z)$. Because $\mathcal{L}^*$ is the adjoint of $\mathcal{L}$ it follows that $e^{\mathcal{L}^* t}$ is the adjoint of $e^{\mathcal{L}t}$ (see Section 7.5). Now let $\mathbb{E}$ denote expectation with respect to initial data distributed according to a random variable on $\mathcal{Z}$ with density $\rho_0(z)$[4], i.e.

$$\mathbb{E}f := \int_{\mathcal{Z}} f(z)\rho_0(z)\,dz.$$

Theorem 4.8. *Assume that the solution of* (4.1.1) *generates a one-parameter group on $\mathcal{Z}$ so that $\varphi^t(\mathcal{Z}) = \mathcal{Z}$ for all $t \in \mathbb{R}$. Assume also that ϕ is sufficiently smooth so that* (4.3.5) *has a classical solution. Finally, assume that the initial data for* (4.1.1) *is distributed according to a random variable on $\mathcal{Z}$ with density $\rho_0(z)$, smooth enough so that* (4.3.9) *has a classical solution. Then, $z(t)$ is a random variable on $\mathcal{Z}$ with density $\rho(z,t)$ satisfying* (4.3.9).

Proof. Note that, by Theorem 4.6,

$$\begin{aligned}\mathbb{E}(\phi(z(t))) &= \int_{\mathcal{Z}} \phi(\varphi^t(z))\rho_0(z)dz\\ &= \int_{\mathcal{Z}} v(z,t)\rho_0(z)dz\\ &= \int_{\mathcal{Z}} (e^{\mathcal{L}t}\phi)(z)\rho_0(z)dz\\ &= \int_{\mathcal{Z}} (e^{\mathcal{L}^* t}\rho_0)(z)\phi(z)dz.\end{aligned}$$

Also, if $\rho(z,t)$ is the density of $z(t)$, then

$$\mathbb{E}(\phi(z(t))) = \int_{\mathcal{Z}} \rho(z,t)\phi(z)dz.$$

Equating these two expressions for the expectation at time t and using the arbitrariness of ϕ, together with a density argument to extend the equality to all ϕ in $L^2(\mathcal{Z})$, shows that

$$\rho(z,t) = (e^{\mathcal{L}^* t}\rho_0)(z)$$

in $L^2(\mathcal{Z})$. Hence, by the assumed smoothness, the density $\rho(z,t)$ satisfies the adjoint Equation (4.3.9).

Remark 4.9. We can extend the domain of definition of $e^{\mathcal{L}^* t}$ so that it maps $L^1(\mathcal{Z})$ into itself by defining it to be the L^2–adjoint of the operator $e^{\mathcal{L}t} : L^\infty(\mathcal{Z}) \to L^\infty(\mathcal{Z})$ defined by (4.3.7). The resulting operator $e^{\mathcal{L}^* t}$ is continuous on $L^1(\mathcal{Z})$.

The L^2–adjoint of the generator $\mathcal{L}^*$ is then defined in the following way. Let $\mathcal{D}(\mathcal{L}^*)$ denote the set of functions ρ for which the following limit

[4] Once again, here z is an independent spatial variable and should not be confused with the function $z(t)$ solving (4.1.1) with initial condition z_0.

$$\mathcal{L}^* \rho = \lim_{t \to 0} \frac{e^{\mathcal{L}^* t} \rho - \rho}{t} \tag{4.3.10}$$

exists, strongly in $L^1(\mathcal{Z})$. This limit then defines $\mathcal{L}^*$ on $\mathcal{D}(\mathcal{L}^*)$. The definition coincides with (4.3.4) on $C^1(\mathcal{Z})$, which is a dense set in $L^1(\mathcal{Z})$. □

4.4 Ergodicity

Roughly speaking, ergodicity is concerned with the existence and uniqueness of a measure invariant under the dynamics—an invariant measure. This heuristic way of thinking about ergodicity is useful in unifying the slightly different presentations of ergodicity in this and the following two chapters. The subject rapidly becomes quite technical, especially in the context of the nonrandom problems considered in this chapter, and having a heuristic understanding of the subject will facilitate an understanding of Part II of this book, where ergodicity plays a central role in many of the perturbation expansions. We will relate this heuristic way of understanding ergodicity to many other related concepts, in particular: to the convergence of time averages to averages with respect to the invariant measure; and to the one-dimensional null spaces of the generator and its adjoint, in appropriate spaces.

In this chapter we consider the measure space $(\mathcal{Z}, \mathcal{A}, \mu)$, where μ is a measure on $\mathcal{Z}$ and $\mathcal{A}$ denotes a σ–algebra of μ-measurable subsets of $\mathcal{Z}$. Let φ^t denote the solution operator for (4.1.1).

Definition 4.10. *The measure μ is* invariant *if*

$$\mu(\varphi^t(A)) = \mu(A) \quad \forall A \in \mathcal{A}, t > 0.$$

Recall that a set $A \in \mathcal{A}$ is *invariant* under φ^t provided that, for all $t \in \mathbb{R}$,

$$\varphi^t(A) = A.$$

Definition 4.11. *Given an invariant measure μ, the ODE* (4.1.1) *is called* ergodic *if every invariant set A of φ^t is such that either $\mu(A) = 0$ or $\mu(A) = 1$.*

Note that the definition of ergodicity is relative to the measure space in question. Two natural questions now present themselves: (i) how do we identify invariant measures μ?; (ii) given an invariant measure μ how do we check for ergodicity? For ODEs these are, in general, very hard questions.

In the remainder of this section we will assume for simplicity that $\mathcal{Z} = \mathbb{T}^d$ and that the operators $\mathcal{L}$ and $\mathcal{L}^*$ given by (4.3.1) and (4.3.4) are equipped with periodic boundary conditions. The following two theorems address the two preceding questions in turn, specialized to this periodic setting. In the following, it is important to distinguish L^p-spaces with respect to Lebesgue measure and with respect to the invariant measure μ which, in general, will be different.

Theorem 4.12. *Let μ be an invariant probability measure.*

(i) If $\phi \in L^1(\mathbb{T}^d;\mu)$ then

$$\phi^*(z) = \lim_{T\to\infty} \frac{1}{T}\int_0^T \phi(\varphi^t(z))\,dt$$

exists for μ–a.e. $z \in \mathbb{T}^d$;

(ii) A measure μ with density $\rho^\infty \in L^1(\mathbb{T}^d; Leb)$ with respect to Lebesgue measure is invariant if and only if

$$e^{\mathcal{L}^* t}\rho^\infty = \rho^\infty,\ \rho^\infty \in L^1(\mathbb{T}^d; Leb), \quad \forall t \geqslant 0;$$

(iii) A measure μ with density $\rho^\infty \in C^1(\mathbb{T}^d)$ with respect to Lebesgue measure is invariant if and only if

$$\mathcal{L}^*\rho^\infty = 0, \quad \rho^\infty \in C^1(\mathbb{T}^d).$$

Note that condition (iii) has the desirable feature that it need only be checked for continuously differentiable ρ^∞ so that the definition (4.3.4) of $\mathcal{L}^*$ as a differential operator may be used. That we may do this essentially follows from the continuity of the operator $e^{\mathcal{L}^* t}$ on or, relatedly, the density of $C^1(\mathbb{T}^d)$ in, $L^1(\mathbb{T}^d; \mathrm{Leb})$. However, when checking for ergodicity it is not possible to work with the characterization of $\mathcal{L}$ as a differential operator: condition (iii) in the next theorem must be checked for all functions in the domain of definition of $\mathcal{L}$, not just those for which (4.3.1) holds.

Theorem 4.13. *Let φ^t have an invariant probability measure μ.*

(i) If φ^t is ergodic and if $\phi \in L^1(\mathbb{T}^d;\mu)$ then

$$\lim_{T\to\infty} \frac{1}{T}\int_0^T \phi(\varphi^t(z))dt = \int_{\mathbb{T}^d} \phi(z)\mu(dz) \tag{4.4.1}$$

for μ–a.e. $z \in \mathbb{T}^d$;

(ii) φ^t is ergodic if and only if the equation

$$e^{\mathcal{L}t}g = g,\ g \in L^\infty(\mathbb{T}^d), \quad \forall t \geqslant 0$$

has only constant solutions μ–a.e.;

(iii) φ^t is ergodic if and only if the equation

$$\mathcal{L}g = 0, \quad g \in \mathcal{D}(\mathcal{L})$$

has only constant solutions μ–a.e..

Notice that an important aspect of ergodicity, encapsulated in the first item of the preceding theorem, is that time averages may be replaced by averages with respect to the invariant measure. It is this viewpoint on ergodicity that will be crucial in much of our exploitation of ergodicity for ODEs in Part II; we will also use time-average representations when studying similar issues for SDEs. Notice also that, by comparing items (i) in the two preceding theorems, time averages always converge when there is an invariant measure, but that a central feature of ergodicity is that time averages converge to a value that is *independent of the initial condition.*

For ergodic systems, dynamical trajectories visit the whole of the phase space, independently of initial conditions, and spend the same proportion of time in any given set. By choosing ϕ to be I_A, the indicator function of Borel set $A \subseteq \mathbb{T}^d$, we deduce from (4.13) that the measure μ is given by the formula

$$\mu(A) = \lim_{T\to\infty} \frac{1}{T} \int_0^T I_A(z(t))\, dt. \tag{4.4.2}$$

Thus the invariant measure μ measures the proportion of time that typical trajectories spend in a given subset of $\mathbb{T}^d$.

Example 4.14. Consider the equation

$$\frac{dz}{dt} = b(z)$$

where $b \in C^\infty(\mathbb{T}, \mathbb{R})$ and

$$\inf_{z\in\mathbb{T}} b(z) > 0.$$

For each fixed $t \in \mathbb{R}$ we view $z(t)$ as an element of the torus (circle) $\mathbb{T}$. Here

$$\mathcal{L}\phi = b\frac{\partial \phi}{\partial z}$$

is equipped with periodic boundary conditions. Theorem 4.12(iii) shows that the equation has invariant measure μ with density

$$\rho^\infty(z) = \frac{C}{b(z)},$$

where C is chosen so that ρ^∞ integrates to 1 over $\mathbb{T}$. It is natural to ask whether the equation is ergodic with respect to this invariant measure.

We first consider the case $b(z) = \omega > 0$, a constant. Thus μ is simply the Lebesgue measure. Note that $\mathbb{T} = \mathbb{S}$, the circle, and that

$$\varphi^\tau(z) = z + \omega\tau \pmod 1.$$

The *rotation* $\varphi^\tau : \mathbb{S} \to \mathbb{S}$ is well known to be an ergodic map for any fixed τ such that $\omega\tau$ is irrational; this means that any set A satisfying $\varphi^\tau(A) = A$ has Lebesgue (and hence $\mu-$) measure 0 or 1. Consider such a τ. Any invariant set of the ODE must satisfy $\varphi^t(A) = A$ for $t = \tau$ in particular, and hence has Lebesgue (and hence $\mu-$) measure 0 or 1. Thus the ODE is ergodic on $\mathbb{T}$.

The case $b(z)$ not constant may be handled by rescaling time to make the vector field constant. This is possible since b is smooth and strictly positive on $\mathbb{S}$. Since μ is absolutely continuous with respect to Lebesgue measure the ergodicity of the rescaled equation implies the same for the original equation. □

Example 4.15. Consider Equation (4.1.1), viewed as a dynamical system on $\mathbb{T}^d$, in the case where the vector field $h(z)$ is divergence-free, i.e., $\nabla \cdot h(z) = 0$. Then $\mathcal{L}, \mathcal{L}^*$ given by (4.3.1), (4.3.4) satisfy

$$\mathcal{L}^* v = -\nabla \cdot (hv) = -(\nabla \cdot h)v - h \cdot \nabla v = -h \cdot \nabla v = -\mathcal{L}v.$$

Both operators are equipped with periodic boundary conditions. Thus the equation $\mathcal{L}^* \rho^\infty = 0$ on $\mathbb{T}^d$ has the solution $\rho^\infty = 1$. Consequently the Lebesgue measure is invariant for the ODE. However, to prove ergodicity it is necessary to show that the equation $\mathcal{L}g = 0$ has only one solution in $\mathcal{D}(\mathcal{L}) \subset L^\infty$. Determining when this happens is a subtle question. We return to it in Chapter 14 when we study averaging for transport equations. □

The ergodic theory described so far is somewhat limited for applications because it focuses on ergodic measures that have a density ρ^∞ (with respect to the Lebesgue measure) on $\mathcal{Z}$. There are many examples that exhibit ergodic behavior, but the underlying measure μ is not absolutely continuous with respect to Lebesgue measure. Nonetheless, these problems exhibit the crucial property that time averages converge to values independent of the initial condition. A simple example is an ODE (4.1.1) that has a unique globally attracting equilibrium solution. If μ is chosen to be a Dirac mass on this equilibrium point, then (4.4.1) will hold. We give two further illustrations of this kind of behavior.

Example 4.16. Consider the Lorenz equations

$$\begin{aligned}
\frac{dy_1}{dt} &= 10(y_2 - y_1), \\
\frac{dy_2}{dt} &= 28y_1 - y_2 - y_1 y_3, \\
\frac{dy_3}{dt} &= y_1 y_2 - \frac{8}{3} y_3.
\end{aligned} \tag{4.4.3}$$

These equations are ergodic with invariant measure supported on a set of zero volume and fractal dimension between 2 and 3. For such problems the concept of *SRB measures*[5] plays the role that the concept of the invariant measure played in the preceding examples. Time averages of the solution converge to averages with respect to the SRB measure. Thus (4.4.1) holds in this generalized sense.

We illustrate these ideas in Figures 4.1 and 4.2. The first shows the attractor for the Lorenz equations, which is the set on which the invariant measure is supported. The fractal nature is manifest in the banded structure of the set, here shown projected onto the first two components of y. In the second figure we show the empirical measure (histogram) generated by the second component of the Lorenz equations, started from initial data on the attractor. □

Example 4.17. Consider the Harmonic oscillator

$$\frac{d^2\eta}{dt^2} + \omega\eta = 0.$$

[5] From Sinai, Ruelle, and Bowen.

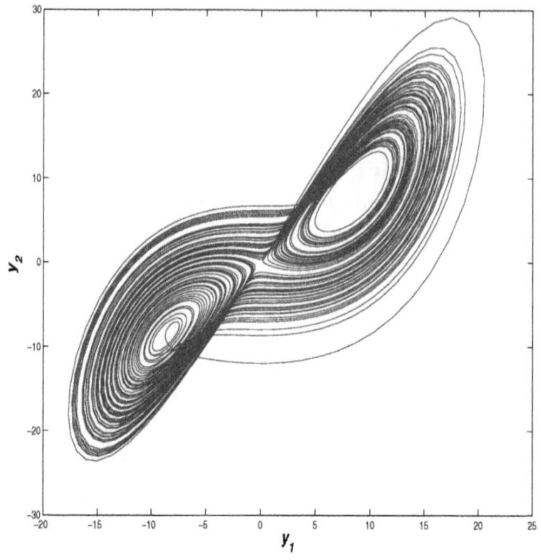

Fig. 4.1. Projection onto (y_1, y_2) of the attractor for the Lorenz equations (4.4.3).

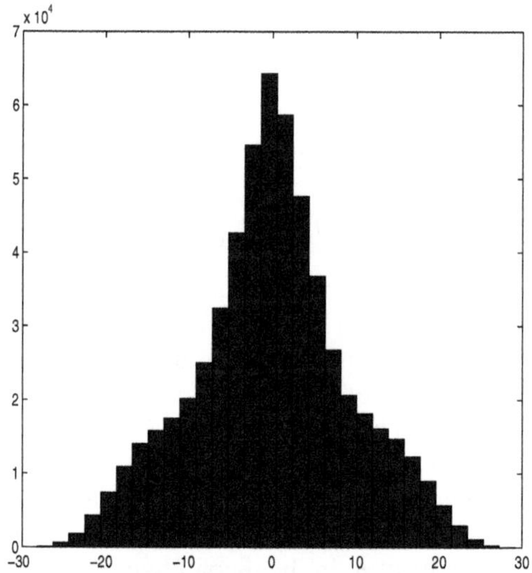

Fig. 4.2. Empirical measure for component y_2 of (4.4.3).

This can be written as the first-order system

$$\frac{d\eta}{dt} = v,$$
$$\frac{dv}{dt} = -\omega\eta.$$

The equation has a divergence-free vector field and thus the Lebesgue measure is invariant (although it is not normalizable and hence not a probability measure); furthermore, the formal differential operators $\mathcal{L}$, $\mathcal{L}^*$ satisfy $\mathcal{L} = -\mathcal{L}^*$. The system conserves the energy

$$E = \frac{1}{2}v^2 + \frac{1}{2}\omega\eta^2 = \frac{1}{2}\left(\frac{d\eta}{dt}\right)^2 + \frac{1}{2}\omega\eta^2.$$

A straightforward calculation (see Exercise 2(c)) shows that any smooth function of the Hamiltonian $H(v, \eta)$ is in the null space of $\mathcal{L}$ and hence the equation cannot be ergodic by Theorem 4.13.

However, the equation satisfies a form of ergodicity on the energy shell $\mathcal{E}$ given by $H(v, \eta) := E =$ constant. This can be understood intuitively by noticing that the solution is given by

$$\eta(t) = A\sin(\omega^{\frac{1}{2}}t) + B\cos(\omega^{\frac{1}{2}}t)$$

with

$$\frac{\omega}{2}\left(A^2 + B^2\right) = E.$$

The solution rotates on $\mathcal{E}$, generating an *empirical measure* μ as it does so. This measure quantifies how much time the trajectory spends in subsets of $\mathcal{E}$. It turns out that this measure depends on A and B only through E. A straightforward calculation shows that

$$\eta(t)^2 = \frac{1}{2}\left(A^2 + B^2\right) + AB\sin(2\omega^{\frac{1}{2}}t) + \frac{1}{2}\left(B^2 - A^2\right)\cos(2\omega^{\frac{1}{2}}t).$$

Thus

$$\lim_{T\to\infty}\frac{1}{T}\int_0^T \frac{1}{2}\omega\eta^2(t)\,dt = \frac{1}{4}\omega\left(A^2 + B^2\right) = \frac{1}{2}E.$$

This is an analogue of (4.4.1) in this Hamiltonian problem. Note, however, that memory of the initial condition remains, through E. Assuming that the system does indeed generate an empirical measure μ on $\mathcal{E}$ we deduce that

$$\int_{\mathcal{E}} \frac{\omega}{2}\eta^2\mu(d\eta, dv) = \frac{1}{2}E.$$

The fact that the average kinetic energy of the oscillator has half the total energy is a consequence of the **Virial theorem**, which states that, on average, kinetic and potential energy share the total energy equally. □

4.5 Discussion and Bibliography

The complete proof of Theorem 4.2 can be found in [15, sec. 31] or [68]. The generator of a system of ODEs and its properties is discussed in [188, sec. 7.6]. The use of the generator to study Lyapunov functions and obtain *a priori* estimates on solutions of ODEs may be found in [305]. Sometimes the generator $\mathcal{L}$, defined in Equation (4.3.1), and its adjoint $\mathcal{L}^*$ are called the generators of the **Koopman** and the **Frobenius–Perron** operators, respectively; the Koopman operator is also called the **transfer operator**. It was introduced by Koopman in 1931 [176], as a tool for studying the ergodic properties of classical dynamical systems. A nice discussion of Koopman's original work and its extensions can be found in [266, sec. II.5, sec. VII.4].

The Liouville equation is the fundamental equation of nonequilibrium statistical mechanics; see, for example, [28]. The Fokker–Planck equation, the fundamental equation for stochastic dynamics which we study in Chapter 6, can be derived from the Liouville equation through an appropriate mode elimination procedure. See [269, ch. 9].

The method of characteristics is discussed in numerous PDE books, including [98].

The discussion of ergodicity is fleshed out in more detail in [188]. In particular, Theorem 4.12 follows from theorems 7.3.1, 7.7.1, and 7.8.2 in that book; Theorem 4.13 follows from theorems 4.2.4, 7.7.2, and 7.8.3 in that book. A more thorough discussion of ergodic theory for discrete and continuous-time systems can be found in [260, 327]. See also [14, 70, 290]. For a discusion of SRB measures and further references on the topic, see [338].

The ergodic properties of the Lorenz equation follow from the work of Tucker [318, 319]; see also [298]. For a literature review concerning statistical properties of dynamical systems, and ODEs in particular, see the references in [224]. The proof of the Virial theorem can be found in many textbooks on classical mechanics.

4.6 Exercises

1. Let $\mathcal{Z} = \mathbb{T}^d$. Show that for all $f \in C^1(\mathcal{Z}, \mathbb{R})$ the formal L^2–adjoint of $\mathcal{L}$ defined in Equation (4.3.1) is $\mathcal{L}^*$ defined in Equation (4.3.4).
2. Let $H(p,q) : \mathbb{R}^{2d} \to \mathbb{R}$ be a smooth function and consider the (Hamiltonian) ODE
$$\dot{q} = \nabla_p H(p,q), \quad \dot{p} = -\nabla_q H(p,q) \tag{4.6.1}$$
   a. Write down the generators and its adjoint for (4.6.1).
   b. Write down the Liouville Equation (4.6.1)
   c. Show that every smooth function of the Hamiltonian $H(p(t), q(t))$ solves the Liouville equation.
   d. Is the Hamiltonian system (4.6.1) ergodic?
3. Carry out the same program as in parts a, b, and d of the previous exercise for the (gradient) ODE
$$\dot{q} = -\nabla_q V(q), \tag{4.6.2}$$
where $V(q) : \mathbb{R}^d \to \mathbb{R}$ is a smooth function.

5

Markov Chains

5.1 Setup

In this section we introduce Markov chains on a countable state space. Without loss of generality we take this state space to be $\mathcal{I}$, a subset of the positive integers $\mathbb{N}$.[1]

In Section 5.2 we introduce ideas by studying discrete-time Markov chains. In Section 5.3 we construct an example of a continuous-time Markov chain from the discrete-time object. In Section 5.4 we study the generator of continuous-time Markov chains on $\mathcal{I}$, using the preceding example as motivation for a general definition. Using this concept we discuss existence and uniqueness of solutions for Markov chains in Section 5.5. Ergodicity for finite state Markov chains is overviewed in Section 5.6. Various extensions of the results presented in this chapter, together with bibliographical remarks, are discussed in Section 5.7.

5.2 Discrete-Time Markov Chains

A matrix P with entries p_{ij} is a *stochastic matrix* if

$$\sum_j p_{ij} = 1 \quad \forall i \in \mathcal{I}$$

and $p_{ij} \in [0,1]$ for all $(i,j) \in \mathcal{I} \times \mathcal{I}$.

Definition 5.1. *The random sequence* $\{z_n\}_{n \geqslant 0}$ *is a* discrete-time Markov chain *with initial distribution* ρ_0*, a vector with number of components given by the cardinality of* $\mathcal{I}$*, and* transition matrix P *if it is a Markov stochastic process with state space* $\mathcal{I}$ *and*

- z_0 *has distribution* ρ_0;

[1] In this chapter all sums are over $\mathcal{I}$, unless stated to the contrary. In later chapters, however, we will sometimes find it convenient to work with a doubly indexed state space found as the product of two subsets of the integers.

- *for every* $n \geqslant 0$ *we have, when* $\mathbb{P}(z_n = i) > 0$,

$$\mathbb{P}(z_{n+1} = j | z_n = i) = p_{ij}.$$

The entries of the transition matrix $\{p_{ij}\}_{i,j \in \mathcal{I}}$ are called the **transition probabilities**. By construction, P is a stochastic matrix. Note that, by using the Chapman-Kolmogorov Equation (3.3.5) and an induction, $(P^k)_{ij} = \mathbb{P}(z_k = j | z_0 = i)$. Thus P^k is also a stochastic matrix. Notice also that $\mathbb{P}(z_n = j) = (\rho_0^T P^n)_j$.

The discrete-time Markov chain has transition probabilities from z_n to z_{n+1}, which do not depend on n. The resulting Markov process, on the discrete set $\mathcal{I}$, is thus homogeneous in the sense of Chapter 3; it is sometimes referred to as a discrete-time homogeneous Markov chain. We will only consider homogeneous Markov chains in what follows and hence refrain from explicitly using the term *homogeneous* in the sequel.

Example 5.2. Let $\alpha,\ \beta \in [0, 1]$ and consider the matrix

$$P = \begin{pmatrix} 1-\alpha & \alpha \\ \beta & 1-\beta \end{pmatrix}.$$

Clearly, this is a stochastic matrix, and it is the transition matrix of a two-state Markov chain. □

Example 5.3. **(Symmetric random walk on $\mathbb{Z}$)** The symmetric random walk on $\mathbb{Z}$ is a Markov chain with initial distribution

$$\rho_0^i = \begin{cases} 1 & \text{if } i = 0 \\ 0 & \text{otherwise} \end{cases}$$

and transition probabilities

$$P = \begin{cases} \frac{1}{2} & \text{if } |i-j| = 1 \\ 0 & \text{otherwise.} \end{cases}$$

Equivalently, we could write that

$$\mathbb{P}(z_0 = 0) = 1$$

and

$$\mathbb{P}(z_{n+1} = j | z_n = i) = \begin{cases} \frac{1}{2} & \text{if } j = i+1 \\ \frac{1}{2} & \text{if } j = i-1 \\ 0 & \text{otherwise.} \end{cases} \quad \square$$

Notice that a stochastic matrix satisfies

$$P\mathbf{1} = \mathbf{1}, \tag{5.2.1}$$

where $\mathbf{1}$ is the vector with unit entries, $\mathbf{1} = (1, \dots, 1)$. Combining this with the fact that P has positive entries implies the following fundamental identity:

$$|P|_\infty = 1. \tag{5.2.2}$$

5.3 Continuous-Time Markov Chains

We now study continuous-time Markov chains on $\mathcal{I}$.

Definition 5.4. *A* continuous-time Markov chain *is a Markov stochastic process $\{z(t)\}_{t\in\mathbb{R}^+}$ with state space $E = \mathcal{I}$.*

To introduce continuous-time Markov chains we start by constructing them using discrete-time Markov chains. Let the i.i.d. sequence $\{\tau_n\}_{n\geqslant 0}$ be distributed as $\exp(\lambda)$ for some $\lambda > 0$ and define $\{t_n\}_{n\geqslant 0}$ by $t_{n+1} = t_n + \tau_n$, $t_0 = 0$. Let $\{z_n\}_{n\geqslant 0}$ be a discrete-time Markov chain on $\mathcal{I}$, independent of the $\{\tau_n\}_{n\geqslant 0}$, and set

$$z(t) = z_n, \; t \in [t_n, t_{n+1}). \tag{5.3.1}$$

We call this a *jump chain*.[2] Notice that $z(t)$ takes values in $\mathcal{I}$ and is a càdlàg process. The fact that $z(t)$ is Markov follows from the Markovian structure of $\{z_n\}_{n\geqslant 0}$, together with the properties of the exponential random variable.

Informally we may write $z(t)$ as the solution of the differential equation

$$\frac{dz}{dt} = \delta(t - t_j)\Big(k(z(t^-)) - z(t^-)\Big) \tag{5.3.2}$$

where $k(z)$ is distributed as $p(z, \cdot)$ and the $\{k(z(t_j^-))\}_{j\geqslant 0}$ are drawn independently of one another, and independently of the $\{\tau_j\}$. This representation follows because, integrating over the jump times t_j, we obtain

$$\begin{aligned} z(t_j^+) - z(t_j^-) &= \lim_{\varepsilon\to 0}\int_{t_j-\varepsilon}^{t_j+\varepsilon} \delta(t - t_j)\Big(k(z(t^-)) - z(t^-)\Big)\,dt \\ &= k(z(t_j^-)) - z(t_j^-), \end{aligned}$$

and so $z(t_j^+) = k(z(t_j^-))$ as desired. Making sense of this random differential equation, and in particular showing that it has a solution for all time $t > 0$, is intimately related to the question of showing that the jump times t_j do not accumulate at a finite time. In Section 5.4 we assume that this failure to accumulate does indeed hold. In Section 5.5 we return to the discussion of existence of solutions to this differential equation.

We now find a representation of the matrix $P(t)$ with entries

$$p_{ij}(t) = \mathbb{P}(z(t) = j | z(0) = i), \tag{5.3.3}$$

for the jump chain. We express $P(t)$ in terms of P and λ, the parameters input into the jump chain. Note that, by properties of exponential random variables encapsulated in (3.2.4),

$$\mathbb{P}(k \text{ jumps in } [0,t]) = \frac{e^{-\lambda t}(\lambda t)^k}{k!}.$$

[2] It is also possible to construct jump chains where λ, the rate of jumps, is dependent on the current state.

Thus

$$p_{ij}(t) = \sum_{k=0}^{\infty} \frac{e^{-\lambda t}(\lambda t)^k}{k!} \left(P^k\right)_{ij},$$

since $\left(P^k\right)_{ij} = \mathbb{P}(z_k = j | z_0 = i)$. Hence

$$\begin{aligned} P(t) &= e^{-\lambda t} \sum_{k=0}^{\infty} \frac{(\lambda t)^k}{k!} P^k \\ &= e^{\lambda t (P - I)} \\ &= e^{Lt} \end{aligned}$$

with $L = \lambda(P - I)$. The matrix L is called the **generator** of the continuous-time Markov chain. Making sense of the matrix e^{Lt} when the state space is infinite requires the theory of semigroups; see Section 7.5. Note that, by (5.2.2),

$$L = \lim_{t \to 0} \frac{P(t) - I}{t},$$

where the limit is in the operator norm induced by the ℓ^∞-norm on $\mathcal{I}$.[3] Thus the characterization of the generator coincides with the abstract Definition 3.11. In the next section we show that the generator may be used as the starting point from which to *define* continuous-time Markov chains.

5.4 The Generator

We now abstract the properties of the generator and then take these properties as the general definition of a continuous-time Markov chain.

Definition 5.5. *A matrix* $L : \mathcal{I} \to \mathcal{I}$ *with entries* l_{ij} *is the* generator of a continuous-time Markov chain *if*

- $\sum_j l_{ij} = 0 \quad \forall i \in \mathcal{I}$;
- $l_{ij} \in [0, \infty) \quad \forall (i, j) \in \mathcal{I} \times \mathcal{I}$ *with* $i \neq j$.

Notice that, since P is a stochastic matrix, the matrix L constructed in the example in the previous section satisfies all of the criteria in the preceding definition.

The notion of *generator* can be related to the abstract definition of the generator for Markov processes as given in Chapter 3. From the discussion in Chapter 3 we note that a Markov chain is characterized by its generator. The definition implies that

$$l(i) := -l_{ii} \in [0, \infty) \quad \forall i \in \mathcal{I}. \tag{5.4.1}$$

Given a generator L, it is possible to find a discrete-time Markov chain and a sequence of independent exponential random variables so that an associated jump

[3] If $\mathcal{I}$ is finite dimensional then any ℓ^p-norm may be used.

chain generates paths of the continuous-time Markov chain with generator L. In the case

$$l^* := \sup_i l(i) < \infty \tag{5.4.2}$$

it is possible to do this by choosing i.i.d. exponential random variables and then reversing the construction in the example presented in the previous section.[4] We now illustrate this construction. If (5.4.2) holds, then, for any $\lambda > l^*$,

$$P = I + \lambda^{-1}L \tag{5.4.3}$$

is a stochastic matrix. Generate τ_j as an i.i.d. sequence with τ_0 distributed as an exponential random variable: $\tau \sim \exp(\lambda)$ Now generate z_n from the Markov chain with transition matrix P and define $z(t)$ by (5.3.1). It is possible to check that

$$\frac{P(t) - I}{t} \to L$$

as before, and thus we have constructed a continuous-time stochastic process on $\mathcal{I}$ from the generator of Definition 5.5. Note that $z(t)$ takes values in $\mathcal{I}$, is càdlàg, and may be described by the differential Equation (5.3.2).

We now give another way to see the relationship between the continuous-time Markov chain with transition matrix $P(t)$ and the generator L. Consider a continuous-time Markov chain $z(t)$, $t \geqslant 0$, taking values in the state space $\mathcal{I} \subseteq \{1, 2, \dots\}$. Let $p_{ij}(t)$ be the transition probability from state i to j given by (5.3.3). The Markov property implies that for all $t, \Delta t \geqslant 0$,

$$p_{ij}(t + \Delta t) = \sum_k p_{ik}(t) p_{kj}(\Delta t).$$

This is the Chapman-Kolmogorov Equation (3.3.6) in the discrete state space setting, so that integrals become sums. From this equation it follows that

$$\frac{p_{ij}(t + \Delta t) - p_{ij}(t)}{\Delta t} = \sum_k p_{ik}(t) \ell_{kj}(\Delta t),$$

where

$$\ell_{kj}(\Delta t) = \frac{1}{\Delta t} \times \begin{cases} p_{kj}(\Delta t) & \text{for } k \neq j \\ p_{jj}(\Delta t) - 1 & \text{for } k = j. \end{cases} \tag{5.4.4}$$

Suppose that the limit $\ell_{kj} = \lim_{\Delta t \to 0} \ell_{kj}(\Delta t)$ exists. We then obtain, formally,

$$\frac{dp_{ij}}{dt} = \sum_k p_{ik} \ell_{kj}. \tag{5.4.5}$$

---

[4] When the condition in (5.4.2) fails, the exponential rates must be chosen to depend on the current state.

Because $\sum_j p_{ij}(\Delta t) = 1$, it follows that $\sum_j \ell_{ij}(\Delta t) = 0$, and assuming that the limit exists,

$$\sum_j \ell_{ij} = 0. \tag{5.4.6}$$

This implies that

$$\sum_j p_{ij} = 1$$

for the limiting Equation (5.4.5).

Introducing the matrices $P(t)$, L with entries $p_{ij}(t)$, ℓ_{ij}, respectively, $i, j \in \mathcal{I}$, Equation (5.4.5) reads, in matrix notation,

$$\frac{dP}{dt} = PL, \quad P(0) = I. \tag{5.4.7}$$

It was shown in (5.4.4) that L is calculated from P via the formula

$$L = \lim_{\Delta t \to 0} \frac{1}{\Delta t}\Big(P(\Delta t) - I\Big) = \frac{dP}{dt}\Big|_{t=0}.$$

(This relationship between L and P was also derived in the reverse construction in the last section.) The generator has constants in its null space by (5.4.6):

$$L\mathbf{1} = 0. \tag{5.4.8}$$

As P is a stochastic matrix we deduce that

$$|e^{Lt}|_\infty = 1. \tag{5.4.9}$$

Furthermore, the nonnegativity of the p_{ij} implies that L has nonnegative off-diagonal entries. The condition (5.4.6) thus implies that diagonal entries of $-L$ are also nonnegative.

Notice that

$$P(t) = \exp(Lt) \tag{5.4.10}$$

solves (5.4.7). Thus we see that $P(t)$ and L commute. Consequently, $P(t)$ also solves

$$\frac{dP}{dt} = LP, \quad P(0) = I. \tag{5.4.11}$$

We refer to both (5.4.7) and (5.4.11) as the **master equation** of the Markov chain.

Example 5.6. Let $\alpha \in (0, +\infty)$ and consider the continuous-time Markov chain with generator

$$L = \begin{pmatrix} -\alpha & \alpha \\ \alpha & -\alpha \end{pmatrix}.$$

By calculating the eigenvalues and eigenfunctions of L, we can decompose it as

$$L = U \begin{pmatrix} -2\alpha & 0 \\ 0 & 0 \end{pmatrix} U^T$$

where

$$U = \frac{1}{\sqrt{2}} \begin{pmatrix} 1 & 1 \\ -1 & 1 \end{pmatrix}.$$

The definition of the exponential of a matrix $L = U\Lambda U^T$ with Λ diagonal and U orthogonal is

$$e^{tL} = Ue^{t\Lambda}U^T.$$

Using this definition we may calculate the semigroup generated by L:

$$e^{tL} = \frac{1}{2} \begin{pmatrix} 1+e^{-2\alpha t} & 1-e^{-2\alpha t} \\ 1-e^{-2\alpha t} & 1+e^{-2\alpha t} \end{pmatrix}. \quad \square$$

Let $\rho(t) = (\rho_0(t), \rho_1(t), \dots)^T$ be the transpose of the ith row of $P(t)$, i.e., a column vector whose entries $\rho_j(t) = p_{ij}(t)$ are the probabilities that a system starting in state i will end up, at time t, in each of the states $j \in \mathcal{I}$. Let e_i denote the ith unit vector, zero in all entries except the ith, in which it is one. Directly from (5.4.7) we obtain the following theorem.

Theorem 5.7. *The probability vector* ρ *satisfies*

$$\frac{d\rho}{dt} = L^T\rho, \qquad \rho(0) = e_i. \tag{5.4.12}$$

If the initial state of the Markov chain is random, with probability vector $\rho(0) = \rho_0$ *chosen independently of the transition probabilities in the Markov chain, then*

$$\frac{d\rho}{dt} = L^T\rho, \qquad \rho(0) = \rho_0. \tag{5.4.13}$$

Proof. The first result follows from (5.4.7). Let $\rho^{(i)}$ denote the solution of (5.4.12). If the initial condition is random with $\rho(0) = \rho_0$, a vector with ith component $\rho_{0,i}$, then

$$\rho(t) = \sum_i \rho_{0,i}\rho^{(i)}(t).$$

Differentiating and using (5.4.12) gives (5.4.13). $\square$

Equation (5.4.13) is the Markov chain analogue of the Liouville and Fokker-Planck equations described in Chapters 4 and 6, respectively. We refer to it as the **forward equation**.

Let $\phi : \mathcal{I} \mapsto \mathbb{R}$ be a real-valued function defined on the state space; it can be represented as a vector with entries ϕ_j, $j \in \mathcal{I}$. Then let $v(t) = (v_0(t), v_1(t), \dots)^T$ denote the vector with ith entry

$$v_i(t) = \mathbb{E}\Big(\phi_{z(t)} | z(0) = i\Big),$$

where $\mathbb{E}$ denotes expectation with respect to the Markov transition probabilities.

Theorem 5.8. *The vector of expectations v satisfies the equation*

$$\frac{dv}{dt} = Lv, \quad v(0) = \phi. \tag{5.4.14}$$

Proof. The function $v_i(t)$ can be written explicitly in terms of the transition probabilities:

$$v_i(t) = \sum_j p_{ij}(t)\phi_j. \tag{5.4.15}$$

If we set $\phi = (\phi_0, \phi_1, \dots)^T$ then this can be written in vector form as $v(t) = P(t)\phi$. Differentiating with respect to time and using the master Equation (5.4.11) gives the desired result.

Equation (5.4.14) is the Markov chain analogue of the method of characteristics and of the backward Kolmogorov equation described in Chapters 4 and 6, respectively. We refer to it as the **backward equation**.

5.5 Existence and Uniqueness

A continuous-time Markov chain will have sample paths that exhibit jumps from one state in $\mathcal{I}$ to another. This was made explicit through the construction of the jump chain, under condition (5.4.2). (When this condition fails, a construction where the jump rates depend on the current state is possible.) With this in mind, the following definition is natural.

Definition 5.9. *A continuous-time Markov chain is* nonexplosive *if, with probability one, the jump times do not accumulate at a finite time.*

It is important to understand conditions that ensure nonexplosion. We show that (5.4.2) is sufficient. We generate a sequence $\{z_n\}_{n\geqslant 0}$ from the discrete-time Markov chain with a transition matrix given by (5.4.3), with $\lambda > l^*$. The jump chain associated with this choice of transition matrix P is then given by (5.3.1), where $t_{n+1} = t_n + \tau_n$ and the τ_n are i.i.d. random variables distributed as $\exp(\lambda)$.

Theorem 5.10. *The Markov chain is nonexplosive if* (5.4.2) *holds.*

Proof. We identify the Markov chain corresponding to generator L with the jump chain. Let

$$\zeta = \sum_{n=0}^{\infty} \tau_n.$$

Set $T_n = \lambda\tau_n$ and notice that the $\{T_n\}$ form an i.i.d. sequence with $T_1 \sim \exp(1)$. By the strong law of large numbers (Example 3.26),

$$\frac{1}{N+1}\sum_{n=0}^{N} T_n \to 1 \quad a.s.$$

Hence

$$\zeta := \lim_{N\to\infty} \frac{1}{\lambda} \sum_{n=0}^{N} T_n = \infty \quad a.s.$$

and the result follows.

It is noteworthy that condition (5.4.2) is satisfied whenever $\mathcal{I}$ is finite.

5.6 Ergodicity

As for ODEs, ergodicity for Markov chains is concerned with the existence and uniqueness of an invariant measure. We will relate this definition of ergodicity to the convergence of time averages to a value independent of initial conditions, determined by a unique invariant measure, and to the properties of the null spaces of the generator and its adjoint. In the context of countable state space Markov chains, the invariant measure will be characterized by a vector, the invariant distribution defined later.

For simplicity, we assume that $\mathcal{I}$ is a finite set. We start by discussing discrete-time Markov chains. By (5.2.1), the matrix $(P - I)$ has a nonempty null space, including constant vectors, and hence its transpose also has a nonempty null space. As a consequence, there exists a vector ρ^∞ such that

$$P^T \rho^\infty = \rho^\infty. \tag{5.6.1}$$

In fact we have the following theorem.

Theorem 5.11. *All eigenvalues of P lie in the closed unit circle. The vector ρ^∞ may be chosen so that all of its entries are nonnegative and $\langle \rho^\infty, 1\rangle = 1$.*

The vector ρ^∞ is known as the **invariant distribution**. Note that it defines a probability measure on $\mathcal{I}$. As in the case of ODEs, ergodicity is associated with making this invariant distribution unique. A straightforward way to ensure this is via the following condition.

Definition 5.12. *The discrete-time Markov chain is said to be* ergodic *if the spectrum of P lies strictly inside the unit circle, with the exception of a simple eigenvalue at* 1, *corresponding to a strictly positive invariant distribution.*

Now consider continuous-time Markov chains on $\mathcal{I}$. Using the properties of the generator L we deduce that

$$\begin{aligned} L1 &= 0, \\ L^T \rho^\infty &= 0. \end{aligned} \tag{5.6.2}$$

In fact, using (5.4.3), we have the following theorem.

Theorem 5.13. *All eigenvalues of L lie in the left half-plane. The vector ρ^∞ may be chosen so that all of its entries are nonnegative and $\langle \rho^\infty, 1\rangle = 1$.*

The vector ρ^∞ is again known as the **invariant distribution**. As in the case of discrete-time Markov chains, it defines a probability measure on $\mathcal{I}$. And again ergodicity is associated with making this invariant distribution unique.

Definition 5.14. *The continuous-time Markov chain is said to be* ergodic *if the spectrum of L lies strictly in the left half-plane, with the exception of a simple eigenvalue at zero, corresponding to a strictly positive invariant distribution.*

The following theorem describes the properties of ergodic continuous-time Markov chains. We will use it in the remainder of the book.

Theorem 5.15. *An ergodic continuous-time Markov chain on finite state space $\mathcal{I}$ satisfies the following five properties:*

i) $\mathcal{N}(L) = span\{1\}$;
ii) $\mathcal{N}(L^T) = span\{\rho^\infty\}$, $\rho^\infty(i) > 0\,\forall i \in \mathcal{I}$;
iii) $\exists C, \lambda > 0$ *such that solution of the forward Equation* (5.4.13) *satisfies*

$$|\rho(t) - \rho^\infty|_1 \leqslant Ce^{-\lambda t} \quad \forall t > 0;$$

iv) $\exists C, \lambda > 0$ *such that solution of the backward Equation* (5.4.14) *satisfies*

$$|v(t) - \langle \rho^\infty, \phi \rangle 1|_\infty \leqslant Ce^{-\lambda t} \quad \forall t > 0; \quad \text{and}$$

v)

$$\lim_{T\to\infty} \frac{1}{T}\int_0^T \phi_{z(t)}dt = \langle \rho^\infty, \phi\rangle \quad a.s.$$

Since the state space is finite-dimensional the convergence results hold in any norm; however, the choices as stated are natural from a probabilistic viewpoint.

Notice that, by choosing $\phi = e_i$, we deduce from the final result that the ith component of ρ^∞ can be found as the proportion of time that an arbitrary trajectory of the Markov chain on $t \in [0, \infty)$ spends in state i. This is analogous to the ODE case in the previous chapter and to formula (4.4.2) in particular. Note also that, as in the ODE case, ergodicity is associated with time averages converging to the average against the invariant distribution and hence independent of the initial condition.

Example 5.16. We continue Example 5.6. Notice that L has one-dimensional null space spanned by the vector $(1, 1)^T$. Relatedly L^T has the same property. Thus $\rho^\infty = (1/2, 1/2)^T$. Clearly

$$\lim_{t\to\infty} e^{tL} = \frac{1}{2}\begin{pmatrix} 1 & 1 \\ 1 & 1 \end{pmatrix}.$$

Furthermore the limit is achieved exponentially fast. Since $\rho(0)$ is a distribution (see Example 3.1), it follows that $\rho(t) = e^{tL^T}\rho(0)$ converges to ρ^∞ exponentially fast. Since L is symmetric, the same exponential convergence holds for $v(t)$. □

5.7 Discussion and Bibliography

An excellent introductory textbook on Markov chains is [235]. A more advanced treatment may be found in [303]. The text [94] has a wealth of material on the general setting for Markov processes, including Markov chains. The paper [122] describes a simple algorithm for simulating continuous-time Markov chains, through sample paths of the jump chain. The book [299] describes Markov chains from a computational linear algebra perspective. The book [227] has a very general treatment of ergodicity for Markov chains.

The existence of a nonnegative invariant vector in the discrete-time case follows from the general theory of nonnegative matrices (matrices with no negative entries). Making the eigenpair $(\rho^\infty, 1)$ simple for matrix P^T, and ensuring that $\rho^\infty(i) > 0$ for all i, is equivalent to asking that the matrix is *irreducible*: for any pair $i, j \in \mathcal{I}$ there exists an $n \in \mathbb{Z}^+$ such that $\left(P^n\right)_{ij} > 0$. Asking that the chain be ergodic requires *aperiodicity* as well as irreducibility. Together, aperiodicity and irreducibility imply that there exists an $n \in \mathbb{Z}^+$ such that P^n has all entries strictly positive. This means that there exists an $n \in \mathbb{Z}^+$ for which

$$\mathbb{P}(z_n = j | z_0 = i) > 0 \quad \forall (i,j) \in \mathcal{I} \times \mathcal{I}.$$

Thus there exists a positive integer n such that any member of $\mathcal{I}$ can be reached in n steps with positive probability from any starting point in $\mathcal{I}$. This is another useful heuristic way to think about ergodicity: dynamical trajectories should be able to visit the whole of the state space.[5] These results concerning nonnegative matrices and their spectra may be found in [115]. Note that some texts refer to the irreducible case as ergodic and the irreducible and aperiodic case as *mixing*; see [327].

Extending these ergodicity results to the continuous-time case is fairly straightforward in the case of finite state space, using the jump chain. Recall that, to every continuous-time Markov chain with generator L on finite state space $\mathcal{I}$, we may associate a jump chain with transition matrix given by (5.4.3), for any $\lambda > l^*$ with l^* given by (5.4.2). The existence of a nonnegative invariant vector for the continuous-time process follows from that for the discrete process with transition matrix P. This is because P and L share the same eigenvectors and their eigenvalues $\eta(P)$ and $\eta(L)$ are related by

$$\eta(P) = 1 + \lambda^{-1}\eta(L). \tag{5.7.1}$$

In particular if $\eta(P) = 1$, as for the invariant vector, then $\eta(L) = 0$. The criterion for ergodicity of the continuous-time process simply becomes the condition that the matrix P given by (5.4.3), with $\lambda > l^*$ and l^* given by (5.4.2), is irreducible. Then the spectrum of P lies inside the closed unit circle, with a simple eigenvalue at 1. By (5.7.1) L has a spectrum contained strictly in the left half-plane, together with a simple eigenvalue at 0. Thus the concept of aperiodicity is not required in the continuous-time case.

[5] This is, in fact, the root of the word ergodic = $\varepsilon\rho\gamma o\ +\ o\delta o\varsigma$ (work + path). The term was introduced by Boltzmann in the 1870s in the course of his work on the kinetic theory of gases.

5.8 Exercises

1. Show that the jump chain (5.3.1) satisfies the Markov property.
2. Give a definition of a symmetric random walk on $\mathbb{Z}^d$, where d is arbitrary.
3. Give an example of an asymmetric random walk on $\mathbb{Z}$.
4. Consider the discrete-time Markov chain with transition matrix

$$P = \begin{pmatrix} 1-\alpha & \alpha \\ \beta & 1-\beta \end{pmatrix}$$

with $\alpha,\ \beta \in [0,1]$. Find the invariant density ρ^∞. Under what conditions is the Markov chain ergodic?
5. A stochastic matrix P and a distribution π are said to be in **detailed balance** provided that

$$p_{ij}\pi_i = p_{ji}\pi_j \quad \forall i,j \in \mathcal{I}.$$

Show that π is an invariant distribution for P. (Enforcing this condition is related to making the Markov process reversible.)
6. Consider the continuous-time Markov chain with generator

$$L = \begin{pmatrix} -a & a \\ b & -b \end{pmatrix}.$$

Find the invariant density ρ^∞. Under what conditions is the Markov chain ergodic?
7. Implement a numerical algorithm to simulate paths of the continuous-time Markov chain from the previous example by using the canonical jump chain.

6

Stochastic Differential Equations

6.1 Setup

Gaussian white noise may be thought of informally as a mean-zero Gaussian process with correlation $\delta(t-s)I$. In practice such processes are an idealization of stochastic processes with very short correlation time. In this chapter we describe the background material concerning stochastic differential equations (SDEs)—ODEs driven by white noise—required for the remainder of the book. Let $W(t)$ denote a standard m-dimensional Brownian motion, $h : \mathcal{Z} \to \mathbb{R}^d$ a smooth vector-valued function and $\gamma : \mathcal{Z} \to \mathbb{R}^{d\times m}$ a smooth matrix-valued function. In the following, we typically take $\mathcal{Z} = \mathbb{T}^d$, $\mathbb{R}^d$ or $\mathbb{R}^l \oplus \mathbb{T}^{d-l}$. Consider the Itô SDE

$$\frac{dz}{dt} = h(z) + \gamma(z)\frac{dW}{dt}, \quad z(0) = z_0. \tag{6.1.1}$$

We think of the term dW/dt as representing Gaussian white noise. Such a process exists only as a distribution, and so the precise interpretation of (6.1.1) is as an integral equation for $z(t) \in C(\mathbb{R}^+, \mathcal{Z})$:

$$z(t) = z_0 + \int_0^t h(z(s))ds + \int_0^t \gamma(z(s))dW(s). \tag{6.1.2}$$

In order to make sense of this equation we need to define the stochastic integral against $dW(s)$. We use the Itô interpretation of the stochastic integral as defined in Chapter 3. Because it is notationally convenient to do so, we will frequently write SDEs in the unintegrated form (6.1.1). Whenever we do this, what is written should be interpreted as shorthand for (6.1.2). The function h in (6.1.1) is sometimes referred to as the *drift* and γ as the *diffusion coefficient*.

In later chapters we will often consider $z = (x^T, y^T)^T$, with $x \in \mathcal{X}$, $y \in \mathcal{Y}$. If $\mathcal{Z} = \mathbb{T}^d$ (resp. $\mathbb{R}^d$) then $\mathcal{X} = \mathbb{T}^l$ (resp. $\mathbb{R}^l$) and $\mathcal{Y} = \mathbb{T}^{d-l}$ (resp. $\mathbb{R}^{d-l}$). If $\mathcal{Z} = \mathbb{R}^l \oplus \mathbb{T}^{d-l}$ then $\mathcal{X} = \mathbb{R}^l$ and $\mathcal{Y} = \mathbb{T}^{d-l}$. When we consider Equation (4.1.1) on the torus, this is, as in Chapter 4, simply a convenient shorthand for the case that h, γ are periodic in z (resp. y) and that we consider z (resp. y) as an element of the torus, by working modulo 1 in all directions. On occasion we will consider differential operators on the torus and, in this setting, the operator automatically has periodic boundary conditions.

Sections 6.2–6.4 consider the existence and uniqueness of solutions, the generator, and the Kolmogorov and Fokker–Planck equations, and ergodicity for SDEs, respectively. In Section 6.5 we present extensions of the basic theory, together with references to the literature.

6.2 Existence and Uniqueness

By a solution of (6.1.1) we mean a $\mathcal{Z}$-valued stochastic process $\{z(t)\}$ on $t \in [0,T]$ with the properties:

i) z is continuous and $\mathcal{F}_t$-adapted, where the filtration is generated by the Brownian motion $W(t)$;
ii) $h(z(\cdot)) \in L^1((0,T)), \gamma(z(\cdot)) \in L^2((0,T))$;
iii) Equation (6.1.1) holds for every $t \in [0,T]$ with probability 1.

The solution is called *unique* if any two solutions $x_i(t), i = 1, 2$ satisfy

$$\mathbb{P}(x_1(t) = x_2(t), \quad \forall t \in [0,T]) = 1.$$

In Theorem 4.2 we proved existence and uniqueness of solutions for ODEs (i.e., when $\gamma \equiv 0$ in (6.1.1)) for globally Lipschitz vector fields h. A very similar theorem holds when $\gamma \neq 0$. As for ODEs the conditions can be weakened, when *a priori* bounds on the solution can be found, but we limit ourselves to the simple setup of the following theorem, for expository purposes.

Theorem 6.1. *Assume that both $h(\cdot)$ and $\gamma(\cdot)$ are globally Lipschitz on $\mathcal{Z}$ and that z_0 is a random variable, independent of the Brownian motion $W(t)$, and satisfying*

$$\mathbb{E}|z_0|^2 < \infty.$$

Then the SDE (6.1.1) *has a unique solution $z \in C(\mathbb{R}^+; \mathcal{Z})$ with*

$$\mathbb{E}\left[\int_0^T |z(t)|^2 \, dt\right] < \infty \quad \forall T < \infty.$$

Furthermore, the solution of the SDE is a Markov process.

We conclude the section with two remarks, both of which will play an important role in future chapters.

Remark 6.2. The Stratonovich analogue of (6.1.1) is

$$\frac{dz}{dt} = h(z) + \gamma(z) \circ \frac{dW}{dt}, \quad z(0) = z_0. \tag{6.2.1}$$

By this we mean that $z \in C(\mathbb{R}^+, \mathcal{Z})$ satisfies the integral equation

$$z(t) = z(0) + \int_0^t h(z(s))ds + \int_0^t \gamma(z(s)) \circ dW(s). \tag{6.2.2}$$

By using Definitions (3.4.6) and (3.4.9) it can be shown that z satisfying the Stratonovich SDE (6.2.1) also satisfies the Itô SDE

$$\frac{dz}{dt} = h(z) + \frac{1}{2}\nabla \cdot \left(\gamma(z)\gamma(z)^T\right) - \frac{1}{2}\gamma(z)\nabla \cdot \left(\gamma(z)^T\right) + \gamma(z)\frac{dW}{dt}, \quad z(0) = z_0, \tag{6.2.3}$$

provided that $\gamma(z)$ is differentiable; see Exercise 1.

White noise is, in most applications, an idealization of a stationary random process with short correlation time. In this context the Stratonovich interpretation of an SDE is particularly important because it sometimes arises as the limit obtained by using smooth approximations to white noise.[1] On the other hand, the martingale machinery that comes with the Itô integral makes it more important as a mathematical object. Hence conversion between the two viewpoints is very useful. □

Remark 6.3. The Definition 3.5 of Brownian motion implies the interesting scaling property

$$\{W(ct) : t \geqslant 0\} = \{\sqrt{c}W(t) : t \geqslant 0\},$$

where the preceding should be interpreted as holding in law on $C([0,\infty), \mathbb{R}^d)$. From this it follows that, if $s = ct$, then

$$\frac{dW}{ds} = \frac{1}{\sqrt{c}}\frac{dW}{dt},$$

again in law.

Hence, if we scale time to $s = ct$ in (6.1.1), then we get the equation

$$\frac{dz}{ds} = \frac{1}{c}h(z) + \frac{1}{\sqrt{c}}\gamma(z)\frac{dW}{ds}, \quad z(0) = z_0.$$

(The precise interpretation is as an integral equation, as always.) Notice that, while the SDE transforms unusually under $s = ct$, the Fokker–Planck equation, defined in the next section, transforms in the standard way, because it sees the quadratic term (6.3.1) formed from the diffusion coefficient. □

[1] Under some mild regularity conditions this is always true in one dimension, but not always true in higher dimensions: an additional drift related to the commutator between the row vectors of the diffusion matrix $\gamma(z)$ can appear in the limit as we remove the regularization, in addition to the Stratonovich stochastic integral. See the example in Section 11.7.7 and the discussion in Section 11.8.

6.3 The Generator

Given the function $\gamma(z)$ in the SDE (6.1.1) we define

$$\Gamma(z) = \gamma(z)\gamma(z)^T. \tag{6.3.1}$$

The **generator** $\mathcal{L}$ is then defined as

$$\mathcal{L}v = h \cdot \nabla v + \frac{1}{2}\Gamma : \nabla\nabla v. \tag{6.3.2}$$

This operator, equipped with a suitable domain of definition, is the generator of the Markov process given by (6.1.1), in the sense defined in Chapter 3. We will also be interested in the formal L^2–adjoint operator $\mathcal{L}^*$

$$\mathcal{L}^* v = -\nabla \cdot (hv) + \frac{1}{2}\nabla \cdot \nabla \cdot (\Gamma v).$$

Example 6.4. (i) Consider the SDE

$$\frac{dx}{dt} = \sigma \frac{dW}{dt}$$

on $\mathbb{R}^d$. Its solution is a Brownian motion on $\mathbb{R}^d$ with covariance matrix $\sigma^2 I$. The generator of the Markov process $x(t)$ is

$$\mathcal{L}\phi = \frac{\sigma^2}{2}\Delta\phi.$$

It is formally self-adjoint on $L^2(\mathbb{R}^d)$:

$$\mathcal{L}^*\phi = \frac{\sigma^2}{2}\Delta\phi.$$

(ii) Consider the Ornstein–Uhlenbeck (OU) SDE

$$\frac{dx}{dt} = -\alpha x + \sqrt{2\lambda}\frac{dW}{dt}$$

on $\mathbb{R}^d$ where α, λ are positive constants. The generator of the OU process is

$$\mathcal{L}\phi = -\alpha x \cdot \nabla\phi + \lambda\Delta\phi.$$

Its $L^2(\mathbb{R}^d)$-adjoint is

$$\mathcal{L}^*\phi = \nabla \cdot (\alpha x\phi) + \lambda\Delta\phi.$$

In fact, $\mathcal{L}$ is self-adjoint in a weighted L^2-space, where integration is performed with respect to the density of the invariant measure for the OU process; see Exercise 4.

(iii) The geometric Brownian motion on $\mathbb{R}$ is defined as the solution of the one-dimensional SDE

$$\frac{dx}{dt} = \mu x + \sigma x \frac{dW}{dt}$$

where $\mu \in \mathbb{R}$ and $\sigma > 0$. The generator of the geometric Brownian motion is

$$\mathcal{L}\phi = \mu x \frac{d\phi}{dx} + \frac{\sigma^2 x^2}{2} \frac{d^2\phi}{dx^2}.$$

Its L^2-adjoint is

$$\mathcal{L}^*\phi = -\frac{d}{dx}(\mu x \phi) + \frac{d^2}{dx^2}\left(\frac{\sigma^2 x^2}{2}\phi\right). \quad \square$$

The **Itô formula** that follows is the basic result concerning the rate of change in time of functions $V : \mathcal{Z} \to \mathbb{R}$ evaluated at the solution of a $\mathcal{Z}$-valued SDE. Heuristically it delivers the following result:

$$\frac{d}{dt}\Big(V(z(t))\Big) = \mathcal{L}V(z(t)) + \left\langle \nabla V(z(t)), \gamma(z(t))\frac{dW}{dt} \right\rangle.$$

This is the analogue of (4.3.2) for ODEs. Note that if W were a smooth time-dependent function, this formula would not be correct: there is an additional term in $\mathcal{L}V$, proportional to Γ, which arises from the lack of smoothness of Brownian motion.

As for the SDE (6.1.1) itself, the precise interpretation of the expression for the rate of change of V is in integrated form.

Lemma 6.5. (Itô Formula) *Assume that the conditions of Theorem 6.1 hold. Let $x(t)$ solve (6.1.1) and let $V \in C^2(\mathcal{Z}, \mathbb{R})$. Then the process $V(z(t))$ satisfies*

$$V(z(t)) = V(z(0)) + \int_0^t \mathcal{L}V(z(s))ds + \int_0^t \langle \nabla V(z(s)), \gamma(z(s))\, dW(s)\rangle.$$

As in the setting for ODEs, the formula is readily extended to vector-valued functions $V : \mathcal{Z} \to \mathbb{R}^n$.

Let $\phi : \mathcal{Z} \mapsto \mathbb{R}$ and consider the function

$$v(z,t) = \mathbb{E}\big(\phi(z(t))|z(0) = z\big), \tag{6.3.3}$$

where the expectation is with respect to all Brownian driving paths. By taking expectation in the Itô formula, which removes the stochastic integral, it is possible to deduce the following important consequence of Lemma 6.5.

Theorem 6.6. *Assume that ϕ is chosen sufficiently smooth so that the* **backward Kolmogorov equation**

$$\begin{aligned} \frac{\partial v}{\partial t} &= \mathcal{L}v \quad \mathit{for}\,(z,t) \in \mathcal{Z} \times (0,\infty), \\ v &= \phi \quad \mathit{for}\,(z,t) \in \mathcal{Z} \times \{0\}\,, \end{aligned} \tag{6.3.4}$$

with $\mathcal{L}$ as given in (6.3.2)*, has a unique bounded classical solution*[2] $v(x,t) \in C^{2,1}(\mathcal{Z} \times (0,\infty), \mathbb{R}) \cap C(\mathcal{Z} \times \mathbb{R}^+, \mathbb{R})$. *Then v is given by* (6.3.3) *where $z(t)$ solves* (6.1.2).

This is the analogue of the backward Equation (5.4.14) for Markov chains. If $\gamma \equiv 0$ in (6.1.1), so that the dynamics are deterministic, and φ^t is the flow on $\mathcal{Z}$ so that $z(t) = \varphi^t(z(0))$, then the Kolmogorov Equation (6.3.4) reduces to the hyperbolic Equation (4.3.5) whose characteristics are the integral curves of the ODE (4.1.1).

Example 6.7. Consider the three SDEs from Example 6.4. The corresponding backward Kolmogorov equations are

$$\frac{\partial v}{\partial t} = \frac{\sigma^2}{2} \Delta v$$

for the Brownian motion,

$$\frac{\partial v}{\partial t} = -\alpha x \cdot \nabla v + \lambda \Delta v$$

for the OU process and

$$\frac{\partial v}{\partial t} = \mu x \frac{\partial v}{\partial x} + \frac{\sigma^2 x^2}{2} \frac{\partial^2 v}{\partial x^2}$$

for the geometric Brownian motion. □

A direct consequence of Theorem 6.6 is an equation for propagation of densities. In deriving this equation in the next theorem we make use of the semigroup notation for the solution to time-dependent PDEs; see Section 7.5.

Theorem 6.8. *Consider Equation* (6.1.2) *with z_0 a random variable with density $\rho_0(z)$, independent of w. Assume that the law of $z(t)$ has a density $\rho(z,t) \in C^{2,1}(\mathcal{Z} \times (0,\infty), \mathbb{R}) \cap C(\mathcal{Z} \times \mathbb{R}^+, \mathbb{R})$. Then ρ satisfies the* **Fokker–Planck equation**[3]

$$\frac{\partial \rho}{\partial t} = \mathcal{L}^* \rho \quad \mathit{for}\,(z,t) \in \mathcal{Z} \times (0,\infty), \tag{6.3.5a}$$

$$\rho = \rho_0 \quad \mathit{for}\, z \in \mathcal{Z} \times \{0\}. \tag{6.3.5b}$$

---

[2] See Chapter 7.

[3] Sometimes called the **forward Kolmogorov equation**.

Proof. Let $\mathbb{E}^\mu$ denote averaging with respect to the product measure induced by the measure μ with density ρ_0 on $z(0)$ and the independent driving Wiener measure on the SDE itself. By the previous result, averaging over random z_0 distributed with density $\rho_0(z_0)$, we find

$$
\begin{aligned}
\mathbb{E}^\mu(\phi(z(t))) &= \int_{\mathcal{Z}} v(z_0,t)\rho_0(z_0)\,dz_0 \\
&= \int_{\mathcal{Z}} v(z,t)\rho_0(z)\,dz \\
&= \int_{\mathcal{Z}} (e^{\mathcal{L}t}\phi)(z)\rho_0(z)\,dz \\
&= \int_{\mathcal{Z}} (e^{\mathcal{L}^*t}\rho_0)(z)\phi(z)\,dz,
\end{aligned}
$$

for all ϕ smooth enough that Theorem 6.6 holds. But if $\rho(z,t)$ is the density of $z(t)$ we also have

$$\mathbb{E}^\mu(\phi(z(t))) = \int_{\mathcal{Z}} \rho(z,t)\phi(z)dz.$$

Equating these two expressions for the expectation at time t we obtain

$$\int_{\mathcal{Z}} (e^{\mathcal{L}^*t}\rho_0)(z)\phi(z)\,dz = \int_{\mathcal{Z}} \rho(z,t)\phi(z)\,dz.$$

We use a density argument so that the identity can be extended to all $\phi \in L^2(\mathcal{Z})$. Hence, from the preceding equation we deduce that

$$\rho(z,t) = \left(e^{\mathcal{L}^*t}\rho_0\right)(z).$$

Differentiation of the equation gives (6.3.5a). Setting $t = 0$ gives the initial condition (6.3.5b). Hence ρ is the solution of the initial value problem (6.3.5). □

The Fokker–Planck equation is the continuous analogue of the forward Equation (5.4.12) for Markov chains and of the Liouville Equation (4.3.9) for ODEs.

Remark 6.9. The fact that the solution of the Fokker–Planck equation is the density of a probability measure means that

$$\int_{\mathcal{Z}} \rho(z,t)\,dz = 1, \quad \rho(z,t) \geqslant 0 \ \ \forall t \in \mathbb{R}^+.$$

The conservation law for the integral follows from (6.3.5) because $\mathcal{L}^*$ is an operator in divergence form. Furthermore, the probabilistic interpretation implies that, if ρ is nonnegative at $t = 0$, then it remains nonnegative for all positive times. □

Remark 6.10. In the case where the initial condition for the SDE (6.1.1) is deterministic, $z = z_0$, the initial condition for the Fokker–Planck equation becomes a delta function (Dirac mass):

$$\rho(z, t = 0) = \delta(z - z_0).$$

The solution of the Fokker–Planck equation can be extended to allow for such irregular data. □

Example 6.11. Consider the three SDEs from Example 6.4. The corresponding Fokker–Planck equations are

$$\frac{\partial \rho}{\partial t} = \frac{\sigma^2}{2} \Delta \rho$$

for the Brownian motion,

$$\frac{\partial \rho}{\partial t} = \nabla \cdot (\alpha x \rho) + \lambda \Delta \rho \tag{6.3.6}$$

for the OU process and

$$\frac{\partial \rho}{\partial t} = -\frac{\partial}{\partial x}(\mu x \rho) + \frac{\partial^2}{\partial x^2}\left(\frac{\sigma^2 x^2}{2}\rho\right)$$

for the geometric Brownian motion. □

Note that constants are in the null space of the generator $\mathcal{L}$ given by (6.3.2), which implies that $\dim(\mathrm{Null}(\mathcal{L})) \geqslant 1$. Assuming that the operator $\mathcal{L}$ satisfies the Fredholm alternative (see Section 7.2.3), this implies that $\dim(\mathrm{Null}(\mathcal{L}^*)) \geqslant 1$, too. Thus we expect that $\mathcal{L}^*$ will have a nontrivial null space.

Now suppose ρ is a function in the null space of $\mathcal{L}^*$, which is positive and integrates to 1 on $\mathcal{Z}$. Then ρ is necessarily a steady solution of the Fokker–Planck equation:

$$\mathcal{L}^* \rho = 0, \tag{6.3.7}$$

equipped with the appropriate boundary conditions. Thus, a *stationary distribution* ρ is the solution of the elliptic PDE (6.3.7). We have the following important result giving the **Dirichlet form** associated with the operator $\mathcal{L}$. For simplicity we state and prove this result in the periodic setting.

Theorem 6.12. *Let $\rho \in C^2_{per}(\mathbb{T}^d)$ be any steady solution of the Fokker–Planck equation on $\mathbb{T}^d$ with periodic boundary conditions. Let $f \in C^2_{per}(\mathbb{T}^d)$. Then*

$$\int_{\mathbb{T}^d} \big(-\mathcal{L}f(z)\big) f(z) \rho(z)\, dz = \frac{1}{2}\int_{\mathbb{T}^d} \Big(\nabla f(z) \cdot \Gamma(z) \nabla f(z)\Big) \rho(z)\, dz \tag{6.3.8}$$

$$= \frac{1}{2}\int_{\mathbb{T}^d} |\gamma(z)^T \nabla f(z)|^2 \rho(z) dz. \tag{6.3.9}$$

Proof. We have

$$\begin{aligned}
\mathcal{L}^*(f\rho) &= -\nabla \cdot (h \rho f) + \frac{1}{2} \nabla \cdot \nabla \cdot \Big(\Gamma \rho f\Big) \\
&= (\mathcal{L}^* \rho) f + (-\mathcal{L} f)\rho + \nabla \rho \cdot \Gamma \nabla f \\
&\qquad + \rho \Gamma : \nabla\nabla f + \nabla f \cdot (\nabla \cdot \Gamma)\rho \\
&= (-\mathcal{L} f)\rho + \nabla \rho \cdot \Gamma \nabla f \\
&\qquad + \rho \Gamma : \nabla\nabla f + \nabla f \cdot (\nabla \cdot \Gamma)\rho.
\end{aligned}$$

Now let $g \in C^2_{per}(\mathbb{T}^d)$. Using the fact that $\mathcal{L}$ and $\mathcal{L}^*$ are adjoint operators in $L^2(\mathbb{T}^d)$, and by using the divergence theorem, we use the previous identity to perform the following calculations:

$$\begin{aligned}\int_{\mathbb{T}^d} (\mathcal{L}g) f\rho\, dz &= \int_{\mathbb{T}^d} g\mathcal{L}^*(f\rho)\, dz \\ &= \int_{\mathbb{T}^d} \Big(g(-\mathcal{L}f)\rho + g\nabla\rho\cdot\Gamma\nabla f\Big)\, dz \\ &\quad + \int_{\mathbb{T}^d} g\Big(\nabla f\cdot(\nabla\cdot\Gamma)\rho + \rho\Gamma : \nabla\nabla f\Big)\, dz \\ &= \int_{\mathbb{T}^d} g(-\mathcal{L}f)\rho\, dz - \int_{\mathbb{T}^d}\Big(\nabla g\cdot\Gamma\nabla f\Big)\rho\, dz. \end{aligned} \tag{6.3.10}$$

Equation (6.3.10) implies

$$\int_{\mathbb{T}^d} (-\mathcal{L}g)\, f\rho\, dz + \int_{\mathbb{T}^d} g(-\mathcal{L}f)\rho\, dz = \int_{\mathbb{T}^d}\Big(\nabla g\cdot\Gamma\nabla f\Big)\rho\, dz. \tag{6.3.11}$$

Equation (6.3.8) follows from the preceding equation upon setting $g = f$. □

Roughly speaking the previous result concerning the Dirichlet form shows that $-\mathcal{L}$ is a positive operator, in an appropriate weighted L^2-space. Indeed, let ρ be strictly positive on $\mathcal{Z}$ and define a measure $\mu(dx) = \rho(x)dx$. We can then introduce the weighted Hilbert space $L^2(\mu)$ with inner product and norm as follows:

$$(a,b)_\rho = \int_{\mathbb{T}^d} a(z)\cdot b(z)\rho(z)\, dz, \quad \|a\|_\rho^2 = (a,a)_\rho^2. \tag{6.3.12}$$

Then the preceding theorem shows that

$$(-\mathcal{L}f, f)_\rho = \frac{1}{2}\|\gamma^T\nabla f\|_\rho^2. \tag{6.3.13}$$

Remark 6.13. In suitable functional settings, the previous result concerning the Dirichlet form also applies to other choices of domain $\mathcal{Z}$, not just on the torus. Specifically, the function space should ensure that $\mathcal{L}^*$ is the adjoint of $\mathcal{L}$ and allow the divergence theorem calculation used to reach (6.3.10). Typically, these conditions are realized on noncompact spaces by means of decay assumptions at infinity. □

Example 6.14. For Example 6.4(i), Brownian motion on $\mathbb{T}$, we have $\rho(x) \equiv 1$, so that Lebesgue measure is invariant. In this case, (6.3.13) simply reduces to the integration by parts formula

$$-\frac{\sigma^2}{2}\int_0^1 \frac{d^2 f}{dx^2} f\, dx = \frac{\sigma^2}{2}\int_0^1 \left|\frac{df}{dx}\right|^2 dx. \quad \square$$

6.4 Ergodicity

As for ODEs and Markov chains, ergodicity is related to the existence of a unique invariant measure. As for continuous-time Markov chains, in the SDE case this is best characterized by asking for a unique (up to normalization) function in the null space of the adjoint of the generator. We refer to this function as the invariant distribution or stationary distribution.

It is necessarily the case that constants are in the null space of $\mathcal{L}$:

$$\mathcal{L}\mathbf{1} = 0.$$

The notation $\mathbf{1}$ is used to denote functions that are constant and equal to one, a.e. in an L^p sense. Roughly speaking, we will say that an SDE is ergodic if the null space of its generator consists *only* of constants. Assuming that $\mathcal{L}$ satisfies the Fredholm alternative, this is equivalent to saying that there exists a unique, everywhere positive normalized solution to the stationary Fokker–Planck equation:

$$\mathcal{L}^*\rho^\infty = 0, \quad \inf_{z\in\mathcal{Z}} \rho^\infty > 0, \quad \int_{\mathcal{Z}} \rho^\infty(z)\,dz = 1.$$

In this case we have that the long-time average of a function of the solution to the SDE is equal to the average of this function with respect to the **invariant distribution** $\rho^\infty(z)$ so that the SDE forgets its initial condition. Equation (6.4.2), which makes this idea precise, is often taken as the definition of ergodicity in the physics literature. As well as relating to single sample paths of the SDE, ergodicity can also be related to ensembles of sample paths, over different noise realizations: the solution to the Fokker–Planck Equation (6.3.5) for an ergodic SDE converges, in the limit as $t \to \infty$, to its invariant distribution; see (6.4.1).

Given an SDE, we would like to know whether it is ergodic. We present a rigorous result in the case where $\mathcal{Z} = \mathbb{T}^d$. Consider the SDE (6.1.1) on $\mathbb{T}^d$. We equip both the generator $\mathcal{L}$ and its adjoint $\mathcal{L}^*$ with periodic boundary conditions. Thus

$$\mathcal{D}(\mathcal{L}) = \mathcal{D}(\mathcal{L}^*) = C^2_{per}(\mathbb{T}^d).$$

The following is a mathematically precise statement of ergodicity for SDEs; it ensures that the heuristics just described do indeed hold in the periodic case. The definition generalizes our definition of ergodicity for continuous-time Markov chains.

Definition 6.15. *The SDE* (6.1.1) *with* $\mathcal{Z} = \mathbb{T}^d$ *is said to be* ergodic *if the spectrum of the generator lies strictly in the left half-plane, with the exception of a simple eigenvalue at the origin, corresponding to a strictly positive eigenfunction* $\rho^\infty(z)$.

In the following we use the shorthand notation $\rho(t)$ and $v(t)$ to denote the function-valued time-dependent solutions of the Fokker–Planck and backward Kolmogorov equations, respectively. Thus we may view $\rho(t)$, $v(t)$ as being in a Banach space, for each fixed t, and measure their size through the L^p-norms.

Theorem 6.16. *Equip $\mathcal{L}$, $\mathcal{L}^*$ on $\mathbb{T}^d$ with periodic boundary conditions and assume that $\Gamma(z)$ is strictly positive-definite, uniformly in $z \in \mathbb{T}^d$:*

$$\exists \bar{\gamma} > 0 : \langle \xi, \Gamma(z)\xi \rangle \geqslant \bar{\gamma}|\xi|^2, \quad \forall \xi \in \mathbb{R}^d, z \in \mathbb{T}^d.$$

Then the SDE (6.1.1) *is ergodic and satisfies the following five properties:*

- $\mathcal{N}(\mathcal{L}) = span\{1\}$*;*
- $\mathcal{N}(\mathcal{L}^*) = span\{\rho^\infty\}$, $\inf_{z \in \mathbb{T}^d} \rho^\infty(z) > 0$*;*
- $\exists C, \lambda > 0$ *such that the solution of the Fokker–Planck equation with initial data a Dirac mass at arbitrary* $z(0) \in \mathbb{T}^d$ *satisfies*

$$\|\rho(t) - \rho^\infty\|_1 \leqslant Ce^{-\lambda t} \quad \forall t > 0; \tag{6.4.1}$$

- $\exists C, \lambda > 0$ *such that the solution of the backward Kolmogorov equation with initial data a continuous function* ϕ *satisfies*

$$\left\| v(t) - \left(\int_{\mathbb{T}^d} \phi(z)\rho^\infty(z)\, dz \right) 1 \right\|_\infty \leqslant Ce^{-\lambda t} \quad \forall t > 0;$$

- *for all* $\phi \in C(\mathbb{T}^d)$

$$\lim_{T \to \infty} \frac{1}{T} \int_0^T \phi(z(t))dt = \overline{\phi} := \int_{\mathbb{T}^d} \phi(z)\rho^\infty(z)\, dz, \quad a.s. \tag{6.4.2}$$

Let I_A denote the indicator function of Borel set $A \subseteq \mathcal{Z}$. This function is not continuous but may be approximated by a sequence of continuous functions. By choosing ϕ to be I_A we deduce from the last result that the measure μ defined by

$$\mu(A) = \lim_{T \to \infty} \frac{1}{T} \int_0^T I_A(z(t))\, dt \tag{6.4.3}$$

has density ρ^∞. Thus $\mu(dz) = \rho^\infty(z)dz$. Furthermore, the invariant distribution μ clearly measures the proportion of time that a sample path of the SDE spends in a given set, as for ODEs and Markov chains.

Remark 6.17. Although we do not prove the results stated in the preceding theorem, a few remarks are in order:

(i) The fact that the null space of $\mathcal{L}$ comprises constants, when the diffusion matrix Γ is uniformly positive definite and when periodic boundary conditions are used, may be seen by means of the strong maximum principle – Example 7.15.
(ii) The same fact also follows directly from (6.3.13) if it is assumed that ρ^∞ is strictly positive.
(iii) Convergence of the time average (6.4.2) may be quantified as follows. Let $\mathcal{L}$ be the generator of an ergodic SDE (6.1.1) on $\mathbb{T}^d$ equipped with periodic boundary conditions and satisfying the assumptions of Theorem 6.16. Let Φ be the solution of the **Poisson equation**

$$\mathcal{L}\Phi = \phi - \overline{\phi},$$

where $\overline{\phi}$ denotes the average of a scalar-valued function ϕ with respect to the invariant measure of the SDE. By the Fredholm alternative, Theorem 2.42, this Poisson equation has a solution that is unique up to constants. Applying the Itô formula to $\Phi(x(t))$ gives

$$\Phi(z(t)) = \Phi(z(0)) + \int_0^t \mathcal{L}\Phi(z(s))ds + \int_0^t \langle \nabla\Phi(z(s)), \gamma(z(s))\, dW(s)\rangle$$

so that

$$\frac{1}{T}\Big(\Phi(z(T)) - \Phi(z(0))\Big) = \frac{1}{T}\int_0^T \phi(z(s))\, ds - \overline{\phi} + M_T,$$
$$M_T := \frac{1}{T}\int_0^T \langle \nabla\Phi(z(s)), \gamma(z(s))\, dW(s)\rangle.$$

By the Itô isometry and our assumptions on Φ we have that $\mathbb{E}|M_T|^2 = \mathcal{O}(1/T)$. Thus we obtain

$$\frac{1}{T}\int_0^T \phi(z(s))ds = \overline{\phi} + \mathcal{O}\Big(\frac{1}{\sqrt{T}}\Big)$$

in L^2. This is essentially the ergodic theorem, describing convergence of the time average of an ergodic SDE.[4] Notice also that we may characterize the corrections to this law via the martingale central limit theorem (Theorem 3.33), which may be used to show that $\sqrt{T}M_T$ converges to a Brownian motion with a covariance that can be expressed in terms of the Dirichlet form associated with the generator $\mathcal{L}$, evaluated at the solution of the Poisson equation Φ. □

Example 6.18. Consider a one-dimensional Brownian motion on $\mathbb{T}$:

$$\frac{dz}{dt} = \sigma\frac{dW}{dt}, \quad z(0) = z_0.$$

The generator $\mathcal{L}$ is the differential operator

$$\mathcal{L} = \frac{\sigma^2}{2}\frac{d^2}{dz^2},$$

equipped with periodic boundary conditions on $[0,1]$. This operator is self-adjoint. The null space of both $\mathcal{L}$ and $\mathcal{L}^*$ comprises constant functions on $[0,1]$. Both the backward Kolmogorov and the Fokker–Planck equation reduce to the heat equation

$$\frac{\partial\rho}{\partial t} = \frac{\sigma^2}{2}\frac{\partial^2\rho}{\partial x^2}$$

with periodic boundary conditions in $[0,1]$. A straightforward application of Fourier analysis shows that the solution converges to a constant at an exponential rate; see Exercise 7. □

[4] It is also a form of the law of large numbers.

Note that, while Brownian motion is ergodic on the torus it is not ergodic on $\mathbb{R}$ as the Lebesgue measure is not normalizable as a probability measure. However, although Theorem 6.16 as stated refers to the case where the state space is $\mathbb{T}^d$, it readily extends to a variety of settings with the appropriate function space choice for $\mathcal{L}$ and $\mathcal{L}^*$. To illustrate this, we include two examples on $\mathbb{R}^d$.

Example 6.19. Consider the OU process from Example 6.4

$$\frac{dz}{dt} = -\alpha z + \sqrt{2\lambda}\frac{dW}{dt}, \quad z(0) = z_0, \tag{6.4.4}$$

with $z(t) \in \mathbb{R}$. Here we assume that z_0 is fixed and nonrandom. An application of Itô's formula gives the solution

$$z(t) = e^{-\alpha t} z_0 + \sqrt{2\lambda} \int_0^t e^{-\alpha(t-s)} dW(s).$$

Hence,

$$\mathbb{E}z(t) = z_0 e^{-\alpha t}$$

and, by the Itô isometry,

$$\begin{aligned}
\mathbb{E}\big(z(t) - \mathbb{E}z(t)\big)^2 &= 2\lambda \mathbb{E}\left(\int_0^t e^{-\alpha(t-s)} dW(s)\right)^2 \\
&= 2\lambda \int_0^t e^{-2\alpha(t-s)}\, ds \\
&= \frac{\lambda}{\alpha}\left(1 - e^{-2\alpha t}\right).
\end{aligned}$$

The OU process $z(t)$ is Gaussian. Hence, the preceding calculations show that

$$z(t) \sim \mathcal{N}\Big(m(t), \sigma^2(t)\Big), \tag{6.4.5}$$

where

$$m(t) = e^{-\alpha t} z_0, \quad \sigma^2(t) = \frac{\lambda}{\alpha}(1 - e^{-2\alpha t}),$$

indicating convergence to the Gaussian invariant measure $\mathcal{N}(0, \lambda/\alpha)$ as $t \to \infty$. This is a manifestation of ergodicity.

The Fokker–Planck equation of the OU process is given by (6.3.6):

$$\frac{\partial \rho}{\partial t} = \frac{\partial}{\partial x}(\alpha x \rho) + \lambda \frac{\partial^2 \rho}{\partial x^2}$$

with initial condition being a Dirac mass centered at z_0. It is readily verified that the density associated with the Gaussian measure for $z(t)$, namely

$$\rho(x,t) = \sqrt{\frac{1}{2\pi\sigma^2(t)}} \exp\Big(-\frac{(x - m(t))^2}{\sigma^2(t)}\Big),$$

is a solution of this linear PDE. Thus the density ρ converges to a density ρ^∞, arising from the Gaussian random variable $\mathcal{N}(0, \lambda/\alpha)$, as $t \to \infty$. That is, the unique invariant distribution of the OU process is

$$\rho^\infty(x) = \sqrt{\frac{\alpha}{2\lambda\pi}} \exp\Big(-\frac{\alpha x^2}{2\lambda}\Big). \quad \square \tag{6.4.6}$$

Example 6.20. Consider the equations

$$\begin{aligned} \frac{dx}{dt} &= -ax + y, \quad x(0) = x_0, \\ \frac{dy}{dt} &= -y + \sqrt{2\sigma}\frac{dW}{dt}, \quad y(0) = y_0, \end{aligned}$$

for $(x(t), y(t)) \in \mathbb{R}^2$. We assume $a > 0$. The Fokker–Planck equation takes the form

$$\frac{\partial \rho}{\partial t} + \frac{\partial}{\partial x}((-ax + y)\rho) + \frac{\partial}{\partial y}(-y\rho) = \sigma \frac{\partial^2 \rho}{\partial y^2}.$$

The previous example shows that y has Gaussian distribution and converges to a Gaussian invariant measure. Since

$$x(t) = e^{-at}x(0) + \int_0^t e^{-a(t-s)} y(s) ds$$

and y is Gaussian we deduce that x too is Gaussian: its mean is e^{-at} and the variance is given by

$$\mathbb{E}\Big(\int_0^t e^{-a(t-s)} y(s) ds\Big)^2.$$

These considerations suggest that we seek a steady solution of the Fokker–Planck equation in the form

$$\rho^\infty(x, y) \propto \exp\big(-\alpha x^2 + \beta xy - \gamma y^2\big).$$

(The constant of proportionality should be chosen so that ρ^∞ integrates to 1 on $\mathbb{R}^2$.) Substitution shows that

$$\alpha = \frac{a(a+1)^2}{2\sigma}, \; \beta = \frac{2a(a+1)}{2\sigma}, \; \gamma = \frac{(a+1)}{2\sigma}.$$

Note that we have thus found the density of a Gaussian invariant measure for (x, y):

$$\rho(x, y) = \frac{1}{Z} \exp\Big(-\frac{a(a+1)^2}{2\sigma}x^2 + \frac{2a(a+1)}{2\sigma}xy - \frac{(a+1)}{2\sigma}y^2\Big),$$

where Z is the normalization constant. $\square$

Example 6.21. Consider the SDE

$$\frac{dz}{dt} = v(z) + \sqrt{2\sigma}\frac{dW}{dt}$$

where z is viewed as being an element of $\mathbb{T}^d$ and $\sigma > 0$. If v is divergence-free and smooth then the unique invariant measure is the Lebesgue measure. To see that the Lebesgue measure is invariant note that $\mathcal{L}^*$ is the generator for the SDE with $v(z)$ replaced by $-v(z)$. Theorem 6.16 ensures uniqueness of the invariant measure. While the Lebesgue measure remains invariant if $\sigma = 0$, uniqueness is not automatic in this case (see Chapter 14). □

Example 6.22. Consider the stochastic integral

$$I(t) = \int_0^t \eta(s)\, dW_1(s),$$

where $\eta(t)$ is the Ornstein–Uhlenbeck process defined in Example 6.19, namely

$$\frac{d\eta}{dt} = -\eta + \sqrt{2\sigma}\frac{dW_2}{dt}, \quad \eta(0) = \eta_0.$$

Here $W_1(t)$ and $W_2(t)$ are independent Brownian motions. The invariant measure for η is an $\mathcal{N}(0,\sigma)$ Gaussian random variable. We assume that the initial condition is distributed according to this invariant measure. Hence, $\eta(t)$ is a stationary ergodic Markov process and the martingale central limit theorem (Corollary 3.34) applies. Hence

$$\lim_{\varepsilon \to 0} \varepsilon I(t/\varepsilon^2) = \sqrt{\sigma} W(t),$$

where $W(t)$ is a standard Brownian motion in one dimension. □

6.5 Discussion and Bibliography

Standard textbooks on SDEs are [236, 12, 283, 114]. For a discussion of SDEs from the viewpoint of the Fokker–Planck equation, see [271, 117, 144, 218, 320]. For a discussion of the generator $\mathcal{L}$ and the backward Kolmogorov equation, see [236]. For a discussion concerning ellipticity, hypoellipticity, and smoothness of solutions to these equations, see [275, 276]. The book [139] has a good discussion of ergodicity. The book [210] has a good overview of stability theory and large time properties of SDEs. The Fokker–Planck equation is sometimes referred to as the forward Kolmogorov equation in the mathematics literature. The use of stochastic methods in applied mathematics is overviewed in [64].

We have only discussed *strong* solutions to SDEs. The definition of a *weak* solution, together with existence and uniqueness theorems for weak solutions can be found in [276, ch. 5]. The weak formulation of an SDE is equivalent to the *martingale formulation*; see [304].

A topic of some interest in applications involves the coupling of SDEs (or ODEs) and Markov chains. Consider the SDE (6.1.1) parameterized by $u \in \mathcal{I} = \{1, \cdots, n\}$:

$$\frac{dz}{dt} = h(z,u) + \gamma(z,u)\frac{dW}{dt}, \quad z(0) = z_0. \tag{6.5.1}$$

Let $\mathcal{L}^u$ denote the generator of this SDE for each fixed u. If we assume that u is governed by a continuous-time Markov chain, with generator $L(z)$, so that the transition rates depend on the state z, then the pair (z,u) form a *Markov switching process.* Such processes are indeed Markov processes and are overviewed in [211]. Noting that, for each fixed z, $L(z) \in \mathbb{R}^{n \times n}$, the generator for the combined process (z,x) has the form

$$\mathcal{L} = \begin{pmatrix} \mathcal{L}^1 & & & \\ & \mathcal{L}^2 & & \\ & & \ddots & \\ & & & \mathcal{L}^n \end{pmatrix} + L(z). \tag{6.5.2}$$

This generator acts on functions of the form $v(z,u) = (v(z,1), \cdots, v(z,n))^T$. Then $v(z,u,t) = \mathbb{E}(\phi(z(t),u(t))|z(t) = z, u(t) = u)$ solves the backward Kolmogorov equation

$$\frac{\partial v}{\partial t} = \mathcal{L}v$$

with $v(z,u,0) = \phi(z,u)$. Similarly the vector of probability densities $\rho(z,u,t)$ is propagated by the adjoint equation

$$\frac{\partial \rho}{\partial t} = \mathcal{L}^* v$$

with $\rho(z,u,0)$ describing the initial probabilities in the system. See [326] for a recent application of SDEs driven by Markov chains.

Ergodic properties of SDEs can be studied by a variety of techniques, both functional analytic PDE techniques (see, e.g., [337, 58, 196]) and probabilistic techniques (see, for example, [192, 227, 139]). Lyapunov functions can be a very useful tool for proving ergodicity for SDEs; see [188, sec. 11.9].

In this book we study SDEs in finite dimensions. There is also a well-developed theory of SDEs in infinite dimensions—stochastic PDEs, for example. See [263] for a semigroup-based framework and a brief discussion of ergodicity in this context.

6.6 Exercises

1. Derive the Itô SDE (6.2.3) from the Stratonovich SDE (6.2.1). Using the Itô form of the Stratonovich SDE, find the Fokker–Planck equation for the Stratonovich SDE.
2. Prove Theorem 6.6. (Hint: Use the Itô formula and the martingale property of the stochastic integral.)

3. Consider the OU process defined in Equation (6.4.4).
   a) Calculate all moments of the process $z(t)$.
   b) Verify the formula given for the solution of the Fokker–Planck equation in this case, showing that $\rho(0, x) = \delta(z - x)$.
   c) Deduce the long-time behavior of the OU process from the preceding formula.
4. Consider the OU process defined in Equation (6.4.4). Show that the generator is self-adjoint in the weighted L^2-space defined by the inner product (6.3.12).
5. Consider the SDE

$$m\ddot{x} = -\nabla V(x) - \gamma\dot{x} + \sqrt{\gamma D}\frac{dW}{dt}, \tag{6.6.1}$$

where m, D, γ are positive constants and $V(x) : \mathbb{R}^d \mapsto \mathbb{R}$ is a smooth function.
   a) Write Equation (6.6.1) as a first-order system of SDEs in the form

$$\frac{dz}{dt} = -\frac{1}{D}K\nabla H(z) + \frac{1}{\sqrt{m}}J\nabla H(z) + \sqrt{K}\frac{dB}{dt},$$

   where $z = (x^T, y^T)^T$, J (resp. K) is a skew (resp. symmetric) matrix that you should define, $y = \sqrt{m}\dot{x}$ and $H(z) = 1/2|y|^2 + V(x)$.
   b) Write the corresponding generator and the Fokker–Planck equation.
   c) Solve the stationary Fokker–Planck equation. (Hint: Use separation of variables.)
   d) Solve Equation (6.6.1) in one dimension for the cases $V(x) \equiv 0$ and $V(x) = 1/2x^2$.
6. Consider a Markov chain u with generator

$$L = \begin{pmatrix} -a & a \\ b & -b \end{pmatrix}.$$

Now let x solve an SDE with coefficients depending on u:

$$\frac{dx}{dt} = f(x, u) + \alpha(x, u)\frac{dW}{dt}.$$

Write down the generator for the Markov process (x, u).
7. Consider the case of a Brownian motion on the unit circle considered in Example 6.18. Use Fourier analysis to show that the process is ergodic and that it becomes stationary exponentially fast.

7

Partial Differential Equations

7.1 Setup

In this chapter we outline the basic theory of elliptic, parabolic, and hyperbolic PDEs, as required for this book. In Section 7.2 we study elliptic problems. We start with the Dirichlet and the periodic boundary value problem for elliptic PDEs in divergence form, introducing the notion of weak solutions, and using the Lax-Milgram theorem to prove existence and uniqueness of such solutions. Then we derive the Fredholm alternative for elliptic PDEs. We finish the section on elliptic PDEs with the maximum principle for nondivergence form operators. In Section 7.3 we study parabolic PDEs. We start by showing how the notion of weak solution may be extended to parabolic problems, particularly for problems on bounded spatial domains. Then we introduce the maximum principle for parabolic operators. We conclude the section by studying the Cauchy problem for parabolic PDEs on unbounded domains. Section 7.4 describes the Cauchy problem for hyperbolic PDEs on unbounded domains. Section 7.5 contains a brief discussion of semigroups and how they may be used to unify the understanding of certain aspects of time-dependent PDEs. In Section 7.6 we present various extensions of the basic theory, and we make some bibliographical remarks.

The function space settings that we use were described in Chapter 2. Throughout we will use the terminology *classical solution* to refer to a function that satisfies a PDE pointwise at every point in the domain where the equation itself holds, in the sense that all the required derivatives exist and are continuous in this domain and balance one another as dictated by the equation, and that the boundary conditions are satisfied; the latter requires extension of the function (and possibly derivatives) into the closure of the domain. This is sometimes also termed a *strong solution*. We will also introduce various notions of weak and mild solutions, which have fewer smoothness requirements on the solution.

7.2 Elliptic PDEs

The (homogeneous)[1] **Dirichlet problem** is to find u, a function defined on an open set $\Omega \subset \mathbb{R}^d$, solving

$$-\nabla \cdot (A\nabla u) = f \quad \text{for } x \in \Omega, \tag{7.2.1a}$$

$$u = 0 \quad \text{for } x \in \partial\Omega, \tag{7.2.1b}$$

where $A = A(x)$ is a positive definite matrix and $f = f(x) \in H^{-1}(\Omega)$.

Recall the space H defined to be the set of mean zero $H^1_{per}(\mathbb{T}^d)$ functions – see Equation (2.4.6). The **periodic problem** is to find u solving

$$-\nabla \cdot (A\nabla u) = f, \quad u \text{ is 1-periodic}, \tag{7.2.2}$$

where $A = A(x)$ is a 1-periodic positive definite matrix and $f = f(x) \in H^*$, the dual of H. We view this problem as a PDE on the torus $\mathbb{T}^d$. Recall from Chapter 2 that unless f integrates to zero on the torus, no solution exists. The space H^* given in (2.4.9) ensures that this condition is satisfied.

The class of coefficients $A(x)$ that we will consider is provided in the following definition.

Definition 7.1. *Let $\alpha, \beta \in \mathbb{R}$ be such that $0 < \alpha \leqslant \beta < \infty$. We define $M(\alpha, \beta, \Omega)$ to be the set of $d \times d$ matrices $A \in L^\infty(\Omega; \mathbb{R}^{d\times d})$ such that, for every vector $\xi \in \mathbb{R}^d$ and every $x \in \Omega$,*

(i) $\langle \xi, A(x)\xi \rangle \geqslant \alpha|\xi|^2$, and
(ii) $|A(x)\xi| \leqslant \beta|\xi|$.

Furthermore, we define $M_{per}(\alpha, \beta, \mathbb{T}^d)$ to be the set of matrices in $M(\alpha, \beta, \mathbb{T}^d)$ with periodic coefficients on $\mathbb{T}^d$.

In the first three subsections we study elliptic operators in the form

$$\mathcal{A} = -\nabla \cdot (A\nabla) + b \cdot \nabla + c. \tag{7.2.3}$$

In turn, we study the Dirichlet problem, the periodic problem, and the Fredholm alternative. When studying the maximum principle the elliptic operators will have the form

$$\mathcal{A} = -A : \nabla\nabla + b \cdot \nabla + c. \tag{7.2.4}$$

If $A \in M(\alpha, \beta, \Omega)$ then the operator $\mathcal{A}$ in (7.2.3) or (7.2.4) is said to be *uniformly elliptic*. Operators of the form (7.2.3), and the corresponding PDE, are said to be in *divergence form*. Operators of the form (7.2.4) are said to be in *nondivergence form*. Note that it is possible to convert between the two forms (7.2.3) and (7.2.4), assuming that the matrix $A \in C^1(\Omega; \mathbb{R}^{d\times d})$; see Exercise 1.

[1] Here refering to the homogeneity of the boundary conditions.

7.2.1 The Dirichlet Problem

First we give the precise definition of a solution. For this we will need to introduce the bilinear form

$$a[\phi, \psi] = \int_\Omega \langle A(x)\nabla\phi(x), \nabla\psi(x)\rangle \, dx, \tag{7.2.5}$$

for $\phi,\ \psi \in H_0^1(\Omega)$. Notice that

$$a[\phi, \psi] = (A\nabla\phi, \nabla\psi)$$

with $(\cdot, \cdot)$ the standard $L^2(\Omega)$ inner product. We will use the notation $\langle \cdot, \cdot \rangle_{H^{-1}, H_0^1}$ for the pairing between $H_0^1(\Omega)$ and its dual $H^{-1}(\Omega)$ (see Chapter 2).

Definition 7.2. *We will say that* $u \in H_0^1(\Omega)$ *is a* weak solution *of the boundary value problem* (7.2.1) *if*

$$a[u, v] = \langle f, v\rangle_{H^{-1}, H_0^1} \quad \forall v \in H_0^1(\Omega). \tag{7.2.6}$$

This solution concept arises from multiplying the PDE (7.2.1) by a test function $v \in H_0^1(\Omega)$ and integrating by parts. Asking that the resulting equality holds for all $v \in H_0^1(\Omega)$ gives the desired concept of weak solution. If $f \in L^2(\Omega)$ then, in place of (7.2.6), we sometimes write

$$a[u, v] = (f, v) \quad \forall v \in H_0^1(\Omega). \tag{7.2.7}$$

The Lax–Milgram Theorem 2.40 enables us to prove existence and uniqueness of weak solutions for the class of matrices $A(x)$ given by Definition 7.1.

Theorem 7.3. *The Dirichlet problem* (7.2.1) *with* $A \in M(\alpha, \beta, \Omega)$ *and* $f \in H^{-1}(\Omega)$ *has a unique weak solution* $u \in H_0^1(\Omega)$. *Moreover, the following estimate holds:*

$$\|u\|_{H_0^1(\Omega)} \leqslant \frac{1}{\alpha}\|f\|_{H^{-1}(\Omega)}. \tag{7.2.8}$$

Proof. We have to verify the conditions of the Lax–Milgram theorem. We start with coercivity. We use the positive definiteness of the matrix A to obtain:

$$\begin{aligned} a[u, u] &= \int_\Omega \langle A\nabla u, \nabla u\rangle \, dx \\ &\geqslant \alpha \int_\Omega |\nabla u|^2 \, dx = \alpha \|u\|_{H_0^1(\Omega)}^2. \end{aligned}$$

Now we proceed with continuity. We use the L^∞ bound on A, together with the Cauchy–Schwarz inequality, to estimate:

$$
\begin{aligned}
a[u,v] &= \int_\Omega \langle A\nabla u, \nabla v\rangle \, dx \\
&\leqslant \beta \int_\Omega |\nabla u||\nabla v| \, dx \\
&\leqslant \beta \|\nabla u\|_{L^2(\Omega)} \|\nabla v\|_{L^2(\Omega)} \\
&= \beta \|u\|_{H_0^1(\Omega)} \|v\|_{H_0^1(\Omega)}.
\end{aligned}
$$

The bilinear form $a[u, v]$ satisfies the conditions of the Lax–Milgram theorem and hence there exists a unique solution $u \in H_0^1(\Omega)$ of (7.2.6).

Finally we prove estimate (7.2.8). We have, using the generalized Cauchy–Schwarz inequality (2.4.5),

$$
\begin{aligned}
\alpha \|u\|^2_{H_0^1(\Omega)} &\leqslant a[u,u] = \langle f, u\rangle \\
&\leqslant \|f\|_{H^{-1}(\Omega)} \|u\|_{H_0^1(\Omega)},
\end{aligned}
$$

from which the estimate follows. □

If $f \in L^2(\Omega)$ then the following bound is also useful.

Remark 7.4. Consider the problem (7.2.1) with $A \in M(\alpha, \beta, \Omega)$ and $f \in L^2(\Omega)$. Then

$$
\|u\|_{H_0^1(\Omega)} \leqslant \frac{C_\Omega}{\alpha} \|f\|_{L^2(\Omega)},
$$

where C_Ω is the Poincaré constant for the domain Ω defined in Theorem 2.21. □

The bound (7.2.8) enables us to obtain information on the solution of one-parameter families of Dirichlet problems, induced by parametric dependence in A.

Theorem 7.5. *Assume that there exist positive constants α, β with $\alpha \leqslant \beta$ such that, for all $\varepsilon > 0$, the one-parameter family of matrices $A^\varepsilon = A^\varepsilon(x)$ belongs to $M(\alpha, \beta, \Omega)$. Consider the Dirichlet problem*

$$
-\nabla \cdot (A^\varepsilon \nabla u^\varepsilon) = f, \quad \textit{for } x \in \Omega \tag{7.2.9a}
$$

$$
u^\varepsilon = 0, \quad \textit{for } x \in \partial\Omega, \tag{7.2.9b}
$$

with $f = f(x) \in H^{-1}(\Omega)$. Then there exists a constant C independent of ε such that

$$
\|u^\varepsilon\|_{H_0^1(\Omega)} \leqslant C; \tag{7.2.10}
$$

furthermore, there exists a subsequence $\{\varepsilon_n\}_{n \geqslant 0}$ and a function $u \in H_0^1(\Omega)$ such that

$$
u^{\varepsilon_n}(x) \to u(x) \quad \textit{strongly in } L^2(\Omega).
$$

Proof. Estimate (7.2.8) implies (7.2.10). The Rellich compactness Theorem 2.19 implies that there exists a function $u \in H_0^1(\Omega)$ and a subsequence $\{\varepsilon_n\} \in \varepsilon$ such that u is the strong L^2-limit of u^{ε_n}. □

Remark 7.6. When studying homogenization for elliptic PDEs, in Chapters 12 and 19, we will study Equation (7.2.9), and we will be interested in finding the equation satisfied by the limit u. □

7.2.2 The Periodic Problem

It is intuitively clear that the solution of the periodic problem can be determined only up to a constant. To ensure uniqueness we need to fix this constant; this is why we work in H, the set of mean zero $H^1_{per}(\mathbb{T}^d)$ functions defined in (2.4.6). Recall that we use the notation $a_1[\cdot,\cdot]$ to denote the bilinear form

$$a_1[u,v] = \int_{\mathbb{T}^d} \langle A(x)\nabla u(x), \nabla v(x)\rangle \, dx \quad \forall u,\, v \in H. \tag{7.2.11}$$

Recall that we denote the pairing between H and its dual H^* by $\langle \cdot,\cdot\rangle_{H^*,H}$ (see Chapter 2).

Definition 7.7. *We will say that* $u \in H$ *is a* weak solution *of the boundary value problem* (7.2.2) *if*

$$a_1[u,v] = \langle f, v\rangle_{H^*,H} \quad \forall v \in H. \tag{7.2.12}$$

The structure of H^* means that (7.2.2) has a unique solution when $f \in H^*$ because it ensures that f then has mean zero. Hence existence and uniqueness of weak solutions to (7.2.2) holds within the space H. We have the following theorem.

Theorem 7.8. *The problem* (7.2.2) *with* $A \in M_{per}(\alpha,\beta,\mathbb{T}^d)$ *and* $f \in H^*$ *has a unique weak solution* $u \in H$*. Moreover, the following estimate holds:*

$$\|u\|_H \leqslant \frac{1}{\alpha}\|f\|_{H^*}. \tag{7.2.13}$$

The proof is almost identical to that of Theorem 7.3, so we omit it. The fact that (7.2.2) has a unique solution only when f has mean zero, a condition ensured by asking that $f \in H^*$, can also be shown by means of the Fredholm alternative – a topic that we now turn to.

7.2.3 The Fredholm Alternative

In this section we prove that elliptic differential operators in the divergence form

$$\mathcal{A} = -\nabla\cdot(A\nabla) + b\cdot\nabla + c, \tag{7.2.14}$$

with periodic coefficients A, b, c and equipped with periodic boundary conditions, satisfy the Fredholm alternative. Notice that Theorem 2.42 does not apply directly to the operator $\mathcal{A}$ because it is an unbounded operator. The main idea will be to study the *resolvent* operator

$$R_{\mathcal{A}}(\lambda) = (\mathcal{A} + \lambda I)^{-1}, \tag{7.2.15}$$

where I stands for the identity operator on $L^2_{per}(\mathbb{T}^d)$ and $\lambda > 0$. We will prove that this operator is compact, for λ sufficiently large; consequently, Fredholm theory can be used.

Our assumptions on the coefficients of $\mathcal{A}$ are

$$A \in M_{per}(\alpha, \beta, \mathbb{T}^d), \tag{7.2.16a}$$

$$A(x) = A(x)^T \quad \forall x \in \overline{\Omega}, \tag{7.2.16b}$$

$$b \in C^1_{per}(\mathbb{T}^d), \tag{7.2.16c}$$

$$c \in L^\infty_{per}(\mathbb{T}^d). \tag{7.2.16d}$$

The L^2-adjoint of $\mathcal{A}$ is $\mathcal{A}^*$ given by

$$\mathcal{A}^* U = -\nabla \cdot (A \nabla U) - \nabla \cdot (bU) + cU \tag{7.2.17}$$

also equipped with periodic boundary conditions. We want to study the PDE

$$\mathcal{A}u = f, \quad u \text{ is 1-periodic} \tag{7.2.18}$$

and its adjoint

$$\mathcal{A}^* U = F, \quad U \text{ is 1-periodic}, \tag{7.2.19}$$

for $f,\ F \in L^2_{per}(\mathbb{T}^d)$.

Let $a[\cdot,\cdot], a^*[\cdot,\cdot] : H^1_{per}(\mathbb{T}^d) \times H^1_{per}(\mathbb{T}^d) \to \mathbb{R}$ denote the bilinear forms associated with the operators $\mathcal{A}$ and $\mathcal{A}^*$, i.e.,

$$a[u,v] = \int_{\mathbb{T}^d} \Big(\langle A \nabla u, \nabla v \rangle + (b \cdot \nabla u)\, v + cuv \Big)\, dx \quad \forall\, u,\, v \in H^1_{per}(\mathbb{T}^d) \tag{7.2.20}$$

and

$$a^*[u,v] = \int_{\mathbb{T}^d} \Big(\langle A \nabla u, \nabla v \rangle - \nabla \cdot (bu) v + cuv \Big)\, dx \quad \forall\, u,\, v \in H^1_{per}(\mathbb{T}^d),$$

respectively. As in the previous subsection, we will say that u and U are weak solutions of the PDE (7.2.18) and (7.2.19) provided that

$$a[u,v] = (f,v) \quad \forall v \in H^1_{per}(\mathbb{T}^d) \tag{7.2.21}$$

and

$$a^*[U,V] = (F,V) \quad \forall V \in H^1_{per}(\mathbb{T}^d), \tag{7.2.22}$$

respectively.

We will use $\mathcal{N}$ to denote the null space of an operator. The main result of this section is contained in the next theorem.

Theorem 7.9. (Fredholm Alternative for Periodic Elliptic PDEs) *Assume conditions* (7.2.16). *Then the following alternative holds.*

i) Either there exists a unique solution of (7.2.18) *for every* $f \in L^2_{per}(\mathbb{T}^d)$*; or*

ii) the homogeneous equation

$$\mathcal{A}u = 0, \quad u \text{ is 1-periodic}, \tag{7.2.23}$$

has at least one nontrivial solution, and

$$1 \leqslant dim(\mathcal{N}(\mathcal{A})) = dim(\mathcal{N}(\mathcal{A}^*)) < \infty.$$

In this case the boundary value problem (7.2.18) *has a weak solution if and only if*

$$(f, v) = 0 \quad \forall v \in \mathcal{N}(\mathcal{A}^*).$$

For the proof of this theorem we will use the next two lemmas.

Lemma 7.10. *Assume conditions* (7.2.16). *Then there exist constants* ν, $\mu > 0$ *such that*

$$|a[u, v]| \leqslant \nu \|u\|_{H^1} \|v\|_{H^1}$$

and

$$\frac{\alpha}{2} \|u\|_{H^1}^2 \leqslant a[u, u] + \mu \|u\|_{L^2}^2$$

for all u, $v \in H^1_{per}(\mathbb{T}^d)$.

Proof. 1. We use the L^∞ bounds on the coefficients A, b, together with the Cauchy–Schwarz inequality to deduce:

$$\begin{aligned}
|a(u, v)| &\leqslant \left| \int_{\mathbb{T}^d} \langle A \nabla u \nabla v \rangle \, dx + \int_{\mathbb{T}^d} (b \cdot \nabla u) v \, dx + \int_{\mathbb{T}^d} cuv \, dx \right| \\
&\leqslant \|A\|_{L^\infty} \int_{\mathbb{T}^d} |\nabla u||\nabla v| \, dx + \|b\|_{L^\infty} \int_{\mathbb{T}^d} |\nabla u||v| \, dx + \|c\|_{L^\infty} \int_{\mathbb{T}^d} |u||v| \, dx \\
&\leqslant C \left(\|\nabla u\|_{L^2} \|\nabla v\|_{L^2} + \|\nabla u\|_{L^2} \|v\|_{L^2} + \|u\|_{L^2} \|v\|_{L^2} \right) \\
&\leqslant C \|u\|_{H^1} \|v\|_{H^1}.
\end{aligned}$$

2. We now use the uniform ellipticity of A to compute:

$$\begin{aligned}
\alpha \|\nabla u\|_{L^2}^2 &\leqslant \int_{\mathbb{T}^d} \langle \nabla u, A \nabla u \rangle \, dx \\
&= a[u, u] - \int_{\mathbb{T}^d} (b \cdot \nabla u) \, u \, dx - \int_{\mathbb{T}^d} c|u|^2 \, dx \\
&\leqslant a[u, u] + \int_{\mathbb{T}^d} \left(\|b\|_{L^\infty} |\nabla u||u| + \|c\|_{L^\infty} |u|^2 \right) dx.
\end{aligned} \tag{7.2.24}$$

Now we make use of the algebraic inequality

$$ab \leqslant \delta a^2 + \frac{1}{4\delta} b^2, \quad \forall \delta > 0.$$

Using this in the second term on the right-hand side of (7.2.24) we obtain

$$\int_{\mathbb{T}^d} |\nabla u||u|\, dx \leqslant \delta \|\nabla u\|_{L^2}^2 + \frac{1}{4\delta}\|u\|_{L^2}^2. \tag{7.2.25}$$

We choose δ so that

$$\alpha - \|b\|_{L^\infty}\delta = \frac{\alpha}{2}.$$

We use inequality (7.2.25) with δ chosen as in (7.2.24) to obtain

$$\frac{\alpha}{2}\|\nabla u\|_{L^2}^2 \leqslant a[u,u] + \frac{1}{4\delta}\|b\|_{L^\infty}\|u\|_{L^2}^2 + \|c\|_{L^\infty}\|u\|_{L^2}^2.$$

We now add $\frac{\alpha}{2}\|u\|_{L^2}^2$ on both sides of the preceding inequality to obtain

$$\frac{\alpha}{2}\|u\|_{H^1}^2 \leqslant a[u,u] + \mu\|u\|_{L^2}^2,$$

with

$$\mu = \frac{1}{4\delta}\|b\|_{L^\infty} + \|c\|_{L^\infty} + \frac{\alpha}{2}. \ \square$$

Lemma 7.11. *Assume conditions* (7.2.16). *Take μ from Lemma 7.10. Then for every $\lambda \geqslant \mu$ and each function $f \in L^2_{per}(\mathbb{T}^d)$ there exists a unique weak solution $u \in H^1_{per}(\mathbb{T}^d)$ of the problem*

$$(\mathcal{A} + \lambda I)u = f, \quad u(x) \textit{ is } 1\textit{-periodic}. \tag{7.2.26}$$

Proof. Let $\lambda \geqslant \mu$. Define the operator

$$\mathcal{A}_\lambda := \mathcal{A} + \lambda I. \tag{7.2.27}$$

The bilinear form associated to $\mathcal{A}_\lambda$ is

$$a_\lambda[u,v] = a[u,v] + \lambda(u,v) \quad \forall\, u,v \in H^1_{per}(\mathbb{T}^d). \tag{7.2.28}$$

Now, Lemma 7.10, together with our assumption that $\lambda \geqslant \mu$, implies that the bilinear form $a_\lambda[u,v]$ is continuous and coercive on $H^1_{per}(\mathbb{T}^d)$. Hence the Lax–Milgram theorem applies[2] and we deduce the existence and uniqueness of a solution $u \in H^1_{per}(\mathbb{T}^d)$ of the equation

$$a_\lambda[u,v] = (f,v) \quad \forall v \in H^1_{per}(\mathbb{T}^d). \tag{7.2.29}$$

This is precisely the weak formulation of the boundary value problem (7.2.26). $\square$

Proof of Theorem 7.9. 1. By Lemma 7.11 there exists, for every $g \in L^2_{per}(\mathbb{T}^d)$, a unique solution $u \in H^1_{per}(\mathbb{T}^d)$ of

$$a_\mu[u,v] = (g,v) \quad \forall v \in H^1_{per}(\mathbb{T}^d). \tag{7.2.30}$$

[2] Let $H = H^1_{per}(\mathbb{T}^d)$. Since $f \in L^2_{per}(\mathbb{T}^d)$, we have that $\langle f, v\rangle_{H^*,H} = (f,v)$, and this defines a bounded linear functional on $H^1_{per}(\mathbb{T}^d)$.

We use the resolvent operator defined in (7.2.15) to write the solution of (7.2.30) in the following form:

$$u = R_{\mathcal{A}}(\mu)g. \tag{7.2.31}$$

Consider now Equation (7.2.18). We add the term μu on both sides of this equation to obtain

$$\mathcal{A}_\mu u = \mu u + f,$$

where $\mathcal{A}_\mu$ is defined in (7.2.27). The weak formulation of this equation is

$$a_\mu[u, v] = (\mu u + f, v) \quad \forall v \in H^1_{per}(\mathbb{T}^d).$$

We can rewrite this as an integral equation (see (7.2.31))

$$u = R_{\mathcal{A}}(\mu)(\mu u + f),$$

or, equivalently,

$$(I - K)u = h,$$

where

$$K := \mu R_{\mathcal{A}}(\mu), \quad h = R_{\mathcal{A}}(\mu)f.$$

2. Now we claim that the operator $K : L^2_{per}(\mathbb{T}^d) \to L^2_{per}(\mathbb{T}^d)$ is compact. Indeed, let u be the solution of (7.2.30), which is given by (7.2.31). We use the second estimate in Lemma 7.10, the definition of the bilinear form (7.2.28), and the Cauchy–Schwarz inequality in (7.2.30) to obtain

$$\begin{aligned} \frac{\alpha}{2}\|u\|^2_{H^1} &\leqslant a_\mu[u, u] = (g, u) \\ &\leqslant \|g\|_{L^2}\|u\|_{L^2} \leqslant \|g\|_{L^2}\|u\|_{H^1}. \end{aligned}$$

Consequently,

$$\|u\|_{H^1} \leqslant \frac{2}{\alpha}\|g\|_{L^2}.$$

We use now (7.2.31), the definition of K, and the preceding estimate to deduce that

$$\|Kg\|_{H^1} \leqslant \mu\|u\|_{H^1} \leqslant \frac{2}{\alpha}\mu\|g\|_{L^2}. \tag{7.2.32}$$

By the Rellich compactness theorem, $H^1_{per}(\mathbb{T}^d)$ is compactly embedded in $L^2_{per}(\mathbb{T}^d)$ and consequently estimate (7.2.32) implies that K maps bounded sets in $L^2_{per}(\mathbb{T}^d)$ into compact ones in $L^2_{per}(\mathbb{T}^d)$. Hence, it is a compact operator.
3. We apply now the Fredholm alternative (Theorem 2.42) to the operator K: either

a. there exists a unique $u \in L^2_{per}(\mathbb{T}^d)$ such that

$$(I - K)u = h \tag{7.2.33}$$

or

b. there exists a nontrivial solution $u \in L^2_{per}(\mathbb{T}^d)$ of the homogeneous equation

$$(I - K)u = 0. \tag{7.2.34}$$

Let us assume that a. holds, giving a unique solution of (7.2.33). From the preceding analysis we deduce that there exists a unique weak solution $u \in H^1_{per}(\mathbb{T}^d)$ of (7.2.18). Assume now that b. holds, so that there is a nontrivial solution of (7.2.34). Let N and N^* denote the dimensions of null spaces of $I - K$ and $I - K^*$, respectively. From Theorem 2.42 we know that $N = N^*$. Moreover, it is straightforward to prove that

$$u \in \mathcal{N}(I - K) = 0 \Leftrightarrow a[u, \phi] = 0 \quad \forall u \in H^1_{per}(\mathbb{T}^d)$$

and

$$v \in \mathcal{N}(I - K^*) = 0 \Leftrightarrow a^*[v, \phi] = 0 \quad \forall \phi \in H^1_{per}(\mathbb{T}^d).$$

Thus, the Fredholm alternative for K implies the Fredholm alternative for $\mathcal{A}$ (within the context of weak solutions).
4. Now we prove the final part of the theorem. Let $v \in \mathcal{N}(I - K^*)$. By Theorem 2.42 we know that (7.2.34) has a solution if and only if

$$(h, v) = 0 \quad \forall v \in \mathcal{N}(I - K^*).$$

We compute

$$\begin{aligned}(h, v) = (R_{\mathcal{A}}(\mu)f, v) &= \frac{1}{\mu}(Kf, v) \\ &= \frac{1}{\mu}(f, K^*v) = \frac{1}{\mu}(f, v).\end{aligned}$$

Hence, problem (7.2.18) has a weak solution if and only if

$$(f, v) = 0 \quad \forall v \in \mathcal{N}(\mathcal{A}^*).$$

This completes the proof of the theorem. □

Example 7.12. Let $f \in L^2_{per}(\mathbb{T}^d)$ and assume that $A(x)$ satisfies assumptions (7.2.16a) and (7.2.16b). Then the problem

$$a_1[u, \phi] = (f, \phi) \quad \forall \phi \in H^1_{per}(\mathbb{T}^d),$$

where $a_1[\cdot, \cdot]$ is defined in (7.2.11), has a unique solution $u \in H$ if and only if

$$(f, 1) = 0. \tag{7.2.35}$$

Indeed, consider the homogeneous adjoint equation

$$\mathcal{A}^* v = 0.$$

Clearly, the constant function (say, $v = 1$) is a solution of this equation. The uniform ellipticity of the matrix $A(x)$ implies that

$$\int_{\mathbb{T}^d} |\nabla v|^2 \, dx = 0,$$

so that v is a constant a.e. with respect to Lebesgue measure. Hence the constant solution is unique. Since assumptions (7.2.16a) and (7.2.16b) are satisfied, Theorem 7.9 applies and the result follows. □

Remark 7.13. In the context of the Fredholm alternative, and in other settings, it will often be useful to employ the identity

$$\int_{\mathbb{T}^d} \nabla \cdot q(y) dy = 0,$$

which holds for any 1-periodic C^1 function q. This follows from the divergence theorem, using periodicity to show that the total flux of q through the boundary of the unit cube (torus) $\mathbb{T}^d$ is zero. □

7.2.4 The Maximum Principle

In this section we consider maximum principles for elliptic differential operators in nondivergence form. Specifically we consider the operator $\mathcal{L}$ from (7.2.4) given by

$$\mathcal{L} = A(x) : \nabla\nabla - b(x) \cdot \nabla - c(x). \tag{7.2.36}$$

Throughout we assume that $A, b, c \in C(\overline{\Omega})$ and that $A \in M(\alpha, \beta, \Omega)$. Thus the operator $\mathcal{L}$ is uniformly elliptic in Ω. We use the notation

$$u^+ = \max\{u, 0\}, \quad u^- = -\min\{u, 0\}. \tag{7.2.37}$$

As usual we use $\partial\Omega$ to denote the boundary of the domain Ω, $\partial\Omega = \overline{\Omega} \backslash \Omega$, which we assume to be smooth. First we state the strong maximum principle for uniformly elliptic operators in nondivergence form.

Theorem 7.14. *Assume that* $u \in C^2(\Omega) \cap C(\overline{\Omega})$ *and that* $c : \Omega \to \mathbb{R}^+$. *Let* $\mathcal{L}$ *be the elliptic operator* (7.2.36) *in nondivergence form.*

- *(i) If* $\mathcal{L}u \geqslant 0$ *in* Ω *then*
$$\max_{\overline{\Omega}} u \leqslant \max_{\partial\Omega} u^+;$$
- *(ii) if* $\mathcal{L}u \leqslant 0$ *in* Ω *then*
$$\min_{\overline{\Omega}} u \geqslant -\max_{\partial\Omega} u^-.$$

Thus if $\mathcal{L}u = 0$ *in* Ω *then*

$$\max_{\overline{\Omega}} |u| = \max_{\partial\Omega} |u|.$$

Example 7.15. Consider the SDE (6.1.2) in the case $\mathcal{Z} = \mathbb{T}^d$. The generator $\mathcal{L}$ is given by (6.3.2) equipped with periodic boundary conditions. Assume that $\Gamma \in M_{per}(\alpha, \beta, \mathbb{T}^d)$. Eigenfunctions of this generator that correspond to eigenvalue zero satisfy $\mathcal{L}\psi = 0$ in $\Omega_0 = [0,1]^d$. Consequently, for any such function ψ, the maximum of $|\psi|$ is attained on the boundary of $[0,1]^d$. However, since ψ is defined on $\mathbb{T}^d$, we deduce that $\mathcal{L}\psi = 0$ in $\Omega_a = a + [0,1]^d$ for any $a \in \mathbb{R}^d$. Since $|\psi|$ has maximum attained on the boundary of Ω_a, for any $a \in \mathbb{R}^d$, $|\psi|$ must be a constant function. Combining with the continuity of ψ we deduce that $\mathcal{L}$ on $\mathbb{T}^d$ has one-dimensional null space spanned by constants. By the Fredholm alternative it follows that $\mathcal{L}^*$ also has a one-dimensional null space, spanned by ρ^∞. The fact that the null space is one-dimensional is a reflection of the ergodicity of the process. □

The maximum principle can be used to obtain *a priori bounds* for solutions of elliptic PDEs. A result of this form is the following.

Corollary 7.16. *Let Ω be a bounded domain of $\mathbb{R}^d$ with smooth boundary and let $u \in C(\overline{\Omega}) \cap C^2(\Omega)$ be a classical solution of the inhomogeneous Dirichlet problem*

$$\mathcal{L}u = f \quad \text{for } x \in \Omega,$$

$$u = g \quad \text{for } x \in \partial\Omega,$$

where $\mathcal{L}$ is given by (7.2.36) *and where the coefficients are smooth, A is such that $-\mathcal{L}$ is uniformly elliptic, b is bounded, and $c \geqslant 0$. Then the solution u satisfies the estimate*

$$\|u\|_{L^\infty(\Omega)} \leqslant \|g\|_{L^\infty(\partial\Omega)} + C\|f\|_{L^\infty(\Omega)}. \tag{7.2.38}$$

The constant C depends on the size of the domain, on $\|b\|_{L^\infty(\Omega)}$, and on the uniform ellipticity constant α.

7.3 Parabolic PDEs

7.3.1 Bounded Domains

Suppose that $\mathcal{L}$ is an elliptic differential operator in divergence form, given by (7.2.3). Let $\Omega \subset \mathbb{R}^d$ be bounded and open, $T > 0$, and define $\Omega_T = \Omega \times (0,T)$. Consider the parabolic PDE given by

$$\frac{\partial u}{\partial t} = \mathcal{L}u + f, \quad \text{for } (x,t) \in \Omega_T, \tag{7.3.1a}$$

$$u = 0, \quad \text{for } (x,t) \in \partial\Omega \times (0,T], \tag{7.3.1b}$$

$$u = g, \quad \text{for } (x,t) \in \overline{\Omega} \times \{0\}. \tag{7.3.1c}$$

We assume that $A, b, c \in L^\infty(\Omega)$, that $f \in L^2(\Omega_T)$, and that $g \in L^2(\Omega)$. We also assume that $A \in M(\alpha, \beta, \Omega)$ so that $-\mathcal{L}$ is uniformly elliptic.

We define the bilinear form $a[u, v]$ as in (7.2.20). We view the solution u as being a function $u : [0, T] \to H_0^1(\Omega)$ and work with the following definition of weak solution.

Definition 7.17. *A* weak solution *of* (7.3.1) *is a* $u \in L^2((0, T); H_0^1(\Omega))$ *satisfying:*

- $du/dt \in L^2((0, T); H^{-1}(\Omega))$*;*
- $\langle du/dt, v\rangle_{H^{-1}, H_0^1} + a[u, v] = (f, v) \quad \forall v \in H_0^1(\Omega), a.e.\ t \in (0, T)$*;*
- $u(0) = g$.

Remark 7.18. Standard properties of Sobolev spaces imply that $u \in C([0, T]; L^2(\mathbb{R}^d))$ and, consequently, the equation $u(0) = g$ makes sense. □

Theorem 7.19. *Under the stated assumptions there is a unique weak solution of Equation* (7.3.1)*. Furthermore the solution satisfies, for* $u' = du/dt$,

$$\|u\|_{L^\infty(0,T);L^2(\Omega))} + \|u\|_{L^2((0,T);H_0^1(\Omega))} + \|u'\|_{L^2((0,T);H^{-1}(\Omega))} \leqslant C\Big(\|f\|_{L^2((0,T);L^2(\Omega))} + \|g\|_{L^2(\Omega)}\Big). \tag{7.3.2}$$

The methodology used to prove this result is known as the *energy method*. Estimates of the form (7.3.2) are usually called *energy estimates* The use of weak solutions may be extended to unbounded domains. We illustrate this in the context of transport equations in Section 7.4.

7.3.2 The Maximum Principle

Maximum principles are very useful for studying parabolic PDEs, and they can be used to obtain (pointwise) a priori estimates. We use the notation

$$\Omega_T = \Omega \times (0, T];$$
$$\partial\Omega_T = \partial\Omega \times \{0\}.$$

Hence $\partial\Omega = \overline{\Omega}\backslash\Omega$ and $\partial\Omega_T = \overline{\Omega}_T\backslash\Omega_T$. As in the elliptic case, we assume that $A, b, c \in C(\Omega)$ and that $A \in M(\alpha, \beta, \Omega)$, so that the operator $\mathcal{L}$ is uniformly elliptic in Ω, and we use the notation

$$u^+ = \max\{u, 0\}, \quad u^- = -\min\{u, 0\}$$

as in (7.2.37).

Theorem 7.20. *Assume that* $u \in C^{2,1}(\Omega_T) \cap C(\overline{\Omega}_T)$ *and that* $c : \Omega \to \mathbb{R}^+$*. Consider* $\mathcal{L}$ *the elliptic operator* (7.2.36) *in nondivergence form.*

- *(i) If* $\partial u/\partial t - \mathcal{L}u \leqslant 0$ *in* Ω_T*, then*

$$\max_{\overline{\Omega}_T} u \leqslant \max_{\partial\Omega_T} u^+;$$

- *(ii) if $\partial u/\partial t - \mathcal{L}u \geqslant 0$ in Ω_T, then*

$$\min_{\overline{\Omega}_T} u \geqslant - \max_{\partial\Omega_T} u^- .$$

Thus if $\partial u/\partial t - \mathcal{L}u = 0$ in Ω_T, then

$$\max_{\overline{\Omega}_T} |u| = \max_{\partial\Omega_T} |u|.$$

An important corollary of the strong maximum principle for parabolic PDEs is the following a priori estimate.

Corollary 7.21. *Let Ω be a bounded domain of $\mathbb{R}^d$ with smooth boundary and let $u \in C^{2,1}(\Omega_T, \mathbb{R}) \cap C(\overline{\Omega}_T, \mathbb{R})$ be a classical solution of the equation*

$$\frac{\partial u}{\partial t} - \mathcal{L}u = f, \quad (x,t) \in \Omega \times \mathbb{R}^+$$

with $u(x,0) = g(x)$ and $c(x) \equiv 0$, where $\mathcal{L}$ is given by (7.2.36) *and the coefficients are smooth, A is such that $-\mathcal{L}$ is uniformly elliptic, b is bounded, and $c \geqslant 0$. Assume that $u = 0$ on $\partial\Omega$. Then u satisfies the estimate*

$$\max_{\Omega_T} |u| \leqslant \|g\|_{L^\infty(\Omega)} + \int_0^T \|f(\cdot,s)\|_{L^\infty(\Omega)} ds. \tag{7.3.3}$$

Proof. Let

$$v(t) = \|g\|_{L^\infty(\Omega)} + \int_0^t \|f(\cdot,s)\|_{L^\infty(\Omega)} ds.$$

Notice that

$$\frac{\partial v}{\partial t} - \mathcal{L}v = \|f(\cdot,t)\|_{L^\infty(\Omega)}$$

and $v(0) = \|g\|_{L^\infty(\Omega)}$. Thus, for $e = u - v$,

$$\frac{\partial e}{\partial t} - \mathcal{L}e \leqslant 0$$

and, since $e \leqslant 0$ on $\partial\Omega_T$, we deduce from the maximum principle (Theorem 7.20) that

$$\max_{\overline{\Omega}_T} e \leqslant 0$$

and hence that $u \leqslant v$ in $\overline{\Omega}_T$. Thus

$$\max_{\overline{\Omega}_T} u \leqslant \|g\|_{L^\infty(\Omega)} + \int_0^T \|f(\cdot,s)\|_{L^\infty(\Omega)} ds.$$

Similarly, for $d = u + v$,

$$\frac{\partial d}{\partial t} - \mathcal{L}d \geqslant 0$$

and, since $d \geqslant 0$ on $\partial\Omega_T$, we deduce from the maximum principle (Theorem 7.20) that

$$\min_{\overline{\Omega}_T} d \geqslant 0$$

and hence that $-u \leqslant v$ in $\overline{\Omega}_T$. Thus

$$\min_{\overline{\Omega}_T}(-u) \leqslant \|g\|_{L^\infty(\Omega)} + \int_0^T \|f(\cdot,s)\|_{L^\infty(\Omega)} ds.$$

Combining the upper and lower bounds gives us (7.3.3). □

7.3.3 Unbounded Domains: The Cauchy Problem

In this section we describe the basic theory of parabolic PDEs in nondivergence form and on unbounded domains. Specifically we study the initial value (Cauchy) problem

$$\frac{\partial u}{\partial t} = \mathcal{L}u + f, \quad \text{for } (x,t) \in \mathbb{R}^d \times (0,T], \tag{7.3.4a}$$

$$u = g \quad \text{for } (x,t) \in \mathbb{R}^d \times \{0\}, \tag{7.3.4b}$$

where

$$\mathcal{L} := b(x) \cdot \nabla + \frac{1}{2} A(x) : \nabla\nabla. \tag{7.3.5}$$

We assume that $A, b \in C_b^\infty(\mathbb{R}^d)$, that $f \in C_0^\infty(\mathbb{R}^d \times [0,T])$, and that $A \in M(\alpha, \beta, \mathbb{R}^d)$, $0 < \alpha \leqslant \beta < \infty$, so that the operator $\mathcal{L}$ is uniformly elliptic. We assume also that the initial function $g \in C_b^\infty(\mathbb{R}^d)$.

Theorem 7.22. *Under the stated assumptions, for any $T > 0$, there exists a unique solution $u(x,t) \in C^{2,1}\big(\mathbb{R}^d \times (0,T)\big) \cap C^\infty\big(\mathbb{R}^d \times (0,T)\big)$ to the Cauchy problem* (7.3.4), *for any $t > 0$. Furthermore the following estimate holds.*

$$\|u\|_{L^\infty(\mathbb{R}^d \times [0,t])} \leqslant \|g\|_{L^\infty(\mathbb{R}^d)} + \int_0^t \|f(\cdot,s)\|_{L^\infty(\mathbb{R}^d)} \, ds. \tag{7.3.6}$$

Remark 7.23. The solution $u(x,t)$ in the Theorem 7.22 is a *classical solution.* Estimate (7.3.6) is a consequence of the maximum principle for parabolic PDE (see Corollary 7.21). The *weak formulation* of the Cauchy problem (7.3.4) is obtained by multiplying the equation by a smooth, compactly supported function in $\mathbb{R}^d$ and integrating by parts; see Exercise 8. It is also possible to obtain energy estimates, within the context of weak solutions, for solutions to the Cauchy problem (7.3.4), when $\mathcal{L}$ is in divergence form (see Exercise 11):

$$\|u\|_{L^\infty((0,T);L^2(\mathbb{R}^d))} + C_1\|\nabla u\|_{L^2((0,T);L^2(\mathbb{R}^d))} \leqslant C_2(T)\Big(\|f\|_{L^2((0,T);L^2(\mathbb{R}^d))} + \|g\|_{L^2(\mathbb{R}^d)}\Big). \tag{7.3.7}$$

We discuss weak solutions for transport equations on $\mathbb{R}^d$ in the next section. □

7.4 Transport PDEs

In this section we investigate some basic properties of solutions to the Cauchy problem for linear transport PDEs. These are hyperbolic PDEs, or advection equations. We study problems in nondivergence form – specifically we study the equation

$$\frac{\partial u}{\partial t} + a(x) \cdot \nabla u = f(x) \quad \text{for } (x,t) \in \mathbb{R}^d \times (0,T), \tag{7.4.1a}$$

$$u = g(x) \quad \text{for } (x,t) \in \mathbb{R}^d \times \{0\}, \tag{7.4.1b}$$

where $a \in C_b^2(\mathbb{R}^d;\mathbb{R}^d)$ and $f,\, g \in H^1(\mathbb{R}^d)$. Notice that, when $f \equiv 0$, Equation (7.4.1) is the backward equation for the differential equation

$$\frac{dx}{dt} = -a(x).$$

We will define an appropriate concept of solution for (7.4.1) and state and prove a basic existence and uniqueness theorem.

As always for definitions of weak solutions, this formulation of (7.4.1) involves multiplication by a test function and integration over $\mathbb{R}^d$. We have the following definition.

Definition 7.24. *A function* $u \in L^2((0,T);H^1(\mathbb{R}^d))$ *is a* weak solution *of* (7.4.1) *provided that*

- $\frac{\partial u}{\partial t} \in L^2((0,T);L^2(\mathbb{R}^d))$.
- $(\frac{\partial u}{\partial t},\phi) + (a(x)\cdot\nabla u,\phi) = (u,f)$ *for every* $\phi(x) \in H^1(\mathbb{R}^d)$, *where* $(\cdot,\cdot)$ *denotes the* $L^2(\mathbb{R}^d)$ *inner product.*
- $u(x,0) = g$.

Remark 7.25. This definition is consistent with the definition of weak solutions for parabolic PDEs on bounded domains, Definition 7.17. That definition can be extended to unbounded domains in the parabolic case, as we have done here for transport equations. □

Remark 7.26. Standard properties of Sobolev spaces imply that $u \in C([0,T];L^2(\mathbb{R}^d))$ and, consequently, the equation $u(x,0) = g$ makes sense. □

Existence and uniqueness of weak solutions holds for the linear transport Equation (7.4.1).

Theorem 7.27. *Assume that* $a(x) \in C_b^2(\mathbb{R}^d;\mathbb{R}^d)$ *and that* $f,\, g \in H^1(\mathbb{R}^d)$. *Then there exists a unique weak solution of* (7.4.1). *Furthermore, the following a priori estimate holds:*

$$\|u\|_{L^\infty((0,T);L^2(\mathbb{R}^d))} \leqslant C(T,f,g). \tag{7.4.2}$$

In the case where the vector field $a(x)$ *is divergence free, so that* $\nabla \cdot a(x) = 0$, *and where* $f \equiv 0$, *the estimate becomes*

$$\|u\|_{L^\infty((0,T);L^2(\mathbb{R}^d))} \leqslant \|g\|_{L^2(\mathbb{R}^d)}. \tag{7.4.3}$$

Proof. The proof, which we will only sketch, is based on the *method of vanishing viscosity*: the idea is to "regularize" the initial value problem (7.4.1) by adding second-order spatial derivatives:

$$\frac{\partial u^\varepsilon}{\partial t} + a(x) \cdot \nabla u^\varepsilon - \varepsilon \Delta u^\varepsilon = f(x) \quad \text{for } (x,t) \in \mathbb{R}^d \times (0,T), \tag{7.4.4a}$$

$$u^\varepsilon = g^\varepsilon(x) \quad \text{for } (x,t) \in \mathbb{R}^d \times \{0\}, \tag{7.4.4b}$$

where $\varepsilon > 0$ and g^ε is a smooth, compactly supported approximation to $g(x)$.[3] Existence and uniqueness of solutions to this initial value problem follow from an application of Banach's fixed point theorem in the space $L^\infty((0,T); H^1(\mathbb{R}^d))$. Our assumptions on a, f, and g imply, by parabolic regularity theory, that

$$u^\varepsilon \in L^2(0,T; H^3(\mathbb{R}^d)), \quad \frac{\partial u^\varepsilon}{\partial t} \in L^2((0,T); H^1(\mathbb{R}^d)).$$

Multiplication of (7.4.4) by appropriate functionals of the solution of u^ε, integration by parts, and standard inequalities lead to the following a priori estimate:

$$\|u^\varepsilon\|_{L^\infty((0,T);H^1(\mathbb{R}^d))} + \left\| \frac{\partial u^\varepsilon}{\partial t} \right\|_{L^\infty((0,T);L^2(\mathbb{R}^d))} \leqslant C(g,f). \tag{7.4.5}$$

Crucially, $C(g,f)$ is independent of ε. This estimate implies that we can extract a weakly convergent subsequence, still denoted by u^ε, such that

$$u^\varepsilon \rightharpoonup u \quad \text{weakly in } L^2((0,T); H^1(\mathbb{R}^d))$$

$$\frac{\partial u^\varepsilon}{\partial t} \rightharpoonup u \quad \text{weakly in } L^2((0,T); L^2(\mathbb{R}^d))$$

We pass to the limit as $\varepsilon \to 0$ in the weak formulation of Equation (7.4.4) to obtain the weak formulation of Equation (7.4.1). We can show that $u(x,0) = g(x)$ by using a test function $v \in C([0,T]; H^1(\mathbb{R}^d))$, $v(x,T) = 0$, integrate the weak formulation of (7.4.1) over $(0,T)$, and integrate by parts with respect to t. To show uniqueness, we assume that there are two solutions u_1, u_2 and form their difference $u = u_1 - u_2$, which solves the homogeneous initial value problem

$$\frac{\partial u}{\partial t} + a(x) \cdot \nabla u = 0 \quad \text{for } (x,t) \in \mathbb{R}^d \times (0,T), \tag{7.4.6}$$

$$u = 0 \quad \text{for } (x,t) \in \mathbb{R}^d \times \{0\}, \tag{7.4.7}$$

which leads to

[3] This approximation is obtained by taking the convolution of g with a smooth compactly supported function. The function $g^\varepsilon(x)$ is called the *mollified* approximation to $g(x)$.

$$\frac{1}{2}\frac{d}{dt}\|u\|^2_{L^2(\mathbb{R}^d)} + \int_{\mathbb{R}^d} (a \cdot \nabla u) u \, dx = 0.$$

Now let $v \in C_0^\infty(\mathbb{R}^d)$. We compute

$$\begin{aligned}\int_{\mathbb{R}^d} (a(x) \cdot \nabla v) v \, dx &= \frac{1}{2}\int_{\mathbb{R}^d} a(x) \cdot \nabla v^2 \, dx - \frac{1}{2}\int_{\mathbb{R}^d} (\nabla \cdot a) v^2 \, dx \\ &\leqslant \frac{1}{2}\|\nabla \cdot a\|_{L^\infty(\mathbb{R}^d)} \|v\|^2_{L^2(\mathbb{R}^d)} \leqslant C\|v\|^2_{L^2(\mathbb{R}^d)}. \end{aligned} \quad (7.4.8)$$

An approximation argument yields that the difference $u(x,t)$ satisfies the estimate

$$\int_{\mathbb{R}^d} (a(x) \cdot \nabla u) u \, dx \leqslant C\|u\|^2_{L^2(\mathbb{R}^d)}.$$

Consequently,

$$\frac{d}{dt}\|u\|^2_{L^2(\mathbb{R}^d)} \leqslant C\|u\|^2_{L^2(\mathbb{R}^d)}.$$

Gronwall's inequality now gives $u \equiv 0$ and uniqueness follows.

Let us proceed now with the proof of estimates (7.4.2) and (7.4.3). We multiply Equation (7.4.1a) by u, integrate over $\mathbb{R}^d$, and use estimate (7.4.8) and the Cauchy-Schwarz inequality to obtain

$$\begin{aligned}\frac{1}{2}\frac{d}{dt}\|u\|^2_{L^2(\mathbb{R}^d)} &\leqslant \|f\|_{L^2(\mathbb{R}^d)}\|u\|_{L^2(\mathbb{R}^d)} + C\|u\|^2_{L^2(\mathbb{R}^d)} \\ &\leqslant \|f\|^2_{L^2(\mathbb{R}^d)} + C\|u\|^2_{L^2(\mathbb{R}^d)}.\end{aligned}$$

Gronwall's lemma now yields

$$\|u\|^2_{L^2(\mathbb{R}^d)} \leqslant e^{Ct}\left(\|g\|^2_{L^2(\mathbb{R}^d)} + \|f\|^2_{L^2(\mathbb{R}^d)}\right),$$

from which estimate (7.4.2) follows upon taking the supremum over $[0,T]$.

To prove estimate (7.4.3), we notice that the fact that $a(x)$ is divergence-free, together with the integration by parts leading to (7.4.8), implies that

$$\int_{\mathbb{R}^d} (a \cdot \nabla u) u \, dx = 0.$$

Consequently, and since we have also assumed that $f \equiv 0$, we obtain

$$\frac{1}{2}\frac{d}{dt}\|u\|^2_{L^2(\mathbb{R}^d)} \leqslant 0,$$

from which the estimate follows. □

Remark 7.28. Let u be the unique weak solution of (7.4.1). We multiply the equation by a test function $\phi \in C_0^\infty(\mathbb{R} \times \mathbb{R}^d)$, integrate over $\mathbb{R}^+ \times \mathbb{R}^d$, and integrate by parts in both x and t to obtain

$$\int_{\mathbb{R}^+}\int_{\mathbb{R}^d}\left(\frac{\partial \phi}{\partial t} + \nabla \cdot (a\phi)\right) u \, dx \, dt + \int_{\mathbb{R}^d} g(x)\phi(x,0) \, dx = 0. \quad (7.4.9)$$

This formulation will be very useful in the proof of the averaging theorem for transport equations in Chapter 21. Indeed we will sometimes refer to u satisfying (7.4.9) as a weak solution. □

7.5 Semigroups

Formally the parabolic and hyperbolic PDEs encountered in the preceding sections may be viewed as ODEs in a function space. (This viewpoint underlies the two Definitions 7.17 and 7.24.) They take the form

$$\frac{du}{dt} = \mathcal{L}u + f \tag{7.5.1}$$

where here $\mathcal{L}$ is viewed as an operator on a function space X. To be precise we assume that $\mathcal{L} : \mathcal{D}(\mathcal{L}) \subset X \to X$. Previously in this chapter we viewed $\mathcal{L}$ as an operator that acts on functions $u(x,t)$ at each point (x,t). Adopting the viewpoint that $\mathcal{L}$ acts on a function space casts the PDE as an ODE in function space. This is the *semigroup* approach to evolutionary PDEs.

Arguing formally, from our knowledge of the situation in which X is finite dimensional, we deduce that (7.5.1) has solution

$$u(t) = e^{\mathcal{L}t}u(0) + \int_0^t e^{\mathcal{L}(t-s)} f(s)ds. \tag{7.5.2}$$

This representation is known as the *variation-of-constants formula.* The integral equation can be viewed as the starting point for the *definition* of what are known as *mild solutions* to (7.5.1). To carry out this program requires definition of the operator $e^{\mathcal{L}t}$. There are different approaches to this problem, depending on the setting, many using the Fourier transform, Fourier series, or generalizations.

To illustrate this idea we consider the siutation where X is a Hilbert space and $\mathcal{L}$ has a set of eigenfunctions that form an orthonormal basis for X :

$$-\mathcal{L}\phi_j = \lambda_j \phi_j, \quad \langle \phi_i, \phi_j \rangle = \delta_{ij}.$$

Here δ_{ij} denotes the Kronecker delta tensor and $\langle \cdot \rangle$ the inner product on H. We can then define $e^{\mathcal{L}t}$ via

$$u = \sum_{j=1}^{\infty} u_j \phi_j,$$

$$e^{\mathcal{L}t}u = \sum_{j=1}^{\infty} e^{\lambda_j t} u_j \phi_j.$$

Convergence of this series will depend on the properties of the spectrum of $\mathcal{L}$ and of the regularity of u manifest in the decay of the u_j with j. Once $e^{\mathcal{L}t}$ is defined and if H is a Hilbert space, then it is possible to make sense of its adjoint via

$$\left(e^{\mathcal{L}t}\right)^* = e^{\mathcal{L}^* t}.$$

7.6 Discussion and Bibliography

The material in this chapter is standard and can be found in many books on partial differential equations and functional analysis, such as [53, 121, 98, 274]. Our treatment of the elliptic Dirichlet problem follows closely [98, ch. 6]. Our discussion of elliptic PDEs in the case of periodic boundary conditions is based on [66, ch. 4]. The Fredholm theory for the Dirichlet problem is developed in [98, sec. 6.2.3].

In the case where the data are regular enough so that the Dirichlet problem (7.2.1) admits a classical solution (i.e., a function $u \in C^2(\Omega) \cap C(\overline{\Omega})$ satisfying (7.2.1)), then the weak and classical solutions coincide. See, for example, [121].

We saw in this chapter that the analysis of operators in divergence form is based on energy methods within appropriate function spaces. On the other hand, for PDEs in nondivergence form techniques based on the maximum principle are more suitable. The maximum principle for elliptic PDEs is studied in [121] and for parabolic PDEs in [264]; see also [98]. Of course, provided that the coefficients are C^1, we can rewrite a divergence form PDE in nondivergence form and vice versa, by introducing terms that involve first-order derivatives. Operators in nondivergence form appear naturally in the probabilistic theory of diffusion, as generators of Markov processes. As mentioned in Section 6.5, dimensionality of the null space of the L^2-adjoint of a nondivergence form second-order uniformly elliptic operator is related to the ergodic theory of Markov processes; see [241, 246].

Turning now to parabolic PDEs, the proof of Theorem 7.22 can be found in [113]; see also [112] and [304]. Similar theorems hold for parabolic PDEs with time-dependent coefficients. One can also introduce the concept of the weak solution for the Cauchy problem. It is possible then to prove existence and uniqueness of solutions by using Banach's fixed-point theorem; see [98, sec 7.3.2b]. Parabolic PDEs can also be studied using probabilistic methods. Indeed, under appropriate assumptions on the coefficients, the solution of (7.3.4) admits a probabilistic interpretation, after noting that it is the backward Kolmogorov equation for an SDE, when $f \equiv 0$. More generally, when the parabolic PDE contains a source term proportional to the solution, the Feynman-Kac formula provides a probabilistic interpretation. Probabilistic proofs of existence and uniqueness of solutions to parabolic PDEs can be found in, for example, [156]. Parabolic PDEs with unbounded coefficients are studied in [58, 196].

Most of Section 7.4 is taken from [98, sec 7.3.2], where the details of the proof of Theorem 7.27 are given. As well as use of the vanishing viscosity method, existence and uniqueness theorems for transport PDEs can be proved by the method of characteristics that was introduced in Chapter 4. The solutions constructed by the method of characteristics are classical. The weak solutions that we studied in Section 7.4 are more appropriate for the averaging problem that we study rigorously in Chapter 21.

The semigroup approach to the study of time-dependent PDEs is overviewed in the text [259]. It is central to the proof of existence and uniqueness theorems for dissipative parabolic PDEs through the Hille-Yosida theorem and for proving limit theorems through the Trotter-Kato approximation theorem. The use of semigroups is central to the rigorous theory of limit theorems for stochastic processes; see [181, 182, 183, 94].

7.7 Exercises

1. Let $\mathcal{A}$ be the divergence form elliptic operator (7.2.3) and assume that $A(x) \in C^1(\Omega; \mathbb{R}^d)$. Convert $\mathcal{A}$ into the nondivergence form (7.2.4).
2. Prove the result stated in Remark 7.4, by starting with the weak formulation (7.2.7).
3. Use the Lax–Milgram theorem to prove Theorem 7.8.
4. Use Theorem 7.9 to derive the existence and uniqueness component of Theorem 7.8.
5. State and prove a result analogous to that discussed in Remark 7.4 for the periodic problem (7.2.2).
6. Prove the Fredholm alternative for operator $\mathcal{A}$ defined in (7.2.14) under assumptions analogous to (7.2.16), but adapted to the case of Dirichlet boundary conditions.
7. Prove Corollary 7.16, using similar techniques to those used to prove Corollary 7.21.
8. Consider the parabolic PDE

$$\frac{\partial u}{\partial t} = \mathcal{L}u + f(x) \quad \text{for } (x,t) \in \Omega \times [0,T], \tag{7.7.1a}$$

$$u = 0 \quad \text{for } (x,t) \in \partial\Omega \times (0,T], \tag{7.7.1b}$$

$$u = g \quad \text{for } (x,t) \in \overline{\Omega} \times \{0\} \tag{7.7.1c}$$

where

$$\mathcal{L} := \Delta \tag{7.7.2}$$

and $\Omega \subset \mathbb{R}^d$ is bounded. Formulate a notion of weak solution and prove that the equation has a unique steady solution $\overline{u}(x)$. Prove that $u(x,t) \to \overline{u}(x)$ as $t \to \infty$ by means of energy estimates.

9. Use the method of characteristics to solve the equation

$$\frac{\partial u}{\partial t} + a\frac{\partial u}{\partial x} = b \quad \text{for } (x,t) \in \mathbb{R} \times \mathbb{R}^+,$$
$$u = 0 \quad \text{for } (x,t) \in \mathbb{R} \times \{0\}$$

in the case where $a = a(x)$ and $b = b(x)$ only.

10. Use the method of characteristics to solve Burger's equation

$$\frac{\partial u}{\partial t} + u\frac{\partial u}{\partial x} = 0 \quad \text{for } (x,t) \in \mathbb{R} \times (0,T),$$
$$u = g \quad \text{for } (x,t) \in \mathbb{R} \times \{0\}.$$

What can you say about how large T can be, as a function of initial data g?

11. Consider the Cauchy problem (7.3.4) where the operator $\mathcal{L}$ is in divergence form. Assume that there exists a unique solution that decays sufficiently fast at infinity. Prove the a priori estimate (7.3.7). How does the constant $C_2(T)$ depend on T?
12. Consider the Cauchy problem

$$\frac{\partial u}{\partial t} = \nabla \cdot (A \nabla u), \quad \text{for } (x,t) \in \mathbb{R}^d \times (0,T],$$

$$u = g \ \text{ for } (x,t) \in \mathbb{R}^d \times \{0\},$$

where $A = A(x)$ satisfies the standard assumptions. Show that the solution satisfies the *uniform in time* estimate:

$$\|u\|_{L^\infty(0,T);L^2(\mathbb{R}^d))} + C_1 \|\nabla u\|_{L^2((0,T);L^2(\mathbb{R}^d))} \leqslant C_2 \|g\|_{L^2(\mathbb{R}^d)}.$$

13. Consider the parabolic PDE (7.7.1) with $\mathcal{L} = \Delta$, the domain $\Omega = [0,1]$ and $F \equiv 0$. Reformulate the equation as an ODE on the Hilbert space $L^2(\Omega)$ and find a representation for the resulting semigroup, by using Fourier sine series.

Part II

Perturbation Expansions

8

Invariant Manifolds for ODEs

8.1 Introduction

Perhaps the simplest situation where variable reduction occurs in dynamical systems is that of attractive invariant manifolds. These manifolds *slave* one subset of the variables to another. In this chapter we describe a situation where attractive invariant manifolds can be constructed in scale-separated systems, by means of perturbation expansions. In Section 8.2 we introduce the system of ODEs that we want to simplify by means of the theory of invariant manifolds, and in Section 8.3 we present the simplified equations. The simplified equations are derived in Section 8.4, and several examples are presented in Section 8.5. In Section 8.6 we describe various extensions of the results presented in this chapter, together with bibliographical remarks. We also discuss the material in this section in relation to averaging, the subject of Chapters 9 and 10.

8.2 Full Equations

We consider a system of ODEs of the form (4.1.1) and write z as $z = (x^T, y^T)^T$, where

$$\frac{dx}{dt} = f(x, y), \tag{8.2.1a}$$

$$\frac{dy}{dt} = \frac{1}{\varepsilon} g(x, y), \tag{8.2.1b}$$

and $\varepsilon \ll 1$. Here $x \in \mathcal{X}$ and $y \in \mathcal{Y}$, in the notation of Chapter 4.

Let $\varphi_x^t(y)$ be the solution operator of the fast dynamics with x viewed as a fixed parameter and $\varepsilon = 1$. To be precise, for any $\xi \in \mathcal{X}$, let

$$\frac{d}{dt}\varphi_\xi^t(y) = g(\xi, \varphi_\xi^t(y)), \qquad \varphi_\xi^0(y) = y. \tag{8.2.2}$$

We assume that

$$\lim_{t\to\infty} \varphi_\xi^t(y) = \eta(\xi) \tag{8.2.3}$$

exists, is independent of y, and the convergence is uniform in ξ. Roughly speaking $y(t)$ solving (8.2.1) is given by $y(t) \approx \varphi_{x(0)}^{t/\varepsilon}(y(0))$ for times t that are small compared with 1 (i.e., $t = o(1)$) so that $x(t)$ has not evolved very much. If we then look at short time scales that are nonetheless large compared with ε, so that y is close to its equilibrium point (for example, if $t = \mathcal{O}(\varepsilon^{1/2})$), we deduce that $y(t) \approx \eta(x(0))$. This is the mechanism by which y becomes slaved to x and we now seek to make the heuristics more precise.

Notice that the generator $\mathcal{L}$ for (8.2.1) has the form

$$\mathcal{L} = \frac{1}{\varepsilon}\mathcal{L}_0 + \mathcal{L}_1 \tag{8.2.4}$$

where

$$\mathcal{L}_0 = g(x,y)\cdot\nabla_y, \quad \mathcal{L}_1 = f(x,y)\cdot\nabla_x.$$

In particular, $\mathcal{L}_0$ is the generator of a process on $\mathcal{Y}$ for each fixed x.

Now consider the following PDE for $v(y,t)$ in which x is viewed as a fixed parameter:

$$\frac{\partial v}{\partial t} = \mathcal{L}_0 v, \quad v(y,0) = \phi(y). \tag{8.2.5}$$

Result 4.6 shows that

$$v(y,t) = \phi(\varphi_x^t(y)).$$

Thus, by (8.2.3),

$$v(y,t) \to \phi(\eta(x)), \text{ as } \quad t \to \infty. \tag{8.2.6}$$

This is related to ergodicity, as Equation (8.2.6) shows that the function $v(y,t)$ exhibits no dependence on initial data, asymptotically as $t \to \infty$, and approaches a constant in y. Compare with the discussion of ergodicity in Chapter 4, Theorems 4.12 and 4.13 in particular.

Recall the Definition 4.3 of an invariant set. If this set is a manifold, then we refer to it as an *invariant manifold*. In this chapter we use the scale-separated form of Equations (8.2.1) to construct an approximate invariant manifold. In fact the manifold will have the structure of a *graph*: it will be represented as a function relating the y-coordinates to the x-coordinates. Invariant manifolds representable as graphs are particulary important in describing the dynamics of ODEs close to equilibria, leading to the concepts of *stable, unstable*, and *center manifolds*.

8.3 Simplified Equations

We now state an approximation result that will be derived by formal perturbation arguments in the next section. Define the vector field $F_0(x)$ by

$$F_0(x) = f(x, \eta(x)). \tag{8.3.1}$$

Result 8.1. *For* $\varepsilon \ll 1$ *and time* t *up to* $\mathcal{O}(1)$, $x(t)$ *solving* (8.2.1) *is approximated by* $X(t)$ *solving*

$$\frac{dX}{dt} = F_0(X), \tag{8.3.2}$$

where $F_0(x)$ *is given by* (8.3.1).

Underlying the derivation of this result is an assumption that $y(0)$ is initalized close to $\eta(x(0))$. When this fails, further arguments are required to deal with what is termed an initial or boundary layer; see Section 8.6 for a discussion of this point.

Result 8.1 gives us the leading-order approximation in ε. Keeping the next order yields the refined approximation

$$\frac{dX}{dt} = F_0(X) + \varepsilon F_1(X), \tag{8.3.3}$$

where

$$F_1(x) = \nabla_y f(x, \eta(x))\Big(\nabla_y g(x, \eta(x))\Big)^{-1} \nabla_x \eta(x) f(x, \eta(x)).$$

This approximation requires that $\nabla_y g(x, \eta(x))$ is invertible.

8.4 Derivation

The method used to find these simplified equations is to seek an approximate invariant manifold for the system. Furthermore, we assume that the manifold can be represented as a graph over x, namely $y = \Psi(x)$. The set determined by such a graph is invariant (see Definition 4.3) under the dynamics if

$$\frac{dy}{dt} = \nabla\Psi(x(t))\frac{dx}{dt},$$

whenever $y = \Psi(x)$. This implies that Ψ must solve the nonlinear PDE

$$\frac{1}{\varepsilon} g(x, \Psi(x)) = \nabla\Psi(x) f(x, \Psi(x)).$$

We seek solutions to this equation as a power series

$$\Psi(x) = \Psi_0(x) + \varepsilon\Psi_1(x) + \mathcal{O}(\varepsilon^2).$$

This is our first example of a perturbation expansion.

Substituting and equating coefficients of successive powers of ε to zero yields the hierarchy

$$\begin{aligned} &\mathcal{O}(\tfrac{1}{\varepsilon}) \quad & g(x, \Psi_0(x)) = 0, \\ &\mathcal{O}(1) \quad & \nabla_y g(x, \Psi_0(x))\Psi_1(x) = \nabla\Psi_0(x) f(x, \Psi_0(x)). \end{aligned}$$

Notice that Equations (8.2.2) and (8.2.3) together imply that $g(\xi, \eta(\xi)) = 0$ for all ξ. Hence the $\mathcal{O}(1/\varepsilon)$ equation may be satisfied by choosing $\Psi_0(x) = \eta(x)$, giving the approximation (8.3.2). Since the rate of convergence in (8.2.3) is assumed to be uniform it is natural to assume that $y = \eta(\xi)$ is a hyperbolic equilibrium point[1] of (8.2.2), so that $\nabla_y g(x, \eta(x))$ is invertible. Setting $\Psi_0(x) = \eta(x)$ in the $\mathcal{O}(1)$ equation and inverting yields

$$\Psi_1(x) = \nabla_y g(x, \eta(x))^{-1} \nabla \eta(x) f(x, \eta(x)).$$

Thus

$$\begin{aligned} f(x, \Psi(x)) &= f\left(x, \Psi_0(x) + \varepsilon \Psi_1(x) + \mathcal{O}(\varepsilon^2)\right) \\ &= f(x, \Psi_0(x)) + \varepsilon \nabla_y f(x, \Psi_0(x)) \Psi_1(x) + \mathcal{O}(\varepsilon^2) \\ &= f(x, \eta(x)) + \varepsilon \nabla_y f(x, \eta(x)) \Psi_1(x) + \mathcal{O}(\varepsilon^2), \end{aligned}$$

and the refined approximation (8.3.3) follows.

8.5 Applications

8.5.1 Linear Fast Dynamics

A structure arising in many applications is where the frozen x dynamics, given by $\varphi_\xi^t(\cdot)$, is linear. As a simple example consider the equations

$$\begin{aligned} \frac{dx}{dt} &= f(x, y), \\ \frac{dy}{dt} &= -\frac{y}{\varepsilon} + \frac{\tilde{g}(x)}{\varepsilon}. \end{aligned} \tag{8.5.1}$$

Here $d = 2$ and $\mathcal{X} = \mathcal{Y} = \mathbb{R}$, $\mathcal{Z} = \mathbb{R}^2$. It is straightforward to show that

$$\begin{aligned} \varphi_\xi^t(y) &= e^{-t} y + \int_0^t e^{s-t} \tilde{g}(\xi) ds \\ &= e^{-t} y + (1 - e^{-t}) \tilde{g}(\xi). \end{aligned}$$

Hence (8.2.3) is satisfied for $\eta(\cdot) = \tilde{g}(\cdot)$

The simplified equation given by Result 8.1 is hence

$$\frac{dX}{dt} = f(X, \tilde{g}(X)).$$

Using the fact that $\nabla_y g(x, y) = -1$ we see that the more refined approximation (8.3.3) is

$$\frac{dX}{dt} = f(X, \tilde{g}(X)) \Big(1 - \varepsilon \frac{df}{dy}(X, \tilde{g}(X)) \frac{d\tilde{g}}{dx}(X)\Big).$$

---

[1] A hyperbolic equilibrium point is one where the linearization of the vector field at the equlibrium point contains no spectrum on the imaginary axis.

8.5.2 Large Time Dynamics

The statement of the result concerning simplified dynamics concerns the approximation of $x(t)$ on $\mathcal{O}(1)$ time intervals with respect to ε. However, in many cases the results extend naturally to the infinite time domain. The following example illustrates this idea.

Consider the equations

$$\frac{dx_1}{dt} = -x_2 - x_3, \tag{8.5.2a}$$

$$\frac{dx_2}{dt} = x_1 + \frac{1}{5}x_2, \tag{8.5.2b}$$

$$\frac{dx_3}{dt} = \frac{1}{5} + y - 5x_3, \tag{8.5.2c}$$

$$\frac{dy}{dt} = -\frac{y}{\varepsilon} + \frac{x_1 x_3}{\varepsilon}, \tag{8.5.2d}$$

so that $\mathcal{X} = \mathbb{R}^3$ and $\mathcal{Y} = \mathbb{R}$. Result 8.1 indicates that x should be well approximated by X solving the Rössler system

$$\frac{dX_1}{dt} = -X_2 - X_3, \tag{8.5.3a}$$

$$\frac{dX_2}{dt} = X_1 + \frac{1}{5}X_2, \tag{8.5.3b}$$

$$\frac{dX_3}{dt} = \frac{1}{5} + X_3(X_1 - 5). \tag{8.5.3c}$$

The Rössler equations are chaotic, and consequently comparison of trajectories over long time intervals is not natural. A more useful object is the attractor. A comparison of the numerically generated attractors for the two systems is shown in Figure 8.1. The left side shows the attractor for Equations (8.5.2), projected into the x-coordinates, for $\varepsilon = 10^{-2}$. The right side shows the attractor for the Rössler equations themselves. The agreement is very strong indicating that the simplified dynamics do indeed capture behavior over long time intervals.

8.5.3 Center Manifold

The center manifold is an invariant manifold containing an equilibrium point whose linearization has neutral directions (subspaces corresponding to eigenvalues with zero real part). Consider the equations

$$\frac{dx}{dt} = \lambda x + \sum_{i=0}^{2} a_i x^i y^{2-i},$$

$$\frac{dy}{dt} = x - y + \sum_{i=0}^{2} b_i x^i y^{2-i}.$$

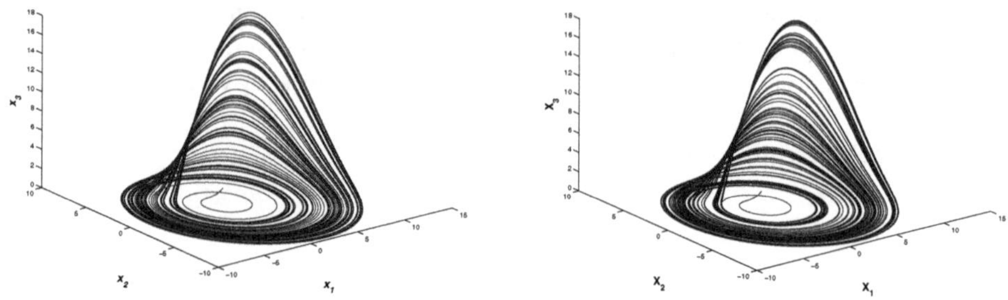

Fig. 8.1. Comparison between the attracting sets for (8.5.2) with $\varepsilon = 0.01$ (left) and (8.5.3) (right), projected on the (x_1, x_2) and (X_1, X_2) planes, respectively.

Here $\lambda \in \mathbb{R}$ and the a_i and b_i are also real numbers. Furthermore, for each t, $x(t) \in \mathbb{R}$ and $y(t) \in \mathbb{R}$. When linearized at the origin, this equation becomes

$$\frac{dx}{dt} = \lambda x,$$
$$\frac{dy}{dt} = x - y.$$

If $z = (x, y)^T$ then

$$\frac{dz}{dt} = Lz$$

with

$$L = \begin{pmatrix} \lambda & 0 \\ 1 & -1 \end{pmatrix}.$$

The eigenvalues of L are λ and -1. As λ passes through 0 the linear stability property of the origin thus changes from stable to unstable. For this reason, studying the equation in the vicinity of $\lambda = 0$ is of interest. In particular we expect to find a center manifold at $\lambda = 0$: an invariant manifold tangent to the eigenspace corresponding to eigenvalue 0 of L.

To construct this manifold rescale the equations as follows. We set

$$x \to \varepsilon x, y \to \varepsilon y, \lambda \to \varepsilon \lambda, t \to \varepsilon^{-1} t.$$

This corresponds to looking for small-amplitude solutions, close to the fixed point at the origin, at parameter values close to the bifurcation values. Such solutions evolve slowly, and hence time is rescaled to capture nontrivial dynamics. The equations become

$$\frac{dx}{dt} = \lambda x + \sum_{i=0}^{2} a_i x^i y^{2-i},$$

$$\frac{dy}{dt} = \frac{1}{\varepsilon}(x - y) + \sum_{i=0}^{2} b_i x^i y^{2-i}.$$

A perturbation expansion gives the invariant manifold $y = x$ and we obtain the following equations for the dynamics on the invariant manifold:

$$\frac{dX}{dt} = \lambda X + AX^2,$$

with $A = \sum_{i=0}^{2} a_i$. The case $\lambda = 0$ gives the center manifold itself, and $\lambda < 0$ the stable manifold.

8.6 Discussion and Bibliography

The topic of invariant manifolds has a long history and is itself the subject of entire books. To do it justice here is impossible, and we provide only brief pointers to the literature. From the perspective of this book, our primary motivation for covering the topic is that it provides a special case of the method of averaging introduced in the next two chapters; furthermore this case can be introduced without appeal to any arguments from ergodic theory or from the theory of stochastic processes. It hence provides a suitable inroad into the topics of this book for readers with a background in dynamical systems; conversely it provides a concrete link between averaging and dynamical systems. We discuss this perspective further in Chapter 10. Note also that the perturbation expansion we use in this chapter is, at a high level, similar to those used in the remainder of Part II. It differs in one significant respect, however: all the remaining chapters involve perturbation expansions for the approximation of linear problems (by working with the backward equation), and rely on repeated use of the Fredholm alternative. In this chapter the strategy underlying the perturbation expansion is somewhat different, as the problem for the graph Ψ is nonlinear and the Fredholm alternative is not used.

Invariant manifolds in general are described in [131] and [331]. These books have considerable emphasis on the construction of unstable, stable, and center manifolds for invariant sets of Equation (4.1.1). In particular, for the case of the simplest invariant set, an equilibrium point, we may change coordinates to a frame in which the origin 0 is an equilibrium point and (4.1.1) takes the form

$$\frac{dz}{dt} = Lz + h_1(z), \quad z(0) = z_0.$$

Here $h_1(z)$ is small compared to $z \to 0$. In the case of a hyperbolic equilibrium point the invariant subspaces of L split into stable and unstable spaces. If we let P denote the orthogonal projection onto the stable space and $Q = I - P$ denote the orthogonal projection onto the unstable space, then introducing $x = Pz, y = Qz$ we obtain the equations

$$\frac{dx}{dt} = L_1 x + f_1(x, y),$$
$$\frac{dy}{dt} = L_2 y + g_1(x, y).$$

The stable manifold is (locally near the origin) representable as a graph $y = \Theta(x)$; likewise the unstable manifold is representable as a graph $x = \Phi(y)$. The center manifold is similar to the stable manifold but occurs when, for example, $P\mathcal{Z}$ comprises neutral directions in L. Center manifolds in particular are discussed in [57]. The special case where the neutral spectrum of L contains a pair of complex conjugate eigenvalues leads to the Hopf bifurcation theorem; see [212].

These special invariant manifold theorems, concerning behavior near fixed points, show the central role of graphs relating one set of variables to another in the construction of invariant manifolds. Such a graph is at the heart of our construction of what is sometimes termed a *slow manifold* for (8.2.1). Early studies of the approximation of ODE with attracting slow manifold by differential-algebraic equations includes the independent work of Levinson and of Tikhonov (see O'Malley [239] and Tikhonov et al. [317]). As mentioned in Section 8.3 the simplest version of the approximation result requires the fast variable y to be initialized close to the invariant manifold. However, even if it is not, an *initial layer* (sometimes termed *boundary layer*) can be introduced to extend the approximation result and studied through the method of matched asymptotic expansions; see [324] and [71, 272].

Our construction of an invariant manifold uses the explicit slaving of y to x through the asymptotically stable fixed points of (8.2.2). More generally, the use of a spectral gap sufficiently large relative to the size of the nonlinear terms is used in the construction of local stable, unstable, and center manifolds (e.g., Carr [57], Wiggins [331]), slow manifolds (Kreiss [178]) and inertial manifolds (Constantin et al. [69]). In particular, the inertial manifold construction shows how ideas from invariant manifold theory extend naturally to infinite dimensions in the context of dissipative PDEs.

References to numerical methods for dynamical systems and for the computation of invariant manifolds in particular may be found in [305]. It is also possible to construct invariant manifolds for stochastic (partial) differential equations. See, for example, [39, 38, 41, 77, 329, 328] and the references therein.

8.7 Exercises

1. Consider the equations

$$\frac{dx}{dt} = \lambda x + a_0 x^3 + a_1 xy,$$
$$\frac{dy}{dt} = -y + \sum_{i=0}^{2} b_i x^i y^{2-i}.$$

Here $\lambda \in \mathbb{R}$ and the a_i and b_i are also real numbers. For each fixed t we have $x(t) \in \mathbb{R}$ and $y(t) \in \mathbb{R}$. Show that the scaling

$$x \to \varepsilon x, y \to \varepsilon^2 y, \lambda \to \varepsilon^2 \lambda, t \to \varepsilon^{-2} t$$

puts this system in a form to which the perturbation techniques of this section apply. Deduce that the center manifold has the form

$$\frac{dX}{dt} = \lambda X + AX^3$$

where $A = a_0 + a_1 b_2$.

2. Assume $\varepsilon > 0$, $A \in \mathbb{R}^{l \times l}$, and $B \in \mathbb{R}^{(d-l)\times(d-l)}$. Consider the equations

$$\begin{aligned}\frac{dx}{dt} &= Ax + \varepsilon f_0(x, y),\\ \frac{dy}{dt} &= -\frac{1}{\varepsilon} By + g_0(x, y),\end{aligned}$$

for $\varepsilon \ll 1$ and $x \in \mathbb{R}^l$, $y \in \mathbb{R}^{d-l}$. Assume that B is symmetric positive definite. Find the first three terms in an expansion for an invariant manifold representing y as a graph over x.

3. Assume $\varepsilon > 0$ and $B \in \mathbb{R}^{(d-l)\times(d-l)}$. Consider the equations

$$\begin{aligned}\frac{dx}{dt} &= f(x, y),\\ \frac{dy}{dt} &= -\frac{1}{\varepsilon}\Big(By - \tilde{g}(x)\Big),\end{aligned}$$

for $\varepsilon \ll 1$ and $x \in \mathbb{R}^l$, $y \in \mathbb{R}^{d-l}$.

a) Assume that B is symmetric positive definite. Find the first term in an expansion for an invariant manifold representing y as a graph over x.

b) Consider the case $d - l = 2$, $\tilde{g} \equiv 0$, and

$$B = \begin{pmatrix} 0 & 1 \\ -1 & 0 \end{pmatrix}.$$

What happens to the solution as $\varepsilon \to 0$?

9

Averaging for Markov Chains

9.1 Introduction

Perhaps the simplest setting in which to expose variable elimination for stochastic dynamical problems is to work in the setting of Markov chains. In this context it is natural to study situations where a subset of the variables evolves rapidly compared with the remainder and can be replaced by their averaged effect. In Section 9.2 we describe the unaveraged Markov chain, and in Section 9.3 we present the averaged equations; the averaged equations are derived in Section 9.4, and an example is given in Section 9.5. In Section 9.6 we discuss various extensions of the results from this chapter and make some bibliographical remarks.

9.2 Full Equations

We work in the setup of Chapter 5 and consider the backward equation

$$\frac{dv}{dt} = Qv. \tag{9.2.1}$$

Recall that this equation, with $v(0) = \phi$, has the property that

$$v_i(t) = \mathbb{E}\Big(\phi_{z(t)} | z(0) = i\Big),$$

where $\mathbb{E}$ denotes expectation with respect to the Markov transition probabilities. We assume that the generator Q[1] takes the form

$$Q = \frac{1}{\varepsilon} Q_0 + Q_1, \tag{9.2.2}$$

with $0 < \varepsilon \ll 1$. We study situations where the state space is indexed by two variables, x and y, and the leading order contribution in Q, namely Q_0, corresponds

[1] In this chapter we denote the generator by Q rather than L because we use index l for the state space; thus we wish to avoid confusion with the components of the generator.

to fast ergodic dynamics in y, with x frozen. Averaging over y then gives the effective reduced dynamics for x.

The precise situation is as follows. Our state space is $\mathcal{I} := \mathcal{I}_x \times \mathcal{I}_y$ with $\mathcal{I}_x,\ \mathcal{I}_y \subseteq \{1, 2, \cdots\}$. We let $q((i,k),(j,l))$ denote the element of the generator associated with transition from $(i,k) \in \mathcal{I}_x \times \mathcal{I}_y$ to $(j,l) \in \mathcal{I}_x \times \mathcal{I}_y$.[2] Consider now a family of Markov chains on $\mathcal{I}_y$, indexed by $i \in \mathcal{I}_x$. We write the generator as $A_0(i)$ with entries as $a_0(k,l;i)$; the indices denote transition from $k \in \mathcal{I}_y$ to $l \in \mathcal{I}_y$ for given fixed $i \in \mathcal{I}_x$. We assume that, for each $i \in \mathcal{I}_x$, $A_0(i)$ generates an ergodic Markov chain on $\mathcal{I}_y$. Hence $A_0(i)$ has a one-dimensional null space for each fixed i, and [3]

$$\begin{aligned} \textstyle\sum_l a_0(k,l;i) &= 0, \quad (i,k) \in \mathcal{I}_x \times \mathcal{I}_y, \\ \textstyle\sum_k \rho^\infty(k;i) a_0(k,l;i) &= 0, \quad (i,l) \in \mathcal{I}_x \times \mathcal{I}_y. \end{aligned} \tag{9.2.3}$$

This is the index form of Equations (5.6.2) with L replaced by $A_0(i)$. Without loss of generality we choose the normalization

$$\sum_k \rho^\infty(k;i) = 1 \quad \forall i \in \mathcal{I}_x.$$

Thus $\rho^\infty(i) = \{\rho^\infty(k;i)\}_{k \in \mathcal{I}_y}$ is the invariant distribution of a Markov chain on $\mathcal{I}_y$, indexed by $i \in \mathcal{I}_x$.

Similarly to the preceding we introduce the generators of a Markov chain on $\mathcal{I}_x$, parameterized by $k \in \mathcal{I}_y$. We denote the generator by $A_1(k)$ with indices $a_1(i,j;k)$; the indices denote transition from $i \in \mathcal{I}_x$ to $j \in \mathcal{I}_x$, for each fixed $k \in \mathcal{I}_y$. With this notation for the A_0, A_1 we introduce generators Q_0, Q_1 of Markov chains on $\mathcal{I}_x \times \mathcal{I}_y$ by

$$\begin{aligned} q_0((i,k),(j,l)) &= a_0(k,l;i)\delta_{ij}, \\ q_1((i,k),(j,l)) &= a_1(i,j;k)\delta_{kl}. \end{aligned} \tag{9.2.4}$$

Here δ_{ij} is the usual Kronecker delta. In the construction of Q_0 (resp. Q_1) the Kronecker delta represents the fact that no transitions are taking place in $\mathcal{I}_x$ (resp. $\mathcal{I}_y$).

To confirm that Q_0, Q_1 as defined are indeed generators, notice that nondiagonal entries $(i,k) \neq (j,l)$ are nonnegative because A_0 and A_1 are generators. Also

$$\begin{aligned} \sum_{j,l} q_0((i,k),(j,l)) &= \sum_{j,l} a_0(k,l;i)\delta_{ij} \\ &= \sum_l a_0(k,l;i) \\ &= 0 \end{aligned}$$

[2] In this chapter and in Chapter 16, we will not use suffices to denote the dependence on the state space as the double-indexing makes this a cluttered notation. Hence we use $q((i,k),(j,l))$ rather than $q_{(i,k),(j,l)}$.

[3] Summation is always over indices in $\mathcal{I}_x$ or $\mathcal{I}_y$ in this chapter. It should be clear from the context which of the two sets is being summed over.

by (9.2.3). A similar calculation shows that

$$\sum_{j,l} q_1((i,k),(j,l)) = 0,$$

using the fact that

$$\sum_{j} a_1(i,j;k) = 0 \quad \forall\, (i,k) \in \mathcal{I}_x \times \mathcal{I}_y,$$

since $A_1(k)$ is a generator for each fixed k. Thus Q_0, Q_1 are also the generators of Markov chains. Finally note that any linear combination of generators, via positive scalar constants, will also be a generator. Hence (9.2.2) defines a generator for any $\varepsilon > 0$.

9.3 Simplified Equations

We define the generator $\bar{Q}_1$ of a Markov chain on $\mathcal{I}_x$ by:

$$\bar{q}_1(i,j) = \sum_{k} \rho^{\infty}(k;i) a_1(i,j;k). \tag{9.3.1}$$

Notice that $\bar{q}_1(i,j) \geqslant 0$ for $i \neq j$ because $\rho^{\infty}(k;i) \geqslant 0$ and $a_1(i,j;k) \geqslant 0$ for $i \neq j$. Furthermore

$$\begin{aligned}\sum_{j} \bar{q}_1(i,j) &= \sum_{k} \rho^{\infty}(k;i) \left(\sum_{j} a_1(i,j;k) \right) \\ &= 0.\end{aligned}$$

Hence $\bar{Q}_1$ is the generator of a Markov chain.

Result 9.1. *Consider Equation (9.2.1) under assumption (9.2.2). Then for $\varepsilon \ll 1$ and times t up to $\mathcal{O}(1)$ the finite-dimensional distributions of $x \in \mathcal{I}_x$ are approximated by a Markov chain X with generator $\bar{Q}_1$.*

We emphasize that x is not itself Markovian: only the pair (x, y) is. As discussed earlier, $\bar{Q}_1$ is the generator of a Markov chain on $\mathcal{I}_x$ alone, and the dynamics in $\mathcal{I}_y$ has been eliminated through averaging. Thus the approximate variable X is Markovian and is governed by the backward equation

$$\frac{dv_0}{dt} = \bar{Q}_1 v_0. \tag{9.3.2}$$

We now provide justification for this elimination of variables, by means of perturbation expansion.

9.4 Derivation

The method used is to show that the backward equation for the full Markov chain in $(x,y) \in \mathcal{I}_x \times \mathcal{I}_y$ can be approximated by the backward Equation (9.3.2) for $x \in \mathcal{I}_x$ alone. We consider Equation (9.2.1) under (9.2.2). We have the backward equation

$$\frac{dv}{dt} = \Big(\frac{1}{\varepsilon}Q_0 + Q_1\Big)v.$$

Unlike in the previous chapter, where we approximated a nonlinear PDE containing a small parameter ε, here the problem is linear. In the following five chapters, all our perturbation expansions are for similar linear equations. The derivation here is hence prototypical of what follows.

We seek solutions $v = v(i,k,t)$ in the form of the multiscale expansion

$$v = v_0 + \varepsilon v_1 + \mathcal{O}(\varepsilon^2). \tag{9.4.1}$$

Substituting and equating coefficients of powers of ε to zero we find

$$\mathcal{O}(\tfrac{1}{\varepsilon}) \; Q_0 v_0 = 0, \tag{9.4.2a}$$

$$\mathcal{O}(1) \; Q_0 v_1 = -Q_1 v_0 + \frac{dv_0}{dt}. \tag{9.4.2b}$$

By (9.2.3) we deduce from (9.4.2a) that v_0 is independent of $k \in \mathcal{I}_y$. Abusing notation, we write

$$v_0(i,k,t) = v_0(i,t)1(k) \tag{9.4.3}$$

where $1(k) = 1$ for all $k \in \mathcal{I}_y$. The operator Q_0 is singular and hence, for (9.4.2b) to have a solution, the Fredholm alternative implies the solvability condition

$$-Q_1 v_0 + \frac{dv_0}{dt} \perp \mathrm{Null}\,\{Q_0^T\}. \tag{9.4.4}$$

From (9.2.3) we deduce that the null space of Q_0^T is characterized by

$$\sum_{k,i} \rho^\infty(k;i)c(i)q_0((i,k),(j,l)) = 0, \tag{9.4.5}$$

for any vector $c = \{c(i)\}$ on $\mathcal{I}_x$. Using (9.4.3) we find that

$$\begin{aligned}\frac{dv_0}{dt} - Q_1 v_0 &= \frac{dv_0}{dt}(i,t)1(k) - \sum_{j,l} a_1(i,j;k)\delta_{kl} v_0(j,t)1(l) \\ &= \Big(\frac{dv_0}{dt}(i,t) - \sum_j a_1(i,j;k)v_0(j,t)\Big)1(k).\end{aligned}$$

Imposing the solvability condition (9.4.4) by means of (9.4.5) we obtain

$$\sum_{k,i} \rho^\infty(k;i)c(i)\Big(\frac{dv_0}{dt}(i,t) - \sum_j a_1(i,j;k)v_0(j,t)\Big) = 0,$$

which implies that

$$\sum_i c(i)\Big(\frac{dv_0}{dt}(i,t) - \sum_j \bar{Q}_1(i,j)v_0(j,t)\Big) = 0.$$

Since c is an arbitrary vector on $\mathcal{I}_x$ we deduce that each component of the sum over i is zero. This yields (9.3.2).

9.5 Application

Consider a simple example where $\mathcal{I}_x = \mathcal{I}_y = \{1, 2\}$. Thus we have a four-state Markov chain on $\mathcal{I} = \mathcal{I}_x \times \mathcal{I}_y$. We assume that the generators of the Markov chains on $\mathcal{I}_y$ and $\mathcal{I}_x$ are given by

$$A_0(i) = \begin{pmatrix} -\theta_i & \theta_i \\ \phi_i & -\phi_i \end{pmatrix}$$

and

$$A_1(k) = \begin{pmatrix} -\alpha_k & \alpha_k \\ \beta_k & -\beta_k \end{pmatrix},$$

respectively. In the first (resp. second) of these Markov chains $i \in \mathcal{I}_x$ (resp. $k \in \mathcal{I}_y$) is a fixed parameter. The parameters $\theta_i, \phi_i, \alpha_k$, and β_k are all nonnegative.

If we order the four states of the Markov chain as $(1, 1)$, $(1, 2)$, $(2, 1)$, and $(2, 2)$, then the generators Q_0 and Q_1 are given by

$$Q_0 = \begin{pmatrix} -\theta_1 & \theta_1 & 0 & 0 \\ \phi_1 & -\phi_1 & 0 & 0 \\ 0 & 0 & -\theta_2 & \theta_2 \\ 0 & 0 & \phi_2 & -\phi_2 \end{pmatrix} \tag{9.5.1}$$

and

$$Q_1 = \begin{pmatrix} -\alpha_1 & 0 & \alpha_1 & 0 \\ 0 & -\alpha_2 & 0 & \alpha_2 \\ \beta_1 & 0 & -\beta_1 & 0 \\ 0 & \beta_2 & 0 & -\beta_2 \end{pmatrix}. \tag{9.5.2}$$

Note that any linear combination of Q_0 and Q_1 will have zeros along the anti-diagonal, and hence the same is true of Q; this reflects the fact that, by construction, transitions in both $\mathcal{I}_x$ and $\mathcal{I}_y$ do not happen simultaneously.

The invariant density of the Markov chain with generator $A_0(i)$ is $\rho^\infty(i) = (\lambda_i, 1 - \lambda_i)^T$ with $\lambda_i = \phi_i/(\theta_i + \phi_i)$. Recall that the averaged Markov chain on $\mathcal{I}_x$ has generator $\bar{Q}_1$ with entries

$$\begin{aligned} q_1(i,j) &= \sum_k \rho^\infty(k;i) a_1(i,j;k) \\ &= \lambda_i a_1(i,j;1) + (1-\lambda_i) a_1(i,j;2). \end{aligned}$$

Thus

$$\bar{Q}_1 = \begin{pmatrix} -\lambda_1\alpha_1 - (1-\lambda_1)\alpha_2 & \lambda_1\alpha_1 + (1-\lambda_1)\alpha_2 \\ \lambda_2\beta_1 + (1-\lambda_2)\beta_2 & -\lambda_2\beta_1 - (1-\lambda_2)\beta_2 \end{pmatrix}. \tag{9.5.3}$$

9.6 Discussion and Bibliography

Two recent monographs where multiscale problems for Markov chains are studied are [335, 336]. See also [291] for a broad discussion of averaging and dimension reduction in stochastic dynamics. Markov chain approximations for SDEs, especially in the large deviation limit, are studied in [111]. Computational methods for multiscale Markov chains are discussed in [85, 86]. Diffusion limits of ODEs driven by Markov chains are studied in [245]. See also [96] for the proof of a related diffusion limit theorem. For a connection between Markov chains and center manifolds, see [262].

In this chapter we have presented averaging for Markov chains. Homogenization (i.e., the central limit theorem) results for Markov chains can be found in [184].

In deriving the approximate equation we implicitly assume that the original Markov chain is prepared in a state that does not depend on the parts of the state space in $\mathcal{I}_y$. If this is not the case then a similar analysis can still be carried out, but an initial layer must be included, over time of order $\mathcal{O}(\varepsilon)$, on which $v(t)$ adjusts from being a function on $\mathcal{I}_x \times \mathcal{I}_y$ to being a function only on $\mathcal{I}_x$, to leading order.

9.7 Exercises

1. Find a multiscale expansion for the invariant measure of the Markov chain with generator $Q = \frac{1}{\varepsilon}Q_0 + Q_1$ when Q_0, Q_1 are given by (9.5.1) and (9.5.2).
2. Find the invariant measure of $\bar{Q}_1$ given by (9.5.3) and interpret your findings in light of your answer to the previous question.
3. Consider the SDE (6.5.1). Assume that u is governed by a two-state Markov chain, with states $\{-1,+1\}$. Write down the generator for the resulting Markov process in (z,u), on the assumption that the generator for the Markov chain has the form
$$L(z) = \begin{pmatrix} -\theta(z) & \theta(z) \\ \phi(z) & -\phi(z) \end{pmatrix}.$$
4. Consider the same setup as in the previous question but where the two-state Markov chain now has generator $\frac{1}{\varepsilon}L(z)$ with $L(z)$ as given in the previous question. Use the method of averaging to find the averaged SDE in z in the limit $\varepsilon \to 0$, where u may be eliminated.

5. Let u be a two-state continuous-time Markov chain with generator as in the previous question. Consider the ODE

$$\frac{dz}{dt} = \lambda(u)z, \quad t \in [0, \infty).$$

Assume that $\lambda(-1) < 0$ and $\lambda(+1) > 0$. Use multiscale analysis to determine conditions under which the trajectories of z do not grow.

6. Let u be a Markov chain on a finite-state space with generator Q taking the form

$$Q = \frac{1}{\varepsilon}Q_0 + Q_1.$$

Assume that the Q_i are generators of Markov chains for $i = 0, 1$ and that Q has a two-dimensional null space:

$$\mathcal{N}(Q_0) = \text{span}\{\phi_0, \phi_1\}.$$

Derive a two-state Markov chain that approximates the dynamics in this null space.

10

Averaging for ODEs and SDEs

10.1 Introduction

Here we take the averaging principle developed in the previous chapter for Markov chains and apply it to ODEs and SDEs. The unifying theme is the approximate solution of the backward equation by means of an appropariate perturbation expansion and consequent elimination of variables.

In Section 10.2 we present the equations we will study, and in Section 10.3 we present the averaged equations. Section 10.4 contains the derivation of the averaged equations; the derivation is carried out in the case where the fast process is stochastic. In Section 10.5 we study how the deterministic situation may be handled. Section 10.6 contains two illustrative examples. Extensions of the results presented in this chapter, together with bibliographical remarks, are given in Section 10.7.

10.2 Full Equations

We write z solving (6.1.1) as $z = (x^T, y^T)^T$ and consider the case where

$$\frac{dx}{dt} = f(x,y), \quad x(0) = x_0, \tag{10.2.1a}$$

$$\frac{dy}{dt} = \frac{1}{\varepsilon}g(x,y) + \frac{1}{\sqrt{\varepsilon}}\beta(x,y)\frac{dV}{dt}, \quad y(0) = y_0, \tag{10.2.1b}$$

with $\varepsilon \ll 1$ and V a standard Brownian motion. Here $x \in \mathcal{X}$, $y \in \mathcal{Y}$, $z \in \mathcal{Z}$, and the notation is as in Sections 4.1 and 6.1.

In Chapter 8 we considered systems in which the fast dynamics converge to an x-dependent fixed point. This gives rise to a situation where the y variables are slaved to the x variables. Averaging generalizes this idea to situations where the dynamics in the y variable, with x fixed, is more complex. As in the previous chapter on Markov chains, we average out the fast variable y over an appropriate invariant measure. We now make these heuristics precise. We define the generators

$$\mathcal{L}_0 = g(x, y) \cdot \nabla_y + \frac{1}{2} B(x, y) : \nabla_y \nabla_y, \tag{10.2.2a}$$

$$\mathcal{L}_1 = f(x, y) \cdot \nabla_x, \tag{10.2.2b}$$

where $B(x, y) = \beta(x, y)\beta(x, y)^T$. To carry out the averaging procedure in this section the most useful way to make an ergodicity assumption is to assume that, for each fixed x, $\mathcal{L}_0$ has one-dimensional null space characterized by

$$\mathcal{L}_0 1(y) = 0, \tag{10.2.3a}$$

$$\mathcal{L}_0^* \rho^\infty(y; x) = 0. \tag{10.2.3b}$$

Here $1(y)$ denotes constants in y. In the case where $\mathcal{Y} = \mathbb{T}^d$ the operators $\mathcal{L}_0$ and $\mathcal{L}_0^*$ are equipped with periodic boundary conditions. In this case these assumptions about the null spaces of $\mathcal{L}_0$ and $\mathcal{L}_0^*$ are known to hold if $B(x, y)$ is strictly positive-definite, uniformly in $(x, y) \in \mathcal{X} \times \mathcal{Y}$, as shown in Theorem 6.16. In more general situations, such as when $\mathcal{Y} = \mathbb{R}^d$ or when the matrix-valued function $B(x, y)$ is degenerate, similar rigorous justifications are possible, but the functional setting is more complicated, typically employing weighted L^p-spaces that characterize the decay of the invariant density at infinity. See the remarks in Section 18.4.

10.3 Simplified Equations

We assume that the generator of the fast process $y(t)$, namely $\mathcal{L}_0$, satisfies (10.2.3) for every $x \in \mathcal{X}$. Define the vector field F by

$$F(x) = \int_{\mathcal{Y}} f(x, y)\, \mu_x(dy) \tag{10.3.1}$$

with $\mu_x(dy) = \rho^\infty(y; x)dy$.

Result 10.1. *For $\varepsilon \ll 1$ and times t up to $\mathcal{O}(1)$, $x(t)$ solving (10.2.1) is approximated by X solving*

$$\frac{dX}{dt} = F(X), \quad X(0) = x_0. \tag{10.3.2}$$

Remark 10.2. A similar result holds even in the case where the equation for the slow variable x is stochastic and has the form

$$\frac{dx}{dt} = f(x, y) + \alpha(x, y)\frac{dU}{dt}, \quad x(0) = x,$$

with U a standard Brownian motion indepenent of V. Under the assumptions of Result 10.1 the averaged equation becomes

$$\frac{dX}{dt} = F(X) + A(X)\frac{dU}{dt}, \quad X(0) = x,$$

where $F(X)$ is the same and

$$A(X)A(X)^T = \int_{\mathcal{Y}} \alpha(x, y)\alpha(x, y)^T\, \mu_x(dy).$$

See Exercise 1. □

10.4 Derivation

As for Markov chains, we derive the averaged equations by working with the backward Kolmogorov equation. Let

$$v(x, y, t) = \mathbb{E}\Big(\phi(x(t), y(t))|x(0) = x, y(0) = y\Big).$$

The backward Equation (6.3.4) for the SDE (10.2.1) is

$$\frac{\partial v}{\partial t} = \frac{1}{\varepsilon}\mathcal{L}_0 v + \mathcal{L}_1 v. \tag{10.4.1}$$

Here $\mathcal{L}_0, \mathcal{L}_1$ are given by (10.2.2) and z in (6.3.4) is (x, y) here. Note that $\mathcal{L}_0$ is a differential operator in y, in which x appears as a parameter. Thus we must equip it with boundary conditions. We simply assume that, with suitable boundary conditions imposed, (10.2.3) holds. In the case where $\mathcal{Y} = \mathbb{T}^d$ and periodic boundary conditions are used, the rigorous results of Chapter 7 apply and the ergodicity assumption on the fast process is satisfied. Note, however, that other functional settings are also possible; the key in what follows is application of the Fredholm alternative to operator equations defined through $\mathcal{L}_0$.

We seek a solution to (10.4.1) in the form of the multiscale expansion

$$v = v_0 + \varepsilon v_1 + \mathcal{O}(\varepsilon^2)$$

and obtain

$$\mathcal{O}(1/\varepsilon)\ \ \mathcal{L}_0 v_0 = 0, \tag{10.4.2a}$$

$$\mathcal{O}(1)\ \ \mathcal{L}_0 v_1 = -\mathcal{L}_1 v_0 + \frac{\partial v_0}{\partial t}. \tag{10.4.2b}$$

Equation (10.4.2a) implies that v_0 is in the null space of $\mathcal{L}_0$ and hence, by (10.2.3) and ergodicity, is a function only of (x, t). Fix x. Then the Fredholm alternative for (10.4.2b), viewed as a differential equation in y, shows that

$$-\mathcal{L}_1 v_0 + \frac{\partial v_0}{\partial t} \perp \text{Null}\,\{\mathcal{L}_0^*\}.$$

By (10.2.3) this implies that

$$\int_{\mathcal{Y}} \rho^\infty(y; x)\Big(\frac{\partial v_0}{\partial t}(x, t) - f(x, y) \cdot \nabla_x v_0(x, t)\Big)dy = 0.$$

Since ρ^∞ is a probability density, we have $\int_{\mathcal{Y}} \rho^\infty(y; x)dy = 1$. Hence

$$\frac{\partial v_0}{\partial t} - \left(\int_{\mathcal{Y}} f(x, y)\mu_x(y)dy\right) \cdot \nabla_x v_0(x, t) = 0$$

so that by (10.3.1),

$$\frac{\partial v_0}{\partial t} - F(x) \cdot \nabla_x v_0 = 0.$$

This is the backward equation for (10.3.2); indeed the method of characteristics as given in Result 4.6 shows that we have the required result.

10.5 Deterministic Problems

In this section we provide a viewpoint on the averaged equation that is useful for two reasons: it applies when Equations (10.2.1) are deterministic, and it forms the basis of numerical methods to compute effective equations in either the deterministic or stochastic context. Our starting point is to analyze the behavior of the fast dynamics in y with x being a fixed parameter.

Let $\varphi_x^t(y)$ be the solution operator of the fast dynamics with x a fixed parameter and $\varepsilon = 1$. To be precise, for fixed ξ,

$$\frac{d}{dt}\varphi_\xi^t(y) = g(\xi, \varphi_\xi^t(y)) + \beta(\xi, \varphi_\xi^t(y))\frac{dV}{dt}, \qquad \varphi_\xi^0(y) = y. \tag{10.5.1}$$

As in Chapter 8, $y(t)$ solving (10.2.1b) is given by $y(t) \approx \varphi_{x(0)}^{t/\varepsilon}(y)$ for times t that are $o(1)$, so that x has not evolved very much. Assume that (10.5.1) is ergodic with invariant measure μ_ξ. On time scales small compared to 1 and large compared to ε, we expect that $x(t)$ is approximately frozen and that $y(t)$ will traverse its (x-dependent) invariant measure on this time scale because it is evolving quickly. Thus it is natural to average $y(t)$ in the $x(t)$ equation, against the invariant measure for (10.5.1) with $\xi = x(t)$.

In the case where $\beta \equiv 0$, $\varphi_\xi^t(y)$ coincides with the solution of (8.2.2). When $\beta \neq 0$, note that $\varphi_\xi^t(y)$ depends on the Brownian motion $\{V(s)\}_{s\in[0,t]}$ and hence is a stochastic process. Rather than assuming convergence to a fixed point, as we did in (8.2.3), we assume here that $\varphi_\xi^t(y)$ is ergodic (see Section 6.4). This implies that the measure defined by

$$\mu_x(A) = \lim_{T\to\infty}\frac{1}{T}\int_0^T I_A(\varphi_x^t(y))\,dt, \quad A \subseteq \mathbb{T}^d, \tag{10.5.2}$$

exists, for I_A the indicator function of arbitrary Borel sets $A \subseteq \mathcal{Y}$. The averaged vector field F in (10.3.1) can be defined using this measure.

When working with an SDE ($\beta \neq 0$), it is natural to assume that $\mu_x(\cdot)$ has a density with respect to the Lebesgue measure so that $\mu_x(dy) = \rho^\infty(y;x)dy$. In fact, under appropriate assumptions on the coefficients $g(x,y)$ and $\beta(x,y)$ it is possible to prove that such a density exists. However, we will illustrate by means of an example arising in Hamiltonian mechanics that this assumption is not necessary. Note also that the situation in Chapter 8 corresponds to the measure $\mu_x(dy)$ being a Dirac mass characterizing the invariant manifold: $\mu_x(dy) = \delta(y - \eta(x))dy$. In this case we obtain

$$F(x) = f(x, \eta(x)).$$

This is precisely the vector field in (8.3.2), and so the simplified equations in Chapter 8 are a special case of those derived here. However, we derived Result 10.1 in the case where β is nonzero and we assumed that the measure μ has a smooth density $\rho^\infty(y;x)$ with respect to Lebesgue measure; that is, we assumed that (10.2.3) holds and we have that $\mu_x(dy) = \rho^\infty(y;x)dy$. It is useful to have an expression for the averaged equation that is also valid for deterministic problems and for the numerical construction of F in either deterministic or random problems. We do this by representing ergodic averages via time averages.

Result 10.3. *An alternative representation of $F(x)$ is via a time average:*

$$F(x) = \lim_{T\to\infty} \frac{1}{T}\int_0^T f\left(x, \varphi_x^s(y)\right) ds. \tag{10.5.3}$$

This representation is found by using (10.5.2) to evaluate (10.3.1). Note that, by ergodicity, the resulting average does not depend on y.

10.6 Applications

We consider two applications of the averaging principle, the first in the context of SDEs and the second in the context of Hamiltonian ODEs.

10.6.1 A Skew-Product SDE

Consider the equations

$$\frac{dx}{dt} = (1-y^2)x,$$
$$\frac{dy}{dt} = -\frac{\alpha}{\varepsilon}y + \sqrt{\frac{2\lambda}{\varepsilon}}\frac{dV}{dt}.$$

Here $\mathcal{X} = \mathcal{Y} = \mathbb{R}$. It is of interest to know whether x will grow in time or remain bounded. We can get insight into this question in the limit $\varepsilon \to 0$ by deriving the averaged equations. Note that y is a time-rescaling of the OU process from Example 6.19. The invariant measure for the ergodic process y is a mean zero Gaussian: $\mathcal{N}(0, \frac{\lambda}{\alpha})$ (see Example 6.19). Note that this measure does not depend on x and hence has density $\rho^\infty(y)$ only. The averaged vector field F is here defined by

$$F(x) = \left(1 - \int_{\mathbb{R}} \rho^\infty(y) y^2 dy\right)x$$

where ρ^∞ is the density associated with Gaussian $\mathcal{N}(0, \lambda/\alpha)$. Thus

$$\int_{\mathbb{R}^d} \rho^\infty(y) y^2 dy = \frac{\lambda}{\alpha}$$

and

$$F(x) = \left(1 - \frac{\lambda}{\alpha}\right)x.$$

Hence the averaged equation is

$$\frac{dX}{dt} = \left(1 - \frac{\lambda}{\alpha}\right)X.$$

From this we deduce that trajectories of x will explode if $\lambda < \alpha$ and will contract if $\lambda > \alpha$. If $\lambda = \alpha$ then the averaged vector field is zero. In this situation we need to rescale time $t \mapsto t/\varepsilon$ to obtain the problem

$$\frac{dx}{dt} = \frac{1}{\varepsilon}(1 - y^2)x,$$
$$\frac{dx}{dt} = -\frac{\alpha}{\varepsilon^2}y + \sqrt{\frac{2\alpha}{\varepsilon^2}}\frac{dv}{dt}.$$

On this longer time scale, nontrivial dynamics occur. SDEs of this form are the topic of Chapter 11, and this specific example is considered in Section 11.7.

10.6.2 Hamiltonian Mechanics[1]

In many applications Hamiltonian systems with strong potential forces, responsible for fast, small-amplitude oscillations around a constraining submanifold, are encountered. It is then of interest to describe the evolution of the slowly evolving degrees of freedom by averaging over the rapidly oscillating variables. We give an example of this. The example is interesting because it shows that the formalism of this chapter can be extended to pure ordinary differential equations with no noise present; it also illustrates that it is possible to deal with situations where the limiting measure μ retains some memory of initial conditions – in this case the total energy of the system.

Consider a two-particle system with Hamiltonian

$$H(x, p, y, v) = \frac{1}{2}(p^2 + v^2) + \Phi(x) + \frac{\omega(x)}{2\,\varepsilon^2}y^2, \tag{10.6.1}$$

where (x, y) are the coordinates and (p, v) are the conjugate momenta of the two particles, $\Phi(x)$ is a nonnegative potential, and $\omega(x)$ is assumed to satisfy $\omega(x) \geqslant \bar{\omega} > 0$ for all x. The corresponding equations of motion are

$$\frac{dx}{dt} = p,$$
$$\frac{dp}{dt} = -\Phi'(x) - \frac{\omega'(x)}{2\varepsilon^2}y^2,$$
$$\frac{dy}{dt} = v,$$
$$\frac{dv}{dt} = -\frac{\omega(x)}{\varepsilon^2}y.$$

We let E denote the value of the Hamiltonian H at time $t = 0$:

[1] This example was developed in collaboration with R. Kupferman.

$$E = H(x(0), p(0), y(0), v(0)).$$

Note that E is the total energy of the two-particle system. We assume that E is bounded independently of ε. Since the Hamiltonian H is conserved in time, Φ is nonnegative, and $\omega \geqslant \bar{\omega}$, Equation (10.6.1) implies that

$$y^2 \leqslant 2\varepsilon^2 E/\bar{\omega}.$$

Hence the solution approaches the submanifold $y = 0$ as $\varepsilon \to 0$. Note, however, that y appears in the combination y/ε in the x equations and in the expression for the energy H. Thus it is natural to make the change of variables $\eta = y/\varepsilon$. The equations then read

$$\begin{aligned} \frac{dx}{dt} &= p, \\ \frac{dp}{dt} &= -\Phi'(x) - \frac{\omega'(x)}{2}\eta^2, \\ \frac{d\eta}{dt} &= \frac{1}{\varepsilon}v, \\ \frac{dv}{dt} &= -\frac{\omega(x)}{\varepsilon}\eta. \end{aligned} \tag{10.6.2}$$

In these variables we recover a system of the form (10.2.1) with "slow" variables, $x \leftarrow (x, p)$, and "fast" variables, $y \leftarrow (\eta, v)$. It is instructive to write the equation in second-order form as

$$\begin{aligned} \frac{d^2x}{dt^2} + \Phi'(x) + \frac{1}{2}\omega'(x)\eta^2 = 0, \\ \frac{d^2\eta}{dt^2} + \frac{1}{\varepsilon^2}\omega(x)\eta = 0. \end{aligned}$$

The fast equations represent a harmonic oscillator whose frequency $\omega^{1/2}(x)$ is modulated by the x variables.

Consider the fast dynamics with (x, p) frozen. The Hamiltonian for this fast dynamics is, for $\varepsilon = 1$ and x frozen,

$$H_{\text{fast}} = \frac{1}{2}v^2 + \frac{\omega(x)}{2}\eta^2.$$

The energy of the fast system, at given (x, p), which is conserved while (x, p) is frozen, is found by subtracting the energy associated with the frozen variables from the total energy of the original system. We denote the result of this calculation by

$$E_{\text{fast}} = E - \frac{1}{2}p^2 - \Phi(x).$$

For fixed x, p the dynamics in η, v is confined to the energy shell $H_{\text{fast}}(v, \eta) = E_{\text{fast}}$. We denote this energy shell by $\mathcal{Y}(x, p)$, noting that it is parameterized by the frozen variables (x, p).

The harmonic oscillator is studied in Example 4.17. Using the calculations therein, it follows that the average of the kinetic energy of the fast oscillator against the ergodic measure $\mu_{x,p}$ on $\mathcal{Y}(x,p)$ is

$$\int_{\mathcal{Y}(x,p)} \frac{\omega(x)}{2}\eta^2 \mu_{x,p}(d\eta, dv) = \frac{1}{2}\left[E - \frac{1}{2}p^2 - \Phi(x)\right].$$

Thus

$$\int_{\mathcal{Y}(x,p)} \frac{1}{2}\eta^2 \mu_{x,p}(d\eta, dv) = \frac{1}{2\omega(x)}\left[E - \frac{1}{2}p^2 - \Phi(x)\right].$$

Here (x,p) are viewed as fixed parameters and the total energy E is specified by the initial data of the whole system. The averaging principle states that the rapidly varying η^2 in Equation (10.6.2) for p can be approximated by its ergodic average, giving rise to a closed system of equations for $(X,P) \approx (x,p)$. These are

$$\begin{aligned}\frac{dX}{dt} &= P,\\ \frac{dP}{dt} &= -\Phi'(X) - \frac{\omega'(X)}{2\omega(X)}\left[E - \frac{1}{2}P^2 - \Phi(X)\right],\end{aligned} \tag{10.6.3}$$

with initial data E, $X(0) = X_0 = x(0)$, and $P(0) = P_0 = p(0)$. We will verify below that (X,P) satisfying (10.6.3) conserve the following *adiabatic invariant*

$$J = \frac{1}{\omega^{1/2}(X)}\left[E - \frac{1}{2}P^2 - \Phi(X)\right].$$

Thus, (10.6.3) reduces to the Hamiltonian form

$$\frac{dX}{dt} = P, \tag{10.6.4a}$$

$$\frac{dP}{dt} = -\Phi'(X) - J_0\left[\omega^{1/2}(X)\right]', \tag{10.6.4b}$$

where J_0 is given by

$$J_0 = \frac{1}{\omega^{1/2}(X_0)}\left[E - \frac{1}{2}P_0^2 - \Phi(X_0)\right].$$

This means that the influence of the stiff potential on the slow variables is to induce a Hamiltonian structure, but to replace the potential $\Phi(x)$ by an effective potential,

$$\Phi_{\text{eff}}(x) = \Phi(x) + J_0\,\omega^{1/2}(x).$$

Note that the limiting equation contains memory of the initial conditions for the fast variables, through the constant J_0. Thus the situation differs slightly from that covered by the conjunction of Results 10.1 and 10.3.

To verify that J is indeed conserved in time, note that, from the definition of J and from Equation (10.6.3),

$$
\begin{aligned}
\frac{d}{dt}\Big(\omega^{\frac{1}{2}}(X)J\Big) &= \frac{d}{dt}\Big(E - \frac{1}{2}P^2 - \Phi(X)\Big) \\
&= -P\frac{dP}{dt} - \Phi'(X)\frac{dX}{dt} \\
&= \frac{P\omega'(X)}{2\omega(X)}\Big(E - \frac{1}{2}P^2 - \Phi(X)\Big) \\
&= \frac{P\omega'(X)}{2\omega^{\frac{1}{2}}(X)}J.
\end{aligned}
$$

But, since $dX/dt = P$, we find the alternate expression,

$$
\begin{aligned}
\frac{d}{dt}\Big(\omega^{\frac{1}{2}}(X)J\Big) &= \frac{1}{2}\frac{\omega'(X)}{\omega^{\frac{1}{2}}(X)}\frac{dX}{dt}J + \omega^{\frac{1}{2}}(X)\frac{dJ}{dt} \\
&= \frac{P\omega'(X)}{2\omega^{\frac{1}{2}}(X)}J + \omega^{\frac{1}{2}}(X)\frac{dJ}{dt}.
\end{aligned}
$$

Equating the two expressions gives

$$
\frac{dJ}{dt} = 0,
$$

since $\omega(X)$ is strictly positive.

10.7 Discussion and Bibliography

Averaging is based on some form or ergodicity of the fast process; whether this process is deterministic or stochastic is not of primary importance. However, it is easier, in general, to establish ergodicity for stochastic problems, and this is why our general developments are confined to this case. The averaging method applied to Equations (10.2.1) is analyzed in an instructive manner in [240], where the Liouville equation is used to construct a rigorous proof of the averaged limit. It is sometimes possible to obtain averaging results in the nonergodic case, when the null space of the fast process is finite-dimensional, rather than one-dimensional; see [246, 326].

A detailed account of the averaging method for ODEs, as well as numerous examples, can be found in [281]; see also [13]. An English-language review of the Russian literature can be found in [193]. An overview of the topic of slow manifolds, especially in the context of Hamiltonian problems, may be found in [199]. The paper [321] provides an overview of variable elimination in a wealth of problems with scale separation.

Anosov's theorem is the name often given to the averaging principle in the context of ODEs – (10.2.1) with $\beta \equiv 0$. This theorem requires the fast dynamics to be ergodic. Often ergodicity fails due to the presence of "resonant zones"–regions in $\mathcal{X}$ for which the fast dynamics is not ergodic. Arnold and Neistadt [193] extended Anosov's result to situations in which the ergodicity assumption fails on a sufficiently small set of $x \in \mathcal{X}$. Those results were further generalized and extended to the stochastic framework by Kifer, who also studied the diffusive and large deviation character of the discrepancy between the effective and exact solutions [169, 170, 171, 172]; see also [111, ch. 7].

The situations in which the fast dynamics tend to fixed points, periodic solutions, or chaotic solutions can be treated in a unified manner through the introduction of *Young measures* (see [29, 309]). Artstein and co-workers considered a class of singularly perturbed system of type (10.2.1), with attention given to the limiting behavior of both slow and fast variables. In all of these cases the pair (x, y) can be shown to converge to (X, μ_X), where X is the solution of

$$\frac{dX}{dt} = \int_{\mathbb{T}^d} f(X, y)\, \mu_X(dy)$$

and μ_X is the ergodic measure on $\mathbb{T}^d$; the convergence of y to μ_X is in the sense of Young measures. (In the case of a fixed point the Young measure is a Dirac mass concentrated at a point.) A general theorem along these lines is proved in [17].

There are many generalizations of this idea. The case of nonautonomous fast dynamics, as well as a case with infinite dimensions are covered in [18]. Moreover, these results still make sense even if there is no unique invariant measure μ_x, in which case the slow variables can be proved to satisfy a (nondeterministic) differential inclusion [19].

In the context of SDE, an interesting generalization of (10.2.1) is to consider systems of the form

$$\frac{dx}{dt} = f(x, y) + \alpha(x, y)\frac{dU}{dt}, \tag{10.7.1a}$$

$$\frac{dy}{dt} = \frac{1}{\varepsilon}g(x, y) + \frac{1}{\sqrt{\varepsilon}}\beta(x, y)\frac{dV}{dt}. \tag{10.7.1b}$$

The simplified equation is then an SDE, not an ODE (see Remark 10.2).This situation is a subcase of the setup we consider in the next chapter. It can be obtained by setting $f_0 = 0$ in that chapter, letting $f_1 = f$ there, and by identifying ε here with ε^2 in that chapter.

In the application section we studied the averaging principle for a two-scale Hamiltonian system. The systematic study of Hamiltonian problems with two time scales was initiated by Rubin and Ungar [277]. More recently the ideas of Neistadt, based on normal form theory, have been applied to such problems [32]; this approach is very powerful, yielding very tight, exponential, error estimates between the original and limiting variables. A different approach to the problem, using the techniques of time-homogenization [43], is the paper [44]. The example presented in Section 10.6.2 is taken from that paper. The heuristic derivation we have given here is made rigorous in [44], using time-homogenization techniques, and it is also generalized to higher dimension. Resonances become increasingly important as the co-dimension, m, increases, limiting the applicability of the averaging approach to such two-scale Hamiltonian systems (Takens [306]).

Numerical work on multiscale ODEs and SDEs is overviewed in the next chapter.

10.8 Exercises

1. Derive the averaged equation resulting from the SDE (10.7.1) under the assumption that U and V are independent, standard Brownian motions (see Remark 10.2).
2. Let $\Phi : \mathcal{X} \times \mathcal{Y} : \mathbb{R}^+$ and consider the equations

$$\frac{dx}{dt} = -\nabla_x \Phi(x,y) + \sqrt{2\sigma}\frac{dU}{dt}$$
$$\frac{dy}{dt} = -\frac{1}{\varepsilon}\nabla_y \Phi(x,y) + \sqrt{\frac{2\sigma}{\varepsilon}}\frac{dV}{dt},$$

where U and V are standard Brownian motions of appropriate dimensions. Under a Fredholm alternative assumption, which you should clearly state, show that the averaged equation for X has the form

$$\frac{dX}{dt} = -\nabla\Psi(X) + \sqrt{2\sigma}\frac{dW}{dt}$$

where the *Fixman potential* Ψ is given by

$$\exp\Bigl(-\frac{1}{\sigma}\Psi(x)\Bigr) = \int_{\mathcal{Y}} \exp\Bigl(-\frac{1}{\sigma}\Phi(x,y)\Bigr)dy.$$

Here W is the Brownian motion of appropriate dimension. (In fact, strong convergence techniques, such as those highlighted in Chapter 17, may be used to show that $X \approx x$ strongly for $W = U$.).
3. Let Φ be as in the previous question. Write the following second-order system as a system of coupled first-order SDEs:

$$\frac{d^2x}{dt^2} + \frac{dx}{dt} = -\nabla_x \Phi(x,y) + \sqrt{2\sigma}\frac{dU}{dt},$$
$$\varepsilon\frac{d^2y}{dt^2} + \frac{dy}{dt} = -\frac{1}{\varepsilon}\nabla_y \Phi(x,y) + \sqrt{\frac{2\sigma}{\varepsilon}}\frac{dV}{dt}.$$

Find the stationary distribution of the fast process y explicitly. Find the averaged equation for X, using the previous question to guide you.
4. Derive the averaged equation from the example in Subsection 10.6.1 by use of formula (10.5.3) from Result 10.3.
5. Let u be a continuous-time Markov chain with generator

$$L = \begin{pmatrix} -a & a \\ b & -b \end{pmatrix}.$$

Without loss of generality label the state-space $\mathcal{I} = \{-1, +1\}$. Define two functions $\omega : \mathcal{I} \to (0,\infty)$ and $m : \mathcal{I} \to (-\infty,\infty)$ by $\omega(\pm 1) = \omega^{\pm}$ and $m(\pm 1) = m^{\pm}$. Now consider the stochastic differential equations, with coefficients depending on u, given by

$$\frac{dx}{dt} = f(x, y) + \sqrt{2\sigma}\frac{dU}{dt},$$
$$\frac{dy}{dt} = -\frac{1}{\varepsilon}\omega(u)(y - m(u)) + \sqrt{\frac{2\sigma}{\varepsilon}}\frac{dV}{dt},$$

with U and V standard Brownian motions of appropriate dimensions. Write the generator for the process (x, y, u) and use multiscale analysis to derive the averaged coupled Markov chain and SDE of the form

$$\frac{dX}{dt} = F(X, u) + \sqrt{2\sigma}\frac{dW}{dt}$$

where W is a standard Brownian motion with the same dimension as U.

6. Generalize the previous exercise to the case where the transition rates of the Markov chain, determined by a and b, depend on x and y.
7. Find a representation for the effective coefficient matrix $A(x)$ in Remark 10.2, using time-averaging.

11

Homogenization for ODEs and SDEs

11.1 Introduction

In this chapter we continue our study of systems of SDEs with two widely separated characteristic time scales. The setting is similar to the one considered in the previous chapter. The difference is that in this chapter we seek to derive an effective equation describing dynamics on the longer, *diffusive time scale*. This is the time scale of interest when the effective drift $F(x)$ defined in Equation (10.3.1) vanishes due, for example, to the symmetries of the problem. The vanishing of the effective drift is captured in the centering condition; see Equation (11.2.5). In contrast to the case considered in the previous chapter, in the diffusive time scale the effective equation is stochastic, even when noise does not act directly on the slow variables, that is, even when $\alpha(x,y) \equiv 0$ in Equation (11.2.1).

In Section 11.2 we present the SDEs that we will analyze in this chapter. Section 11.3 contains the simplified equations, which we derive in Section 11.4. In Section 11.5 we describe various properties of the simplified equations. The derivation assumes that the fast process to be eliminated is stochastic. In Section 11.6 we show how the deterministic case can be handled. In Section 11.7 we present various applications of the theory developed in this chapter: the case where the fast process is of Ornstein–Uhlenbeck type is in Section 11.7.1 and the case where the fast process is a chaotic deterministic process is in Section 11.7.2. Deriving the Stratonovich stochastic integral as the limit of smooth approximations to white noise is considered in Section 11.7.3; Stokes' law is studied in Section 11.7.4. The Green–Kubo formula from statistical mechanics is derived in Section 11.7.5. The case where the stochastic integral in the limiting equation can be interpreted in neither the Itô nor the Stratonovich sense in considered in Section 11.7.6. Lévy area corrections are studied in Section 11.7.7. Various extensions of the results presented in this chapter, together with bibliographical remarks, are presented in Section 11.8.

11.2 Full Equations

Consider the SDEs

$$\frac{dx}{dt} = \frac{1}{\varepsilon} f_0(x, y) + f_1(x, y) + \alpha(x, y)\frac{dU}{dt}, \quad x(0) = x_0, \tag{11.2.1a}$$

$$\frac{dy}{dt} = \frac{1}{\varepsilon^2} g(x, y) + \frac{1}{\varepsilon}\beta(x, y)\frac{dV}{dt}, \quad y(0) = y_0. \tag{11.2.1b}$$

Here U and V are indepenent standard Brownian motions. Both the x and y equations contain fast dynamics, but the dynamics in y is an order of magnitude faster than in x. As discussed in Sections 4.1 and 6.1 $x \in \mathcal{X}$, $y \in \mathcal{Y}$, and $\mathcal{X} \oplus \mathcal{Y} = \mathcal{Z}$.

For Equation (11.2.1), the backward Kolmogorov Equation (6.3.4) with $\phi = \phi(x)$ is[1]

$$\frac{\partial v}{\partial t} = \frac{1}{\varepsilon^2}\mathcal{L}_0 v + \frac{1}{\varepsilon}\mathcal{L}_1 v + \mathcal{L}_2 v, \quad \text{for } (x, y, t) \in \mathcal{X} \times \mathcal{Y} \times \mathbb{R}^+, \tag{11.2.2a}$$

$$v = \phi(x), \quad \text{for } (x, y, t) \in \mathcal{X} \times \mathcal{Y} \times \{0\}, \tag{11.2.2b}$$

where

$$\mathcal{L}_0 = g \cdot \nabla_y + \frac{1}{2} B : \nabla_y \nabla_y, \tag{11.2.3a}$$

$$\mathcal{L}_1 = f_0 \cdot \nabla_x, \tag{11.2.3b}$$

$$\mathcal{L}_2 = f_1 \cdot \nabla_x + \frac{1}{2} A : \nabla_x \nabla_x, \tag{11.2.3c}$$

with

$$A(x, y) := \alpha(x, y)\alpha(x, y)^T,$$
$$B(x, y) := \beta(x, y)\beta(x, y)^T.$$

By using the method of multiple scales we eliminate the y dependence in this Kolmogorov equation, to identify a simplified equation for the dynamics of x alone.

In terms of the generator $\mathcal{L}_0$, which is viewed as a differential operator in y, in which x appears as a parameter, the natural ergodicity assumption to make for variable elimination is the statement that $\mathcal{L}_0$ has one-dimensional null space characterized by

$$\mathcal{L}_0 1(y) = 0, \tag{11.2.4a}$$

$$\mathcal{L}_0^* \rho^\infty(y; x) = 0. \tag{11.2.4b}$$

Here $1(y)$ denotes constants in y and $\rho^\infty(y; x)$ is the density of an ergodic measure $\mu_x(dy) = \rho^\infty(y; x)dy$. We also assume that $f_0(x, y)$ averages to zero under this measure, so that the **centering condition**

[1] For simplicity we will take the initial condition of the backward Kolmogorov equation to be independent of y. This is not necessary. See the discussion in Section 11.8

$$\int_{\mathcal{Y}} f_0(x,y)\mu_x(dy) = 0 \quad \forall x \in \mathcal{X} \tag{11.2.5}$$

holds. It can then be shown that the term involving f_0 in the x equation will, in the limit $\varepsilon \to 0$, give rise to $\mathcal{O}(1)$ effective drift and noise contributions in an approximate equation for x.

As in the previous chapter, in the case where $\mathcal{Y} = \mathbb{T}^d$, the operators $\mathcal{L}_0$ and $\mathcal{L}_0^*$ are equipped with periodic boundary conditions. Then, assuming that $B(x,y)$ is strictly positive definite, uniformly in $(x,y) \in \mathcal{X} \times \mathbb{T}^d$, Theorem 6.16 justifies the statement that the null space of $\mathcal{L}_0^*$ is one-dimensional. In more general situations, such as when $\mathcal{Y} = \mathbb{R}^d$, or $B(x,y)$ is degenerate, similar rigorous justifications are possible, but the functional setting is more complicated, typically employing weighted L^p-spaces that characterize the decay of the invariant density at infinity.

When $\mathcal{Y} = \mathbb{T}^d$ and $B(x,y)$ is strictly positive definite, Theorem 7.9 also applies, and we have a solvability theory for Poisson equations of the form

$$-\mathcal{L}_0\phi = h. \tag{11.2.6}$$

In particular, the equation has a solution if and only if the right-hand side of the preceding equation is centered with respect to the invariant measure of the fast process $\mu_x(dy)$:

$$\int_{\mathbb{T}^d} h(x,y)\,\mu_x(dy) = 0 \quad \forall x \in \mathcal{X}. \tag{11.2.7}$$

When (11.2.7) is satisfied, the solution of (11.2.6) is unique up to a constant in the null space of $\mathcal{L}_0$. We can fix this constant by requiring that

$$\int_{\mathbb{T}^d} \phi(x,y)\mu_x(dy) = 0 \quad \forall x \in \mathcal{X}.$$

In more general situations, such as when $\mathcal{Y} = \mathbb{R}^d$ or $B(x,y)$ is degenerate, the question of existence and uniqueness of solutions to the Poisson Equation (11.2.6) becomes more complicated; however, analogous results are posible in function space settings that enforce appropriate decay properties at infinity. See the remarks and references to the literature in Section 11.8.

11.3 Simplified Equations

We assume that the operator $\mathcal{L}_0$ satisfies the Fredholm alternative, Theorem 2.42, and has one-dimensional null space characterized by (11.2.4). We define the **cell problem** [2] as follows:

$$-\mathcal{L}_0\Phi(x,y) = f_0(x,y), \quad \int_{\mathcal{Y}} \Phi(x,y)\rho^\infty(y;x)dy = 0. \tag{11.3.1}$$

---

[2] The word "cell" here refers to the periodic unit cell, which sets the scale for the fast variable in the case $\mathcal{Y} = \mathbb{T}^d$. The terminology comes from the theory of periodic homogenization for PDEs.

This is viewed as a PDE in y, with x a parameter. By the Fredholm alternative, (11.3.1) has a unique solution, since f_0 satisfies (11.2.5). We may then define a vector field F by

$$\begin{aligned} F(x) &= \int_{\mathcal{Y}} \big(f_1(x,y) + (\nabla_x \Phi(x,y)) f_0(x,y)\big) \rho^\infty(y;x) dy \\ &= F_1(x) + F_0(x) \end{aligned} \tag{11.3.2}$$

and a diffusion matrix $A(x)$ by

$$A(x)A(x)^T = A_1(x) + \frac{1}{2}\Big(A_0(x) + A_0(x)^T\Big), \tag{11.3.3}$$

where

$$A_0(x) := 2\int_{\mathcal{Y}} f_0(x,y) \otimes \Phi(x,y) \rho^\infty(y;x) dy, \tag{11.3.4}$$

$$A_1(x) := \int_{\mathcal{Y}} A(x,y) \rho^\infty(y;x) dy. \tag{11.3.5}$$

To make sure that $A(x)$ is well defined, it is necessary to prove that the sum of $A_1(x)$ and the symmetric part of $A_0(x)$ are positive semidefinite. This is done in Section 11.5.

Result 11.1. *For $\varepsilon \ll 1$ and times t up to $\mathcal{O}(1)$, the process $x(t)$, the solution of (11.2.1), is approximated by the process $X(t)$, the solution of*

$$\frac{dX}{dt} = F(X) + A(X)\frac{dW}{dt}, \quad X(0) = x_0. \tag{11.3.6}$$

Remark 11.2. Notice that knowledge of AA^T is not sufficient to determine A uniquely. As a result, Equation (11.3.3) does not determine the limiting SDE (11.3.6) uniquely. This is a consequence of the fact that there may be many SDEs that have the same generator. This in turn relates to the fact that the approximation of the solution to (11.2.1) by the solution to (11.3.6) is only valid in the sense of weak convergence of probability measures; see Chapter 18. □

11.4 Derivation

We seek a multiscale expansion for the solution of (11.2.2) with the form

$$v = v_0 + \varepsilon v_1 + \varepsilon^2 v_2 + \cdots. \tag{11.4.1}$$

Here $v_j = v_j(x,y,t)$. Substituting this expansion into (11.2.2) and equating powers of ε gives a hierarchy of equations, the first three of which are

$$\mathcal{O}(1/\varepsilon^2) \qquad -\mathcal{L}_0 v_0 = 0, \tag{11.4.2a}$$

$$\mathcal{O}(1/\varepsilon) \qquad -\mathcal{L}_0 v_1 = \mathcal{L}_1 v_0, \tag{11.4.2b}$$

$$\mathcal{O}(1) \qquad -\mathcal{L}_0 v_2 = -\frac{\partial v_0}{\partial t} + \mathcal{L}_1 v_1 + \mathcal{L}_2 v_0. \tag{11.4.2c}$$

By (11.2.4) Equation (11.4.2a) implies that the first term in the expansion is independent of y, $v_0 = v_0(x,t)$. We proceed now with Equation (11.4.2b). The solvability condition is satisfied for this equation since, by assumption (11.2.5), $f_0(x,y)$ is centered with respect to the invariant measure for $\varphi_x^t(\cdot)$ and, from (11.2.3b),

$$\mathcal{L}_1 v_0 = f_0(x,y) \cdot \nabla_x v_0(x,t).$$

Equation (11.4.2b) becomes

$$-\mathcal{L}_0 v_1 = f_0(x,y) \cdot \nabla_x v_0(x,t). \tag{11.4.3}$$

Since $\mathcal{L}_0$ is a differential operator in y alone with x appearing as a parameter, the general solution of (11.4.3) has the form

$$v_1(x,y,t) = \Phi(x,y) \cdot \nabla_x v_0(x,t) + \Phi_1(x,t). \tag{11.4.4}$$

The function Φ_1 plays no role in what follows so we set it to zero. Thus we represent the solution v_1 as a linear operator acting on v_0. As our aim is to find a closed equation for v_0, this form for v_1 is a useful representation of the solution. Substituting for v_1 in (11.4.3) shows that Φ solves the cell problem (11.3.1). Condition (11.2.5) ensures that there is a solution to the cell problem and the normalization condition makes it unique. Turning now to Equation (11.4.2c) we see that the right-hand side takes the form

$$-\Big(\frac{\partial v_0}{\partial t} - \mathcal{L}_2 v_0 - \mathcal{L}_1\big(\Phi \cdot \nabla_x v_0\big)\Big).$$

Hence solvability of (11.4.2c) for each fixed x requires

$$\begin{aligned}\frac{\partial v_0}{\partial t} &= \int_{\mathcal{Y}} \rho^\infty(y;x)\mathcal{L}_2 v_0(x,t)dy + \int_{\mathcal{Y}} \rho^\infty(y;x)\mathcal{L}_1\big(\Phi(x,y) \cdot \nabla_x v_0(x,t)\big)\,dy \\ &= I_1 + I_2.\end{aligned} \tag{11.4.5}$$

We consider the two terms on the right-hand side separately. The first is

$$\begin{aligned}I_1 &= \int_{\mathcal{Y}} \rho^\infty(y;x)\big(f_1(x,y) \cdot \nabla_x + \frac{1}{2}A(x,y) : \nabla_x \nabla_x\big) v_0(x,t)dy \\ &= F_1(x) \cdot \nabla_x v_0(x,t) + \frac{1}{2}A_1(x) : \nabla_x \nabla_x v_0(x,t).\end{aligned}$$

Now for the second term I_2, note that

$$\mathcal{L}_1(\Phi \cdot \nabla_x v_0) = f_0 \otimes \Phi : \nabla_x \nabla_x v_0 + (\nabla_x \Phi f_0) \cdot \nabla_x v_0.$$

Hence $I_2 = I_3 + I_4$ where

$$I_3 = \int_{\mathcal{Y}} \rho^\infty(y;x)\big(\nabla_x \Phi(x,y) f_0(x,y)\big) \cdot \nabla_x v_0(x,t)\, dy$$

and

$$I_4 = \int_{\mathcal{Y}} \rho^\infty(y;x)\big(f_0(x,y) \otimes \Phi(x,y) : \nabla_x \nabla_x v_0(x,t)\big)\, dy.$$

Thus

$$I_2 = F_0(x) \cdot \nabla_x v_0(x,t) + \frac{1}{2} A_0(x) : \nabla_x \nabla_x v_0(x,t).$$

Combining our simplifications of the right-hand side of (11.4.5) we obtain, since by (2.2.2) only the symmetric part of A_0 is required to calculate the Frobenius inner product with another symmetric matrix, the following expression:

$$\frac{\partial v_0}{\partial t} = F(x) \cdot \nabla_x v_0 + \frac{1}{2} A(x)A(x)^T : \nabla_x \nabla_x v_0.$$

This is the backward equation corresponding to the reduced dynamics given in (11.3.6).

11.5 Properties of the Simplified Equations

The effective SDE (11.3.6) is only well defined if $A(x)A(x)^T$ given by (11.3.3), (11.3.5) is nonnegative definite. We now prove that this is indeed the case.

Theorem 11.3. *Consider the case where $\mathcal{Y} = \mathbb{T}^d$ and $\mathcal{L}_0$ is equipped with periodic boundary conditions. Then*

$$\langle \xi, A_1(x)\xi + A_0(x)\xi \rangle \geqslant 0 \quad \forall x \in \mathcal{X}, \xi \in \mathbb{R}^l.$$

Hence the real-valued matrix function $A(x)$ is well defined by (11.3.3) *since $A(x)A(x)^T$ is nonnegative definite.*

Proof. Let $\phi(x,y) = \xi \cdot \Phi(x,y)$. Then ϕ solves

$$-\mathcal{L}_0 \phi = \xi \cdot f_0.$$

By Theorem 6.12 we have

$$\begin{aligned}
&\langle \xi, A_1(x)\xi + A_0(x)\xi \rangle \\
&\quad = \int_{\mathcal{Y}} \Big(|\alpha(x,y)^T \xi|^2 - 2(\mathcal{L}_0 \phi(x,y))\phi(x,y) \Big) \rho^\infty(y;x) dy \\
&\quad = \int_{\mathcal{Y}} \Big(|\alpha(x,y)^T \xi|^2 + |\beta(x,y)^T \nabla_y \phi(x,y)|^2 \Big) \rho^\infty(y;x) dy \\
&\quad \geqslant 0.
\end{aligned}$$

Thus

$$\begin{aligned}\langle \xi, AA^T\xi\rangle &= \langle \xi, A_1\xi\rangle + \frac{1}{2}\langle \xi, (A_0 + A_0^T)\xi\rangle \\ &= \langle \xi, (A_1 + A_0)\xi\rangle \geqslant 0.\square\end{aligned}$$

Two important remarks are in order.

Remark 11.4. Techniques similar to those used in the proof of the previous theorem, using (6.3.11) instead of the Dirichlet form itself, show that

$$\frac{1}{2}\Big(A_0(x)+A_0(x)^T\Big) = \int_{\mathcal{Y}} \Big(\nabla_y \Phi(x,y)\beta(x,y)\otimes\nabla_y\Phi(x,y)\beta(x,y)\Big)\rho^\infty(y;x)dy. \ \square \tag{11.5.1}$$

Remark 11.5. By virtue of Remark 6.13 we see that the proceeding theorem can be extended to settings other than $\mathcal{Y} = \mathbb{T}^d$. □

11.6 Deterministic Problems

As in the previous chapter, it is useful to have representations of the effective equation in terms of time averages, both for numerical purposes and for deterministic problems. To this end, a second representation of $A_0(x)$ and $F_0(x)$ is as follows. Let $\varphi_\xi^t(y)$ solve (10.5.1) and let $\mathbb{E}^{\mu_x}$ be the product measure formed from use of $\mu_x(\cdot)$ on initial data and standard independent Wiener measure on driving Brownian motions. Using this notation we may now employ a time integral to represent the solution of the cell problem, leading to the following representation formulae. Derivation is given at the end of the section.

Result 11.6. *Alternative representations of the vector field $F_0(x)$ and diffusion matrix $A_0(x)$ can be found through the following integrals over time and $\mathbb{E}^{\mu_x}$:*

$$A_0(x) = 2\int_0^\infty \mathbb{E}^{\mu_x}\Big(f_0(x,y)\otimes f_0(x,\varphi_x^t(y))\Big)\,dt \tag{11.6.1}$$

and, if the generator $\mathcal{L}_0$ is independent of x, then

$$F_0(x) = \int_0^\infty \mathbb{E}^{\mu_x}\Big(\nabla_x f_0(x,\varphi_x^t(y))f_0(x,y)\Big)\,dt. \tag{11.6.2}$$

All these representations hold for any y, by ergodicity.

The integral over t in this result enables us to express the effective equations without explicit reference to the solution of the cell problem Φ and requires sufficiently fast *decay of correlations* in order to be well-defined.

Another pair of alternative representations of $F(x)$ and $A(x)A(x)^T$ may be found by using time averaging (over s) to replace the expectations in the previous result. The expressions for A_0 and F_0 then involve two time integrals: the integral over

s is an ergodic average, replacing averaging with respect to the stationary measure on path space; the integral over t expresses the effective equations without reference to the solution of the cell problem Φ and, again, requires sufficiently fast *decay of correlations* in order to be well-defined. In fact the well posedness of the cell problem (11.3.1) implies the decay of correlations property.

Result 11.7. *Alternative representations of the vector field F and diffusion matrix A can be found through the following integrals over time:*

$$F_1(x) = \lim_{T\to\infty} \frac{1}{T} \int_0^T f_1(x, \varphi_x^s(y))\, ds,$$

$$A_1(x) = \lim_{T\to\infty} \frac{1}{T} \int_0^T A(x, \varphi_x^s(y))\, ds;$$

and

$$A_0(x) = 2\int_0^\infty \Big(\lim_{T\to\infty} \frac{1}{T} \int_0^T f_0(x, \varphi_x^s(y)) \otimes f_0(x, \varphi_x^{t+s}(y)) ds\Big)\, dt, \qquad (11.6.3)$$

where $\varphi_x^t(y)$ solves (10.5.1). *Furthermore, if the generator $\mathcal{L}_0$ is independent of x, then*

$$F_0(x) = \int_0^\infty \Big(\lim_{T\to\infty} \frac{1}{T} \int_0^T \nabla_x f_0(x, \varphi_x^{t+s}(y)) f_0(x, \varphi_x^s(y)) ds\Big) dt.$$

All these representations hold for any y, by ergodicity.

The following result will be useful to us in deriving the alternate representations of $A_0(x)$ and $F_0(x)$ in the two preceding results. It uses ergodicity to represent the solution of the cell problem, and related Poisson equations, as time integrals.

Result 11.8. *Let $\mathcal{L}$ be the generator of the ergodic Markov process $y(t)$ on $\mathcal{Y}$ which satisfies the SDE*

$$\frac{dy}{dt} = g(y) + \beta(y)\frac{dV}{dt}, \quad y(t) = y \qquad (11.6.4)$$

and let $\mu(dy)$ denote the unique invariant measure. Assume that h is centered with respect to μ:

$$\int_{\mathcal{Y}} h(y)\mu(dy) = 0.$$

Then the solution $f(y)$ of the Poisson equation

$$-\mathcal{L}f = h, \quad \int_{\mathcal{Y}} f(y)\mu(dy) = 0$$

admits the representation formula

$$f(y) = \int_0^\infty \big(e^{\mathcal{L}t}h\big)(y)\, dt. \qquad (11.6.5)$$

Proof. We apply the Itô formula to $f(y(t))$ to obtain

$$\begin{aligned} f(y(t)) - f(y) &= \int_0^t \mathcal{L}f(y(s))\,ds + \int_0^t \langle \nabla_y f(y(s)), \beta(y(s))\,dW(s)\rangle \\ &= \int_0^t -h(y(s))\,ds + \int_0^t \langle \nabla_y f(y(s)), \beta(y(s))\,dW(s)\rangle. \end{aligned}$$

We take expectation with respect to the Wiener measure and use the martingale property of stochastic integrals and the fact that $\mathbb{E}h(y(s)|y(0) = y)$ solves the backward Kolmogorov equation to conclude that

$$f(y) = \mathbb{E}f(y(t)) + \int_0^t (e^{\mathcal{L}s}h)(y)\,ds.$$

We take the limit $t \to \infty$ and use the ergodicity of the process $y(t)$, together with the fact that $f(y)$ is centered with respect to the invariant measure with density $\rho^\infty(y;x)$, to deduce that

$$\begin{aligned} f(y) &= \lim_{t\to\infty} \mathbb{E}f(y(t)) + \int_0^\infty (e^{\mathcal{L}t}h)(y)\,dt \\ &= \int_{\mathcal{Y}} f(y)\mu(dy) + \int_0^\infty (e^{\mathcal{L}t}h)(y)\,dt \\ &= \int_0^\infty (e^{\mathcal{L}t}h)(y)\,dt \end{aligned}$$

and the proof is complete. □

Remark 11.9. Notice that the preceding result implies that we can write, at least formally,

$$\mathcal{L}^{-1} = -\int_0^\infty e^{\mathcal{L}t}\,dt$$

when applied to functions centered with respect to μ. Furthermore, the result is also valid for the case where the coefficients in (11.6.4) depend on a parameter x. □

We complete the section by deriving the alternative expressions for $A(x)$ and $F(x)$ through time integration, given in Results 11.7 and 11.6. The expressions for $F_1(x)$ and $A_1(x)$ in Result 11.7 are immediate from ergodicity, simply using the fact that the time average equals the average against ρ^∞. By use of Result 11.8, the solution to the cell problem can be written as

$$\Phi(x,y) = \int_0^\infty (e^{\mathcal{L}_0 t}f_0)(x,y)\,dt = \int_0^\infty \mathbb{E}f_0(x, \varphi_x^t(y))\,dt \tag{11.6.6}$$

where $\mathbb{E}$ denotes expectation with respect to the Wiener measure. Now

$$F_0(x) = \int_{\mathcal{Y}} \rho^\infty(y;x)\nabla_x\Phi(x,y)f_0(x,y)\,dy.$$

In the case where $\mathcal{L}_0$ is x-independent so that $\varphi_x^t(\cdot) = \varphi^t(\cdot)$ is also x-independent, as are $\mu_x = \mu$ and $\rho^\infty(\,\cdot\,; x) = \rho^\infty(\cdot)$, we may use (11.6.6) to see that

$$F_0(x) = \int_{\mathcal{Y}} \rho^\infty(y;x) \int_0^\infty \mathbb{E}\nabla_x f_0(x, \varphi^t(y)) f_0(x,y)\, dt\, dy,$$

where $\mathbb{E}$ is expectation with respect to Wiener measure. Recal that $\mathbb{E}^{\mu_x}$ denotes the product measure formed from distributing y in its invariant measure, together with the Brownian motion driving the equation for $\varphi^t(y)$. Changing the order of integration we find that

$$F_0(x) = \int_0^\infty \mathbb{E}^{\mu_x}\Big(\nabla_x f_0(x, \varphi_x^t(y)) f_0(x,y)\Big)\, dt \tag{11.6.7}$$

as required for the expression in Result 11.6. Now we replace averages over $\mathbb{E}^{\mu_x}$ by time averaging to obtain, for all y,

$$F_0(x) = \int_0^\infty \Big(\lim_{T\to\infty} \frac{1}{T}\int_0^T \nabla_x f_0(x, \varphi_x^{t+s}(y)) f_0(x, \varphi_x^s(y))\, ds\Big)\, dt,$$

and so we obtain the desired formula for Result 11.7.

A similar calculation to that yielding (11.6.7) gives (11.6.1) for $A_0(x)$ in Result 11.6. Replacing the average against $\mathbb{E}^{\mu_x}$ by time average we arrive at the desired formula for $A_0(x)$ in Result 11.7.

11.7 Applications

We give a number of examples illustrating the wide applicability of the ideas in this chapter.

11.7.1 Fast Ornstein-Uhlenbeck Noise

Consider the equations

$$\frac{dx}{dt} = \frac{1}{\varepsilon}(1-y^2)x, \tag{11.7.1}$$

$$\frac{dy}{dt} = -\frac{\alpha}{\varepsilon^2} y + \sqrt{\frac{2\alpha}{\varepsilon^2}}\frac{dV}{dt}, \tag{11.7.2}$$

where $V(t)$ is a standard one-dimensional Brownian motion. Here

$$f_0(x,y) = (1-y^2)x \text{ and } f_1(x,y) = 0.$$

Recall that the equation for y is a time-rescaling of the OU process from Example 6.19, with $\lambda = \alpha$. Furthermore, these equations arise from the first application in Section 10.6, in the case where $\lambda = \alpha$ and after time rescaling to produce nonzero effects.

We have that

$$\int_{-\infty}^{\infty}(1-y^2)x\rho^{\infty}(y)\,dy=0,$$

where $\rho^{\infty}(y)$ is the invariant density of the Ornstein–Uhlenbeck process, namely a standard unit normal distribution. Thus the theory put forward in this chapter applies.

The generator of the process $\varphi_{\xi}^{t}(\cdot)=\varphi^{t}(\cdot)$ is

$$\mathcal{L}_0=-\alpha y\frac{\partial}{\partial y}+\alpha\frac{\partial^2}{\partial y^2} \tag{11.7.3}$$

and the cell problem (Poisson equation) (11.3.1) becomes

$$\alpha y\frac{\partial \Phi}{\partial y}-\alpha\frac{\partial^2\Phi}{\partial y^2}=(1-y^2)x.$$

The unique centered solution to this equation is

$$\Phi(y,x)=\frac{1}{2\alpha}(1-y^2)x.$$

Under the standard normal distribution, the fourth and second moments take values 3 and 1, respectively. Hence, the coefficients in the limiting Equation (11.3.6) are

$$F(x)=\int_{-\infty}^{\infty}\left(-\frac{1}{2\alpha}y^2(1-y^2)x\right)\rho^{\infty}(y)\,dy=\frac{1}{\alpha}x$$

and

$$A^2(x)=2\int_{-\infty}^{\infty}\left(-\frac{1}{2\alpha}y^2x(1-y^2)x\right)\rho^{\infty}(y)\,dy=\frac{2}{\alpha}x^2.$$

The homogenized SDE is thus

$$\frac{dX}{dt}=\frac{X}{\alpha}+\sqrt{\frac{2}{\alpha}}X\frac{dW}{dt}. \tag{11.7.4}$$

This is the geometric Brownian motion studied in Example 6.4. The solution is

$$X(t)=X(0)\exp\left(\sqrt{\frac{2}{\alpha}}W(t)\right).$$

It converges neither to 0 nor to ∞, but subsequences in time attain both limits. This should be compared with the behavior found in the first example in Section 10.6, which gives rise to decay (resp. growth) if $\lambda>\alpha$ (resp. $\lambda<\alpha$). Our example corresponds to the case $\lambda=\alpha$ with time rescaled to see nontrivial dynamics. It thus lies between decay and growth. Notice that we could have also taken the function in front of the white noise with a minus sign; see Remark 11.2.

Let us now obtain the coefficients of the homogenized equation by using the alternative representations (11.6.1) and (11.6.2). To this end we need to study the variable $\varphi^t(y)$ solving (10.5.1). From the calculations presented in Example 6.19 we have that

$$\varphi^t(y) = e^{-\alpha t}y + \sqrt{2\alpha}\int_0^t e^{-\alpha(t-s)}dV(s),$$
$$\varphi^t(y)^2 = e^{-2\alpha t}y^2 + \sqrt{2\alpha}ye^{-\alpha t}\int_0^t e^{-\alpha(t-s)}dV(s) + 2\alpha\Big(\int_0^t e^{-\alpha(t-s)}dV(s)\Big)^2. \tag{11.7.5}$$

In addition, by the Itô isometry,

$$\mathbb{E}\Big(\int_0^t e^{-\alpha(t-s)}dV(s)\Big)^2 = \int_0^t e^{-2\alpha(t-s)}ds, = \frac{1}{2\alpha}\Big(1 - e^{-2\alpha t}\Big).$$

To construct the measure $\mathbb{E}^{\mu^x}$ we take the initial condition y to be a standard unit Gaussian distribution and an independent driving Brownian motion V. (The measure is, in fact, independent of x in this particular example.) Thus, by stationarity under this initial Gaussian distribution,

$$\int \rho^\infty(y)y^2\,dy = 1, \quad \mathbb{E}^{\mu^x}\varphi^t(y)^2 = 1.$$

Furthermore

$$\begin{aligned}\mathbb{E}^{\mu^x}\Big(\int \rho^\infty(y)y^2\varphi^t(y)^2\,dy\Big) &= e^{-2\alpha t}\int \rho^\infty(y)y^4\,dy \\ &\quad +2\alpha\mathbb{E}^{\mu^x}\Big(\int_0^t e^{-\alpha(t-s)}\,dV(s)\Big)^2 \\ &= 3e^{-2\alpha t} + 1 - e^{-2\alpha t} \\ &= 1 + 2e^{-2\alpha t}.\end{aligned}$$

Since $f_0(x,y) = (1-y^2)x$, combining these calculations in (11.6.2) gives

$$\begin{aligned}F_0(x) &= x\int_0^\infty \mathbb{E}^{\mu_x}\Big((1-\varphi^t(y)^2)(1-y^2)\Big)dt \\ &= x\int_0^\infty 2e^{-2\alpha t}dt \\ &= \frac{x}{\alpha}.\end{aligned} \tag{11.7.6}$$

Similarly from (11.6.1) we obtain

$$A_0(x) = \frac{2x^2}{\alpha}.$$

This confirms that the effective equation is (11.7.4).

11.7.2 Fast Chaotic Noise

We now consider an example that is entirely deterministic but behaves stochastically when we eliminate a fast chaotic variable. In this context it is essential to use the representation of the effective diffusion coefficient given in Result 11.7. This representation uses time integrals, and makes no reference to averaging over the invariant measure (which does not have a density with respect to Lebesgue measure in this example; see Example 4.16). Consider the equations

$$\frac{dx}{dt} = x - x^3 + \frac{\lambda}{\varepsilon}y_2, \tag{11.7.7}$$

$$\begin{aligned}\frac{dy_1}{dt} &= \frac{10}{\varepsilon^2}(y_2 - y_1),\\ \frac{dy_2}{dt} &= \frac{1}{\varepsilon^2}(28y_1 - y_2 - y_1y_3),\\ \frac{dy_3}{dt} &= \frac{1}{\varepsilon^2}(y_1y_2 - \frac{8}{3}y_3).\end{aligned} \tag{11.7.8}$$

The vector $y = (y_1, y_2, y_3)^T$ solves the Lorenz equations, at parameter values where the solution is ergodic (see Example 4.16). In the invariant measure the component y_2 has mean zero. Thus the centering condition holds. The equation for x is a scalar ODE driven by a chaotic signal with characteristic time ε^2. Because $f_0(x,y) \propto y_2$, with invariant measure shown in Figure 4.2, and because $f_1 = (x,y) = f_1(x)$ only, the candidate equation for the approximate dynamics is

$$\frac{dX}{dt} = X - X^3 + \sigma\frac{dW}{dt}, \tag{11.7.9}$$

where σ is a constant. Now let $\psi^t(y) = e_2 \cdot \varphi^t(y)$. Then the constant σ can be found by use of (11.6.3) giving

$$\sigma^2 = 2\lambda^2 \int_0^\infty \frac{1}{T}\Big(\lim_{T\to\infty}\int_0^T \psi^s(y)\psi^{t+s}(y)ds\Big)dt.$$

This is the integrated autocorrelation function of y_2. By ergodicity we expect the value of σ^2 to be independent of y and to be determined by the SRB measure for the Lorenz equations alone. Notice that the formula is expected to make sense, even though the cell problem is not well-posed in this case because the generator of the fast process is not elliptic.

Another way to derive this result is as follows. Gaussian white noise $\sigma\dot{W}$, the time derivative of Brownian motion, may be thought of as a delta-correlated stationary process. The integral of its autocorrelation function on $[0,\infty)$ gives $\sigma^2/2$. On the assumption that y_2 has a correlation function that decays in time, and noting that this has time scale ε^2, the autocorrelation of $\frac{\lambda}{\varepsilon}\psi^{s/\varepsilon^2}(y)$ at timelag t may be calculated and integrated from 0 to ∞; matching this with the known result for Gaussian white noise gives the desired result for σ^2.

11.7.3 Stratonovich Corrections

When white noise is approximated by a smooth process this often leads to Stratonovich interpretations of stochastic integrals, at least in one dimension. We use multiscale analysis to illustrate this phenomenon by means of a simple example. Consider the equations

$$\frac{dx}{dt} = \frac{1}{\varepsilon} f(x) y,$$
$$\frac{dy}{dt} = -\frac{\alpha y}{\varepsilon^2} + \sqrt{\frac{2\alpha}{\varepsilon^2}} \frac{dV}{dt}, \tag{11.7.10}$$

with V being a standard one-dimensional Brownian motion.

Assume for simplicity that $y(0) = 0$. Then

$$\mathbb{E}(y(t)y(s)) = e^{-\frac{\alpha}{\varepsilon^2}|t-s|}$$

and, consequently,

$$\lim_{\varepsilon \to 0} \mathbb{E}\left(\frac{y(t)}{\varepsilon} \frac{y(s)}{\varepsilon}\right) = \frac{2}{\alpha} \delta(t-s),$$

which implies the heuristic

$$\lim_{\varepsilon \to 0} \frac{y(t)}{\varepsilon} = \sqrt{\frac{2}{\alpha}} \frac{dV}{dt}. \tag{11.7.11}$$

Another way of seeing this is by solving (11.7.10) for y/ε:

$$\frac{y}{\varepsilon} = \sqrt{\frac{2}{\alpha}} \frac{dV}{dt} - \frac{\varepsilon}{\alpha} \frac{dy}{dt}. \tag{11.7.12}$$

If we neglect the $\mathcal{O}(\varepsilon)$ term on the right-hand side, then we arrive, again, at the heuristic (11.7.11).

Both of these arguments lead us to conjecture a limiting equation of the form

$$\frac{dX}{dt} = \sqrt{\frac{2}{\alpha}} f(X) \frac{dV}{dt}. \tag{11.7.13}$$

We will show that, as applied, *the heuristic gives the incorrect limit:* this is because, in one dimension, whenever white noise is approximated by a smooth process, the limiting equation should be interpreted in the Stratonovich sense, giving

$$\frac{dX}{dt} = \sqrt{\frac{2}{\alpha}} f(X) \circ \frac{dV}{dt} \tag{11.7.14}$$

in this case. We now derive this limit equation by the techniques introduced in this chapter.

The cell problem is

$$-\mathcal{L}_0 \Phi(x, y) = f(x) y$$

with $\mathcal{L}_0$ given by (11.7.3). The solution is readily seen to be

$$\Phi(x,y) = \frac{1}{\alpha} f(x) y, \quad \nabla_x \Phi(x,y) = \frac{1}{\alpha} f'(x) y.$$

The invariant density is

$$\rho^\infty(y) = \frac{1}{\sqrt{2\pi}} \exp\Big(-\frac{y^2}{2}\Big),$$

which is in the null space of $\mathcal{L}_0^*$ and corresponds to a standard unit Gaussian $\mathcal{N}(0,1)$ random variable.

From Equation (11.3.2) we have

$$\begin{aligned} F(x) &= \int_{\mathbb{R}} \frac{1}{\alpha} f'(x) f(x) y^2 \rho^\infty(y) dy \\ &= \frac{1}{\alpha} f'(x) f(x). \end{aligned}$$

Also (11.3.3) gives

$$\begin{aligned} A(x)^2 &= \int_{\mathbb{R}} \frac{2}{\alpha} f(x)^2 y^2 \rho^\infty(y) dy \\ &= \frac{2}{\alpha} f(x)^2. \end{aligned}$$

The limiting equation is therefore the Itô SDE

$$\frac{dX}{dt} = \frac{1}{\alpha} f'(X) f(X) + \sqrt{\frac{2}{\alpha}} f(X) \frac{dV}{dt}.$$

This is the Itô form of (11.7.14), by Remark 6.2. Hence, the desired result is established.

11.7.4 Stokes' Law

The previous example may be viewed as describing the motion of a massless particle with position x in a velocity field proportional to $f(x)y$, with y an OU process. If the particle has mass m then it is natural to study the generalized equation

$$m\frac{d^2x}{dt^2} = \frac{1}{\varepsilon} f(x) y - \frac{dx}{dt}, \qquad (11.7.15a)$$

$$\frac{dy}{dt} = -\frac{\alpha y}{\varepsilon^2} + \sqrt{\frac{2\alpha}{\varepsilon^2}} \frac{dV}{dt}. \qquad (11.7.15b)$$

(Note that setting $m = 0$ gives the previous example.) Equation (11.7.15a) is *Stokes' law*, stating that the force on the particle is proportional to a drag force,

$$\frac{1}{\varepsilon} f(x) y - \frac{dx}{dt},$$

which is equal to the difference between the fluid velocity and the particle velocity. As in the previous example, y is a fluctuating OU process. For simplicity we consider the case of unit mass, $m = 1$.

Using the heuristic argument from the previous section it is natural to conjecture the limiting equation

$$\frac{d^2X}{dt^2} = \sqrt{\frac{2}{\alpha}} f(X)\frac{dV}{dt} - \frac{dX}{dt}. \tag{11.7.16}$$

In contrast to the previous application, the conjecture that this is the limiting equation turns out to be correct. The reason is that, here, x is smoother and the Itô and Stratonovich integrals coincide; there is no Itô correction to the Stratonovich integral. (To see this it is necessary to first write (11.7.16) as a first-order system; see Exercise 2a). We verify the result by using the multiscale techniques introduced in this chapter.

We first write (11.7.15) as the first-order system

$$\begin{aligned}
\frac{dx}{dt} &= r,\\
\frac{dr}{dt} &= -r + \frac{1}{\varepsilon} f(x) y,\\
\frac{dy}{dt} &= -\frac{1}{\varepsilon^2}\alpha y + \frac{1}{\varepsilon}\sqrt{2\alpha}\frac{dV}{dt}.
\end{aligned}$$

Here (x, r) are slow variables (x in (11.2.1)) and y the fast variables (y in (11.2.1)). The cell problem is now given by

$$\mathcal{L}_0\Phi(x,r,y) = -f_0(x,r,y) = \begin{pmatrix} 0 \\ -f(x)y \end{pmatrix},$$

with $\mathcal{L}_0$ given by (11.7.3). The solution is

$$\Phi(x,r,y) = \begin{pmatrix} 0 \\ \frac{1}{\alpha} f(x) y \end{pmatrix}, \quad \nabla_{(x,r)}\Phi(x,y) = \begin{pmatrix} 0 & 0 \\ \frac{1}{\alpha} f'(x) y & 0 \end{pmatrix}.$$

Notice that f_0 is in the null space of $\nabla_{(x,r)}\Phi$, and hence (11.3.2) gives

$$F(X,R) = F_1(X,R) = \begin{pmatrix} R \\ -R \end{pmatrix}. \tag{11.7.17}$$

From (11.3.3) we have

$$A(X,R)A(X,R)^T = \int_{\mathbb{R}} 2\begin{pmatrix} 0 & 0 \\ 0 & \frac{1}{\alpha} f(X)^2 y^2 \end{pmatrix} \rho^\infty(y)dy.$$

Recall that $\rho^\infty(y)$ is the density of an $\mathcal{N}(0,1)$ Gaussian random variable. Evaluating the integral gives

$$A(X,R)A(X,R)^T = \begin{pmatrix} 0 & 0 \\ 0 & \frac{2}{\alpha} f(X)^2 \end{pmatrix}.$$

Hence a natural choice for $A(x)$ is

$$A(X,R) = \begin{pmatrix} 0 \\ \sqrt{\frac{2}{\alpha}} f(X) \end{pmatrix}.$$

Thus from (11.7.17) and (11.7.18) we obtain the limiting equation

$$\begin{aligned} \frac{dX}{dt} &= R, \\ \frac{dR}{dt} &= -R + \sqrt{\frac{2}{\alpha}} f(X) \frac{dW}{dt}, \end{aligned}$$

which, upon elimination of R, is seen to coincide with the conjectured limit (11.7.16).

11.7.5 Green–Kubo Formula

In the previous application we encountered the equation of motion for a particle with significant mass, subject to Stokes drag. Here we study the same equation of motion, but where the velocity field is steady. We also assume that the particle is subject to molecular diffusion. The equation of motion is thus

$$\frac{d^2x}{dt^2} = f(x) - \frac{dx}{dt} + \sigma \frac{dU}{dt}. \tag{11.7.18}$$

Here U is a standard unit Brownian motion. We will study the effective diffusive behavior of the particle x on large length and time scales, under the assumption that $f(x)$ is a mean zero periodic function. We show that, on appropriate large length and time scales, the particle performs an effective Brownian motion, and we calculate its diffusion coefficient.

To this end we rescale the equation of motion by setting $x \to x/\varepsilon$ and $t \to t/\varepsilon^2$ to obtain

$$\varepsilon^2 \frac{d^2x}{dt^2} = \frac{1}{\varepsilon} f\left(\frac{x}{\varepsilon}\right) - \frac{dx}{dt} + \sigma \frac{dU}{dt}.$$

Introducing the variables $y = \varepsilon dx/dt$ and $z = x/\varepsilon$ we obtain the system

$$\begin{aligned} \frac{dx}{dt} &= \frac{1}{\varepsilon} y, \\ \frac{dy}{dt} &= -\frac{1}{\varepsilon^2} y + \frac{1}{\varepsilon^2} f(z) + \frac{\sigma}{\varepsilon} \frac{dW}{dt}, \\ \frac{dz}{dt} &= \frac{1}{\varepsilon^2} y. \end{aligned}$$

The process (y,z) is ergodic, with characteristic time scale ε^2, and plays the role of y in (11.2.1); x plays the role of x in (11.2.1). The operator $\mathcal{L}_0$ is the generator of the process (y,z). Furthermore

$$f_1(x,y,z) = 0, \quad f_0(x,y,z) = y.$$

Thus, since the evolution of (y, z) is independent of x, $\Phi(x, y, z)$, the solution of the cell problem, is also x−independent. Hence (11.3.2) gives $F(x) = 0$. Turning now to the effective diffusivity we find that, since $\alpha(x, y) = A(x, y) = 0$, (11.3.3) gives $A(x)^2 = A_0(x)$. Now define $\psi^t(y, z)$ to be the component of $\varphi^t(y, z)$ projected onto the y coordinate. By Result 11.7 we have that

$$A_0(x) = 2\int_0^\infty \Big(\lim_{T\to\infty} \frac{1}{T}\int_0^T \psi^s(y)\psi^{s+t}(y)ds\Big)dt.$$

The expression

$$C(t) = \lim_{T\to\infty} \frac{1}{T}\int_0^T \psi^s(y)\psi^{s+t}(y)ds$$

is the *velocity autocorrelation function*. Thus the effective equation is

$$\frac{dX}{dt} = \sqrt{2D}\frac{dW}{dt},$$

a Brownian motion with diffusion coefficient

$$D = \int_0^\infty C(t)dt.$$

Thus, the effective diffusion coefficient is given by the *integrated velocity autocorrelation*. This is an example of the **Green–Kubo formula**.

11.7.6 Neither Itô nor Stratonovich

We again use Stokes' law (11.7.15a), now for a particle of small mass $m = \tau_0\varepsilon^2$ where $\tau_0 = \mathcal{O}(1)$, and neglecting molecular diffusion. If we also assume that the velocity field of the underlying fluid is of the form $\frac{1}{\varepsilon}f(x)\eta$ where η solves an SDE, then we obtain

$$\tau_0\varepsilon^2\frac{d^2x}{dt^2} = -\frac{dx}{dt} + \frac{1}{\varepsilon}f(x)\eta, \tag{11.7.19a}$$

$$\frac{d\eta}{dt} = \frac{1}{\varepsilon^2}g(\eta) + \frac{1}{\varepsilon}\sqrt{2\sigma(\eta)}\frac{dW}{dt}. \tag{11.7.19b}$$

We interpret equations (11.7.19b) in the Itô sense. We assume that $g(\eta)$, $\sigma(\eta)$ are such that there exists a unique stationary solution of the Fokker-Planck equation for (11.7.19b), so that η is ergodic.

We write (11.7.19) as a first-order system,

$$\begin{aligned}\frac{dx}{dt} &= \frac{1}{\varepsilon\sqrt{\tau_0}}v,\\ \frac{dv}{dt} &= \frac{f(x)\eta}{\varepsilon^2\sqrt{\tau_0}} - \frac{v}{\tau_0\varepsilon^2},\\ \frac{d\eta}{dt} &= \frac{g(\eta)}{\varepsilon^2} + \frac{\sqrt{2\sigma(\eta)}}{\varepsilon}\frac{dW}{dt}.\end{aligned} \tag{11.7.20}$$

Equations (11.7.20) are of the form (11.2.1) and, under the assumption that the fast process (v, η) is ergodic, the theory developed in this chapter applies. In order to calculate the effective coefficients we need to solve the stationary Fokker–Planck equation

$$\mathcal{L}_0^* \rho(x, v, \eta) = 0$$

and the cell problem

$$-\mathcal{L}_0 h = \frac{v}{\sqrt{\tau_0}}, \tag{11.7.21}$$

where

$$\mathcal{L}_0 = g(\eta)\frac{\partial}{\partial \eta} + \sigma(\eta)\frac{\partial^2}{\partial \eta^2} + \left(\frac{f(x)\eta}{\sqrt{\tau_0}} - \frac{v}{\tau_0}\right)\frac{\partial}{\partial v}.$$

Equation (11.7.21) can be simplified considerably: we look for a solution of the form

$$h(x, v, \eta) = \left(\sqrt{\tau_0}\, v + f(x)\widehat{h}(\eta)\right). \tag{11.7.22}$$

Substituting this expression in the cell problem we obtain, after some algebra, the equation

$$-\mathcal{L}_\eta \widehat{h} = \eta.$$

Here $\mathcal{L}_\eta$ denotes the generator of η. We assume that the unique invariant measure for $\eta(t)$ has density $\rho_\eta(\eta)$ with respect to Lebesgue measure; the centering condition that ensures the well-posedness of the Poisson equation for $\widehat{h}$ is

$$\int_{\mathbb{R}} \eta \rho_\eta(\eta)\, d\eta = 0.$$

We assume that this holds. The homogenized SDE is

$$\frac{dX}{dt} = F(X) + \sqrt{D(X)}\frac{dW}{dt}, \tag{11.7.23}$$

where

$$F(x) := \int_{\mathbb{R}^2} \left(\frac{v}{\sqrt{\tau_0}}\widehat{h}(\eta) f'(x)\right) \rho(x, v, \eta)\, dv d\eta$$

and

$$D(x) := 2\int_{\mathbb{R}^2} \left(v^2 + \frac{v}{\sqrt{\tau_0}}\widehat{h}(\eta) f(x)\right) \rho(x, v, \eta)\, dv d\eta.$$

In the case where $\eta(t)$ is the Ornstein–Uhlenbeck process,

$$\frac{d\eta}{dt} = -\frac{\alpha}{\varepsilon^2}\eta + \sqrt{\frac{2\lambda}{\varepsilon^2}}\frac{dW}{dt}, \tag{11.7.24}$$

we can compute the homogenized coefficients $D(X)$ and $B(X)$ explicitly. The effective SDE is

$$\frac{dX}{dt} = \frac{\lambda}{\alpha^2(1 + \tau_0\alpha)} f(X) f'(X) + \sqrt{\frac{2\lambda}{\alpha^2}} f(X)\frac{dW}{dt}. \tag{11.7.25}$$

Note that in the limit $\tau_0 \to \infty$ we recover the Itô stochastic integral, as in Subsection 11.7.4, whereas in the limit $\tau_0 \to 0$ we recover the Itô interpretation of the Stratonovich stochastic integral as in Subsection 11.7.3. For $\tau_0 \in (0, \infty)$ the limiting equation is of neither the Itô nor the Stratonovich form. In fact, Equation (11.7.25) can be written in the form

$$X(t) = x_0 + \int_0^t \frac{2\lambda}{\alpha^2} f(X) \widehat{\circ} dW(t),$$

where the definition of the stochastic integral through Riemann sums depends on the value of τ_0. The fact that we recover this interesting limit is very much tied to the scaling of the mass as $\mathcal{O}(\varepsilon^2)$. This scaling ensures that the time scale of the ergodic process η and the relaxation time of the particle are the same. Resonance between these time scales gives the desired effect.

11.7.7 The Lévy Area Correction[3]

In Section 11.7.3 we saw that smooth approximation to white noise in one dimension leads to the Stratonovich stochastic integral. This is not true in general, however, in the multidimensional case: an additional drift can appear in the limit. This extra drift contribution is related to the properties of the Lévy area of the limit process (see the discussion in Section 11.8).

Consider the fast–slow system

$$\dot{x}_1 = \frac{1}{\varepsilon} y_1, \tag{11.7.26a}$$

$$\dot{x}_2 = \frac{1}{\varepsilon} y_2, \tag{11.7.26b}$$

$$\dot{x}_3 = \frac{1}{\varepsilon} \left(x_1 y_2 - x_2 y_1 \right), \tag{11.7.26c}$$

$$\dot{y}_1 = -\frac{1}{\varepsilon^2} y_1 - \alpha \frac{1}{\varepsilon^2} y_2 + \frac{1}{\varepsilon} \dot{W}_1, \tag{11.7.26d}$$

$$\dot{y}_2 = -\frac{1}{\varepsilon^2} y_2 + \alpha \frac{1}{\varepsilon^2} y_1 + \frac{1}{\varepsilon} \dot{W}_2, \tag{11.7.26e}$$

where $\alpha > 0$. Here W_1, W_2 are standard independent Brownian motions.

Notice that Equations (11.7.26d) and (11.7.26e) may be written in the form

$$\dot{y} = -\frac{1}{\varepsilon^2} y + \frac{1}{\varepsilon^2} \alpha J y + \frac{1}{\varepsilon} \dot{W},$$

where $y = (y_1, y_2)$, $W = (W_1, W_2)$, and J is the antisymmetric (symplectic) matrix

$$J = \begin{pmatrix} 0 & -1 \\ 1 & 0. \end{pmatrix}.$$

[3] This section was written in collaboration with M. Hairer.

Applying the heuristic that

$$y \approx \varepsilon (I - \alpha J)^{-1} \frac{dW}{dt}$$

leads to the conjectured limiting equations

$$\dot{x}_1 = \frac{1}{1+\alpha^2}\left(\dot{W}_1 - \alpha \dot{W}_2\right), \tag{11.7.27a}$$

$$\dot{x}_2 = \frac{1}{1+\alpha^2}\left(\dot{W}_2 + \alpha \dot{W}_1\right), \tag{11.7.27b}$$

$$\dot{x}_3 = \frac{1}{1+\alpha^2}\left((\alpha x_1 - x_2)\dot{W}_1 + (\alpha x_2 + x_1)\dot{W}_2\right). \tag{11.7.27c}$$

We know from Subsections 11.7.3 and 11.7.6 that we must take care in conjecturing such a limit as typically smooth approximations of white noise give rise to the Stratonovich stochastic integral. However, in this case Itô and Stratonovich coincide so this issue does not arise. Nonetheless, the conjectured limit equation is wrong.

Multiscale techniques, as described in this chapter, lead to the correct homogenized system:

$$\dot{x}_1 = \frac{1}{1+\alpha^2}\left(\dot{W}_1 - \alpha \dot{W}_2\right), \tag{11.7.28a}$$

$$\dot{x}_2 = \frac{1}{1+\alpha^2}\left(\dot{W}_2 + \alpha \dot{W}_1\right), \tag{11.7.28b}$$

$$\dot{x}_3 = \frac{1}{1+\alpha^2}\left((\alpha x_1 - x_2)\dot{W}_1 + (\alpha x_2 + x_1)\dot{W}_2\right) + \frac{\alpha}{1+\alpha^2}. \tag{11.7.28c}$$

Notice the additional constant drift that appears in Equation (11.7.28c). It is the antisymmetric part in the equation for the fast process y that is responsible for the presence of the additional drift in the homogenized equation. In particular, when $\alpha = 0$ the homogenized equation becomes

$$\dot{x}_1 = \dot{W}_1,$$

$$\dot{x}_2 = \dot{W}_2,$$

$$\dot{x}_3 = -x_2\dot{W}_1 + x_1\dot{W}_2,$$

which agrees with the original (in general incorrect) conjectured limit (11.7.27).

11.8 Discussion and Bibliography

The perturbation approach adopted in this chapter, and more general related ones, is covered in a series of papers by Papanicolaou and co-workers – see [244, 241, 242, 240], building on original work of Khasminkii [165, 166]. See [155, 154, 31, 205, 244, 242, 240, 155, 154] for further material. We adapted the general analysis to the simple case where $\mathcal{Y} = \mathbb{T}^d$. This may be extended to, for example $\mathbb{R}^d$, by working in the appropriate functional setting; see [249, 250, 251].

The basic perturbation expansion outlined in this chapter can be rigorously justified and weak convergence of x to X proved as $\varepsilon \to 0$; see Kurtz [181] and Chapter 18. The perturbation expansion that underlies the approach is clearly exposed in [241]; see also [117; ch. 6; 321; 291]. Similar problems are analyzed in [27, ch. 8], by using eigenfunction expansions for the Fokker–Planck operator of the fast process. Projection operator techniques are also often employed in the physics literature as a method for eliminating fast variables. See [117, ch. 6] and the references therein.

Studying the derivation of effective stochastic models when the original system is an ODE is a subject investigated in some generality in [242]. The specific example in Section 11.7.2 relies on the ergodicity of the Lorenz equations, something established in [318, 319]. Use of the integrated autocorrelation function to calculate the effective diffusion coefficient numerically is highlighted in [322]; a different approach to finding the effective diffusion coefficient is described in [125]. The program described is carried out in discrete time by Beck [31], who uses a skew-product structure to facilitate an analysis; the ideas can then be rigorously justified in some cases. A skew-product setup is also employed in [322] and [125]. A rigorous limit theorem for ODEs driven by a fast mixing system is proved in [225], using the large deviation principle for dynamical systems developed in [224]. In the paper [208], the idea that fast chaotic motion can introduce noise in slow variables is pursued for an interesting physically motivated problem where the fast chaotic behavior arises from the Burgers bath of [204]. Further numerical experiments on the Burgers bath are reported in [209].

Related work can be found in [124], and similar ideas in continuous time are addressed in [155, 154] for differential equations; however, rather than developing a systematic expansion in powers of ε, they find the exact solution of the Fokker–Planck equation, projected into the space $\mathcal{X}$, by use of the Mori-Zwanzig formalism [65], and then make power series expansions in ε of the resulting problem.

In Section 11.7.5 we derived a formula for the effective diffusion coefficient in terms of the integral of the velocity autocorrelation function, giving the Green–Kubo formula. This calculates a *transport coefficient* via the time integral of an autocorrelation function. The Green–Kubo formula, and other transport coefficients, are studied in many books on statistical mechanics; see, for example, [28, ch. 11, 269].

Applications of multiscale analysis to climate models, where the atmosphere evolves quickly relative to the slow oceanic variations, are surveyed in Majda et al. [205, 202]. Further applications to the atmospheric sciences may be found in [206, 207]; see also [78]. Stokes' law, Equation (11.7.15a) is a phenomenological model for the motion of inertial particles in fluids; see [217]. Models of the form (11.7.15), where the velocity field of the fluid in which the particles are immersed is taken to be a Gaussian Markovian random field, were developed in [288, 289] and analyzed further in [254]. Similar Gaussian models for passive tracers were studied in [55, 56].

The fact that smooth approximations to white noise in one dimension lead, in the limit as we remove the regularization, to Stratonovich stochastic integrals (see Section 11.7.3) is often called the Wong–Zakai theorem after [332]. Whether one should interpret the stochastic integral in the sense of Itô or Stratonovich is usually called the Itô-versus-Stratonovich problem. In cases where more than one fast time scale is present, as in the example considered in Section 11.7.6, the correct interpretation of the stochastic integral in the limiting SDE depends on the order with which we take the limits; see [109, 280]. As was shown in Section 11.7.6, there are instances where the stochastic integral in the limiting SDE can be interpreted in neither the Itô nor the Stratonovich sense; see [129, 180, 255]. A similar phenomenon for the case where the fast process is a discrete deterministic chaotic map was observed in [124]. An interesting setup to consider in this context is Stokes' law (11.7.15) in the case where the mass is small:

$$\varepsilon^a \frac{d^2x}{dt^2} = \frac{1}{\varepsilon} f(x) y - \frac{dx}{dt} + \sigma \frac{dU}{dt},$$
$$\frac{dy}{dt} = -\frac{\alpha y}{\varepsilon^2} + \sqrt{\frac{2\alpha}{\varepsilon^2}} \frac{dV}{dt}.$$

Setting $\varepsilon = 0$ in the first equation, and invoking a white noise approximation for y/ε leads to the conjecture that the limit X of x satisfies a first-order SDE. The question then becomes the interpretation of the stochastic integral. In [180] multiscale expansions are used to derive the limiting equation satisfied by x in the cases $a = 1, 2$, and 3. The case $a = 1$ leads to the Itô equation in the limit, the case $a = 3$ to the Stratonovich equation, and $a = 2$ to an intermediate limit between the two.

In higher dimensions smooth approximations to white noise result (in general, and depending of the type of regularization) in an additional drift – apart from the Stratonovich stochastic integral - which is related to the commutator between the row vectors of the diffusion matrix; see [151]. A rigorous framework for understanding examples such as that presented in Section 11.7.7, based on the theory of rough paths, can be found in [198].

In this chapter we have considered equations of the form (11.2.1), where U and V are independent Brownian motions. Frequently applications arise where the noise in the two processes are correlated. We will cover such situations in Chapter 13, where we study homogenization for parabolic PDEs. The structure of the linear equations considered will be general enough to subsume the form of the backward Kolmogorov equation, which arises from (11.2.1) when U and V are correlated – in fact they are identical. The main change over the derivation in this chapter is that the operator $\mathcal{L}_1$ has additional terms arising from the correlation in the noises; see Exercises 5 and 1.

When writing the backward Kolmogorov equation for the full system, Equation (11.2.2), we assumed that the initial conditions depended only on the slow variable x. This assumption simplifies the analysis but is not necessary. If the initial condition is a function of both x and y, then an initial (or boundary) layer appears that has to be resolved. This can be achieved by adding appropriate terms in the two–scale expansion that decay exponentially fast in time. This is done in [336] for continuous-time Markov chains and in [167] for SDEs. In this case the initial conditions for the homogenized SDE are obtained by averaging the initial conditions of the original SDE with respect to the invariant measure of the fast process.

In this chapter we have studied homogenization for finite-dimensional stochastic systems. Similar results can be proved for infinite-dimensional stochastic systems, SPDEs. See [40] for an application of the techniques developed in this chapter to the stochastic Burgers equation.

The use of the representations in Result 11.1 is discussed in [241]. The representations in Results 11.7 and 11.6 for the effective drift and diffusion can be used in the design of coarse time-stepping algorithms; see [322]. In general, the presence of two widely separated characteristic time scales in the SDEs (11.2.1) renders their numerical solution a formidable task. New numerical methods have been developed that aim at the efficient numerical solution of such problems. In the context of averaging for Hamiltonian systems the subject is described in [116]; the subject is revisited, in a more general setting, in [93]. Many of these methods exploit the fact that for ε sufficiently small the solution of (11.2.1a) can be approximated by the solution of the homogenized Equation (11.3.6). The homogenized coefficients are computed through formulae of the form (11.6.3) or (11.6.1), integrating Equation (11.2.1b) over short time intervals; see [322, 81, 84, 123]. An ambitious program to numerically compute a subset of variables from a (possibly stochastic) dynamical system is outlined in [162]; this approach does not use scale separation explicitly and finds application in a range of different problems; see [163, 164, 149, 30, 190, 278, 334]. Numerical methods for multiscale problems are overviewed in [83]. For work on parameter estimation for multiscale SDEs see [258]. For other (partly computational) work on dimension reduction in stochastic systems see [59, 148, 273].

11.9 Exercises

1. Find the homogenized equation for the SDEs

$$\frac{dx}{dt} = \frac{1}{\varepsilon} f_0(x,y) + f_1(x,y) + \alpha_0(x,y)\frac{dU}{dt} + \alpha_1(x,y)\frac{dV}{dt}, \quad x(0) = x_0,$$
$$\frac{dy}{dt} = \frac{1}{\varepsilon^2} g(x,y) + \frac{1}{\varepsilon} g_1(x,y) + \frac{1}{\varepsilon}\beta(x,y)\frac{dV}{dt}, \quad y(0) = y_0,$$

   assuming that f_0 satisfies the centering condition and that U and V are independent Brownian motions.

2. a. Let $\mathcal{Y}$ denote either $\mathbb{T}^d$ or $\mathbb{R}^d$. What is the generator $\mathcal{L}$ for the process $y \in \mathcal{Y}$ given by

$$\frac{dy}{dt} = g(y) + \frac{dV}{dt}?$$

   In the case where $g(y) = -\nabla\Psi(y)$ find a function in the null space of $\mathcal{L}^*$.

b. Find the homogenized SDE arising from the system

$$\frac{dx}{dt} = \frac{1}{\varepsilon} f(x, y),$$
$$\frac{dy}{dt} = \frac{1}{\varepsilon^2} g(y) + \frac{1}{\varepsilon} \frac{dV}{dt},$$

in the case where $g = -\nabla \Psi(y)$.

c. Define the cell problem, giving appropriate conditions to make the solution unique in the case $\mathcal{Y} = \mathbb{T}^d$. State clearly any assumptions on f that are required in the preceding derivation.

3. Use the Itô formula to derive the solution to the SDE (11.7.4). Convert this SDE into Stratonovich form. What do you observe?

4. a. Let $\mathcal{Y}$ be either $\mathbb{T}^d$ or $\mathbb{R}^d$. Write down the generator $\mathcal{L}_0$ for the process $y \in \mathcal{Y}$ given by:

$$\frac{dy}{dt} = g(y) + \frac{dV}{dt}.$$

In the case where g is divergence-free, find a function in the null space of $\mathcal{L}_0^*$.

b. Find the averaged SDE arising from the system

$$\frac{dx}{dt} = f(x, y),$$
$$\frac{dy}{dt} = \frac{1}{\varepsilon} g(y) + \frac{1}{\sqrt{\varepsilon}} \frac{dV}{dt},$$

in the case where g is divergence-free.

c. Find the homogenized SDE arising from the system

$$\frac{dx}{dt} = \frac{1}{\varepsilon} f(x, y),$$
$$\frac{dy}{dt} = \frac{1}{\varepsilon^2} g(y) + \frac{1}{\varepsilon} \frac{dV}{dt},$$

in the case where g is divergence-free.

d. Define the cell problem, giving appropriate conditions to make the solution unique in the case $\mathcal{Y} = \mathbb{T}^d$. Clearly state any assumptions on f that are required in the preceding derivation.

5. Consider the equation of motion

$$\frac{dx}{dt} = f(x) + \sigma \frac{dW}{dt},$$

where $f(x)$ is divergence-free and periodic with mean zero. It is of interest to understand how x behaves on large length and time scales. To this end, rescale the equation of motion by setting $x \to x/\varepsilon$ and $t \to t/\varepsilon^2$ and introduce $y = x/\varepsilon$. Write down a pair of coupled SDEs for x and y. Use the methods developed in Exercise 1 to enable elimination of y to obtain an effective equation for x.

6. Carry out the analysis presented in Section 11.7.6 in arbitrary dimensions. Does the limiting equation have the same structure as in the one-dimensional case?
7. Derive Equation (11.7.25) from (11.7.23) when $\eta(t)$ is given by (11.7.24).
8. (The Kramers to Smoluchowski limit) Consider the Langevin equation

$$\varepsilon^2 \frac{d^2 x}{dt^2} = b(x) - \frac{dx}{dt} + \sqrt{2\sigma}\frac{dW}{dt}, \tag{11.9.1}$$

where the particle mass is assumed to be small, $m = \varepsilon^2$.

a. Write (11.9.1) as a first-order system by introducing the variable $y = \varepsilon \dot{x}$.
b. Use multiscale analysis to show that, when $\varepsilon \ll 1$ the solution of (11.9.1) is well approximated by the solution of the Smoluchowski equation

$$\frac{dX}{dt} = b(X) + \sqrt{2\sigma}\frac{dW}{dt}.$$

c. Calculate the first correction to the Smoluchowski equation.

9. Write Equations (11.7.16) as a first-order system and show that the Itô and Stratonovich forms of the equation coincide.

12

Homogenization for Elliptic PDEs

12.1 Introduction

In this chapter we use multiscale expansions to study the problem of homogenization for second-order uniformly elliptic PDEs in divergence form. At a purely formal level the calculations used to derive the homogenized equations are very similar to those used in the previous chapter to study homogenization for SDEs. The primary difference is that there is no time dependence in the linear equations that we study.

In Section 12.2 we present the boundary value problem studied in this chapter. Section 12.3 contains the simplified (homogenized) equations, and their derivation is given in Section 12.4. Section 12.5 studies the structure of the simplified equation, showing that it inherits ellipticity from the original equation. In Section 12.6 we describe two applications of the theory, both explicitly solvable, a one-dimensional example and a two-dimensional layered material.

12.2 Full Equations

We study uniformly elliptic PDEs in divergence form, with Dirichlet boundary conditions:

$$-\nabla \cdot \left(A^{\varepsilon} \nabla u^{\varepsilon}\right) = f \quad \text{for } x \in \Omega, \tag{12.2.1a}$$

$$u^{\varepsilon} = 0 \quad \text{for } x \in \partial\Omega. \tag{12.2.1b}$$

Here $u^{\varepsilon} = u^{\varepsilon}(x)$ is an unknown scalar field, to be determined, $A^{\varepsilon} = A(x/\varepsilon)$ a given matrix field and $f = f(x)$ a given scalar field. Unlike the problems in the previous four chapters, there are not two different explicit variables x and y. We will introduce $y = x/\varepsilon$ to create a setting similar to that in the previous chapters. Our goal is then to derive a homogenized equation in which y is eliminated, in the limit $\varepsilon \to 0$. Furthermore, we study various properties of the homogenized coefficients.

We take $\Omega \subset \mathbb{R}^d$, open, bounded with smooth boundary. We will assume that the matrix-valued function $A(y)$ is smooth, 1–periodic, and uniformly positive definite.

This assumption implies that the differential operator that appears on the left-hand side of (12.2.1a) is uniformly elliptic (see Chapter 7). Furthermore, we take the function $f(x)$ to be smooth and independent of ε. To summarize, we make the following assumptions:

$$f \in C^\infty(\mathbb{R}^d, \mathbb{R}); \tag{12.2.2a}$$

$$A \in C^\infty_{per}(\mathbb{T}^d, \mathbb{R}^{d\times d}); \tag{12.2.2b}$$

$$\exists \alpha > 0 : \langle \xi, A(y)\xi \rangle \geqslant \alpha|\xi|^2, \quad \forall y \in \mathbb{T}^d \; \forall \xi \in \mathbb{R}^d. \tag{12.2.2c}$$

Notice that our assumptions on A imply that $A^\varepsilon \in M(\alpha, \beta, \Omega)$ for some appropriate β and α independent of ε. The regularity assumptions are more stringent than is necessary; we make them at this point in order to carry out the formal calculations that follow. Allowing minimal regularity assumptions is an important issue, however: in many applications one expects that the coefficient $A(y)$ will have jumps when passing from one material phase to the other. Our proofs of homogenization theorems in Chapter 19 will weaken the regularity assumptions we make here.

Let $\mathcal{A}_0 = -\nabla_y \cdot (A\nabla_y)$ equipped with periodic boundary conditions on the unit torus and with $A = A(y)$. This operator will play a central role in the following. It was studied in Example 7.12; there it was shown that it has a one-dimensional null space, comprising constants in y. Furthermore, use of the Fredholm alternative shows that the Poisson equation

$$\mathcal{A}_0 v = h, \quad v \text{ is 1-periodic}, \tag{12.2.3}$$

has a solution if and only if

$$\int_{\mathbb{T}^d} h(y)dy = 0. \tag{12.2.4}$$

The solution is unique up to an additive constant. Among all solutions of (12.2.3) which satisfy the solvability condition we will choose the unique solution whose integral over $\mathbb{T}^d$ vanishes:

$$\mathcal{A}_0 v = h, \quad v \text{ is 1-periodic}, \int_{\mathbb{T}^d} v(y)\, dy = 0.$$

Equations of the form (12.2.3) will play a central role in what follows.

12.3 Simplified Equations

Define the **effective diffusion tensor** by the formula

$$\overline{A} = \int_{\mathbb{T}^d} \left(A(y) + A(y)\nabla\chi(y)^T \right) dy \tag{12.3.1}$$

where the vector field $\chi : \mathbb{T}^d \to \mathbb{R}^d$ satisfies the **cell problem**

$$-\nabla_y \cdot \left(\nabla_y \chi A^T\right) = \nabla_y \cdot A^T, \quad \chi \text{ is 1-periodic}. \tag{12.3.2}$$

Result 12.1. *For $0 < \varepsilon \ll 1$ the solution u^ε of Equation* (12.2.1) *is approximately given by the solution u of the homogenized equation*

$$-\nabla \cdot \left(\overline{A}\nabla u\right) = f \ \textit{ for } x \in \Omega, \tag{12.3.3a}$$

$$u = 0 \ \textit{ for } x \in \partial\Omega. \tag{12.3.3b}$$

Notice that the field χ is determined up to a constant vector. However, since only $\nabla_y \chi$ enters into the formula for the homogenized matrix $\overline{A}$ appearing in the homogenized equation, the value of this constant is irrelevant. For definiteness, however, we work with the unique solution χ found by imposing the normalization

$$\int_{\mathbb{T}^d} \chi(y)dy = 0. \tag{12.3.4}$$

The cell problem can be written in an alternative, sometimes useful, form by writing an equation for each component of χ:

$$\mathcal{A}_0 \chi_\ell = \nabla_y \cdot a_\ell, \quad \ell = 1, \ldots, d\,, \tag{12.3.5}$$

where $a_\ell = Ae_\ell, \ell = 1, \ldots, d$, and $\{e_\ell\}_{\ell=1}^d$ is the standard basis on $\mathbb{R}^d$. Thus a_ℓ is the ℓth column of A.

Remark 12.2. Since the Hessian $\nabla_x \nabla_x u$ is symmetric, it follows from property (2.2.2) applied to (12.3.1) that the following expression for $\overline{A}$ is equally valid:

$$\overline{A} = \int_{\mathbb{T}^d} \left(A(y)^T + \nabla_y \chi(y) A(y)^T\right) dy.$$

Indeed this expression and (12.3.1) may be combined (for example, averaged) to obtain other equally valid expressions for $\overline{A}$ (for example, symmetric). □

12.4 Derivation

Since a small parameter ε appears in Equation (12.2.1), it is natural to look for a solution in the form of a power series expansion in ε:

$$u^\varepsilon = u_0 + \varepsilon u_1 + \varepsilon^2 u_2 + \ldots.$$

The basic idea behind the method of multiple scales is to assume that all terms in the expansion depend explicitly on *both* x and $y = x/\varepsilon$. Furthermore, since the coefficients of our PDE are periodic functions of x/ε it is reasonable to require that all terms in the expansion are periodic functions of x/ε. Hence, we assume the following ansatz for the solution u^ε:

$$u^\varepsilon(x) = u_0\left(x, \frac{x}{\varepsilon}\right) + \varepsilon\, u_1\left(x, \frac{x}{\varepsilon}\right) + \varepsilon^2\, u_2\left(x, \frac{x}{\varepsilon}\right) + \ldots, \tag{12.4.1}$$

where $u_j(x, y)$, $j = 0, 1, \ldots$, are periodic in y.

The variables x and $y = x/\varepsilon$ represent the "slow" (macroscopic) and "fast" (microscopic) scales of the problem, respectively. For $\varepsilon \ll 1$ the variable y changes much more rapidly than x, and we can think of x as being a constant when looking at the problem at the microscopic scale. This is where scale separation is exploited: *we will treat x and y as independent variables*. Justifying the validity of this assumption as $\varepsilon \to 0$ is one of the main issues in the rigorous theory of homogenization; see Chapter 19.

The fact that $y = x/\varepsilon$ implies that the partial derivatives with respect to x become

$$\nabla \to \nabla_x + \frac{1}{\varepsilon}\nabla_y.$$

In other words, the *total derivative* of a function $g^\varepsilon(x) := g\left(x, \frac{x}{\varepsilon}\right)$ can be expressed as

$$\nabla g^\varepsilon(x) = \nabla_x g(x, y)\Big|_{y=\frac{x}{\varepsilon}} + \frac{1}{\varepsilon}\nabla_y g(x, y)\Big|_{y=\frac{x}{\varepsilon}},$$

where the notation $h(x, y)|_{y=z}$ means that the function $h(x, y)$ is evaluated at $y = z$.

We use the preceding to rewrite the differential operator

$$\mathcal{A}^\varepsilon := -\nabla \cdot (A(y)\nabla)$$

in the form

$$\mathcal{A}^\varepsilon = \frac{1}{\varepsilon^2}\mathcal{A}_0 + \frac{1}{\varepsilon}\mathcal{A}_1 + \mathcal{A}_2, \tag{12.4.2}$$

where

$$\mathcal{A}_0 := -\nabla_y \cdot (A(y)\nabla_y), \tag{12.4.3a}$$
$$\mathcal{A}_1 := -\nabla_y \cdot (A(y)\nabla_x) - \nabla_x \cdot (A(y)\nabla_y), \tag{12.4.3b}$$
$$\mathcal{A}_2 := -\nabla_x \cdot (A(y)\nabla_x). \tag{12.4.3c}$$

Notice that the coefficients in all the operators defined here are periodic functions of y. We equip $\mathcal{A}_0$ with periodic boundary conditions on $\mathbb{T}^d$.

Equation (12.2.1) becomes, on account of (12.4.2),

$$\left(\frac{1}{\varepsilon^2}\mathcal{A}_0 + \frac{1}{\varepsilon}\mathcal{A}_1 + \mathcal{A}_2\right) u^\varepsilon = f \quad \text{for } (x, y) \in \Omega \times \mathbb{T}^d, \tag{12.4.4a}$$

$$u^\varepsilon = 0 \quad \text{for } (x, y) \in \partial\Omega \times \mathbb{T}^d. \tag{12.4.4b}$$

We substitute (12.4.1) into (12.4.4) to deduce:

$$\frac{1}{\varepsilon^2}\mathcal{A}_0 u_0 + \frac{1}{\varepsilon}(\mathcal{A}_0 u_1 + \mathcal{A}_1 u_0) + (\mathcal{A}_0 u_2 + \mathcal{A}_1 u_1 + \mathcal{A}_2 u_0) + \mathcal{O}(\varepsilon) = f. \tag{12.4.5}$$

We equate coefficients of equal powers of ε to zero in the preceding equation and disregard all terms of order higher than 1 to obtain the following sequence of problems:

$$\mathcal{O}(1/\varepsilon^2) \quad \mathcal{A}_0 u_0 = 0, \tag{12.4.6a}$$

$$\mathcal{O}(1/\varepsilon) \quad \mathcal{A}_0 u_1 = -\mathcal{A}_1 u_0, \tag{12.4.6b}$$

$$\mathcal{O}(1) \quad \mathcal{A}_0 u_2 = -\mathcal{A}_1 u_1 - \mathcal{A}_2 u_0 + f. \tag{12.4.6c}$$

Here $u_j(x, y)$ are 1-periodic in their second argument.

Notice that $\mathcal{A}_0$ is a differential operator in y and that x appears in Equations (12.4.6b) and (12.4.6c) merely as a parameter. From (12.4.6a) we deduce that $u_0(x, y) = u(x)$; thus the first term in the multiscale expansion is independent of y. The remaining two equations are of the form (12.2.3) with $v = v(x, y)$ and similarly $h = h(x, y)$; thus x enters as a parameter.

Let us proceed now with (12.4.6b), which becomes

$$\mathcal{A}_0 u_1 = \left(\nabla_y \cdot A^T\right) \cdot \nabla_x u, \quad u_1(x, \cdot) \text{ is 1-periodic}, \int_{\mathbb{T}^d} u_1 \, dy = 0. \tag{12.4.7}$$

The solvability condition (12.2.4) is satisfied because

$$\begin{aligned} \int_{\mathbb{T}^d} \left(\nabla_y \cdot A^T\right) \cdot \nabla_x u \, dy &= \nabla_x u \cdot \int_{\mathbb{T}^d} \nabla_y \cdot A^T \, dy \\ &= 0, \end{aligned}$$

by the divergence theorem and periodicity of $A(\cdot)$; see Remark 7.13. We seek a solution of (12.4.7) using separation of variables:

$$u_1(x, y) = \chi(y) \cdot \nabla_x u(x). \tag{12.4.8}$$

Upon substituting (12.4.8) into (12.4.7) we obtain the cell problem (12.3.2) for the vector field $\chi(y)$. The field $\chi(y)$ is called the **first-order corrector**. Notice that the periodicity of the coefficients implies that the right-hand side of Equation (12.3.2) averages to zero over the unit cell, and consequently the cell problem is well posed. We ensure the uniqueness of solutions to (12.3.2) by requiring the corrector field to have zero average – condition (12.3.4).

Now we consider Equation (12.4.6c). By (12.2.4) we see that, in order for this equation to be well posed, it is necessary and sufficient for the right-hand side to average to zero over $\mathbb{T}^d$. Since we have assumed that the function $f(x)$ is independent of y, the solvability condition implies:

$$\int_{\mathbb{T}^d} \left(\mathcal{A}_2 u_0 + \mathcal{A}_1 u_1\right) \, dy = f. \tag{12.4.9}$$

The first term on the left-hand side of Equation 12.4.9 is

$$\begin{aligned} \int_{\mathbb{T}^d} \mathcal{A}_2 u_0 \, dy &= \int_{\mathbb{T}^d} -\nabla_x \cdot \left(A(y) \nabla_x u\right) \, dy \\ &= -\nabla_x \cdot \left[\left(\int_{\mathbb{T}^d} A(y) \, dy\right) \nabla_x u(x)\right] \\ &= -\left(\int_{\mathbb{T}^d} A(y) \, dy\right) : \nabla_x \nabla_x u(x). \end{aligned} \tag{12.4.10}$$

Moreover

$$\begin{aligned}\int_{\mathbb{T}^d} \mathcal{A}_1 u_1 \, dy &= \int_{\mathbb{T}^d} \left(-\nabla_y \cdot (A(y)\nabla_x u_1) - \nabla_x \cdot (A(y)\nabla_y u_1)\right) dy \\ &=: I_1 + I_2.\end{aligned}$$

The first term $I_1 = 0$ by periodicity and Remark 7.13. Now we consider I_2:

$$\begin{aligned}I_2 &= \int_{\mathbb{T}^d} -\nabla_x \cdot (A(y)\nabla_y u_1) \, dy \\ &= -\int_{\mathbb{T}^d} A(y) : \nabla_x \nabla_y \, (\chi \cdot \nabla_x u) \, dy \\ &= -\left(\int_{\mathbb{T}^d} (A(y)\nabla_y \chi(y)^T) \, dy\right) : \nabla_x \nabla_x u. \qquad (12.4.11)\end{aligned}$$

We substitute (12.4.11) and (12.4.10) in (12.4.9) to obtain the homogenized equation of Result 12.1, where the homogenized coefficient $\overline{A}$ is given by the formula (12.3.1). This completes the derivation.

12.5 Properties of the Simplified Equations

In this section we study some basic properties of the effective coefficients. In particular, we show that the matrix of homogenized coefficients $\overline{A}$ is positive definite, which implies that the homogenized differential operator is uniformly elliptic and that, consequently, the homogenized equation is well posed. Furthermore, we show that symmetry is preserved under homogenization: the homogenized matrix is symmetric if $A(y)$ is. We also show that the homogenization process can create anisotropies: even if the matrix $A(y)$ is diagonal, the matrix of homogenized coefficients $\overline{A}$ need not be.

In order to study the matrix of homogenized coefficients, it is useful to find an alternative representation for $\overline{A}$. To this end, we introduce the bilinear form

$$a_1(\psi, \phi) = \int_{\mathbb{T}^d} \langle \nabla_y \phi, A(y) \nabla_y \psi \rangle \, dy, \qquad (12.5.1)$$

defined for all functions $\phi,\ \psi \in C^1(\mathbb{T}^d)$. Notice that this is the bilinear form associated with the operator $\mathcal{A}_0$, in the sense that

$$\int_{\mathbb{T}^d} \phi \mathcal{A}_0 \psi \, dy = a_1(\phi, \psi) \quad \forall\, \phi, \psi \in C^1_{per}(\mathbb{T}^d). \qquad (12.5.2)$$

Note that, whenever A is symmetric, so is the bilinear form $a_1(\cdot, \cdot)$. We start by obtaining an alternative, equivalent formulation for the cell problem. The formulation is closely related to the weak formulation of elliptic PDEs in divergence form, introduced in Chapter 7. In the rest of this section we will assume that the solution of the cell problem is smooth enough to justify the calculations that follow. It will be enough to assume that each component of the corrector field $\chi(y)$ is continuously differentiable and periodic: $\chi_\ell(y) \in C^1_{per}(\mathbb{T}^d),\ \ell = 1, \dots, d$.

Recall that e_ℓ denotes the unit vector with ith entry δ_{il}. Also let y_ℓ denote the ℓth component of the vector y. Note that $e_\ell = \nabla_y y_\ell$ and recall that $a_\ell = Ae_\ell$, the ℓth column of A. Using these two elementary facts we can obtain the following useful lemma.

Lemma 12.3. *The cell problem* (12.3.2) *can be written in the form*

$$a_1(\phi, \chi_\ell + y_\ell) = 0 \quad \forall \phi \in C^1_{per}(\mathbb{T}^d),\ \ell = 1, \dots d. \tag{12.5.3}$$

Proof. From (12.3.5) we deduce that

$$\mathcal{A}_0 \chi_\ell = \nabla_y \cdot (Ae_\ell) = \nabla_y \cdot (A\nabla_y y_\ell) = -\mathcal{A}_0 y_\ell.$$

Consequently, the cell problem can be written in the form

$$\mathcal{A}_0(\chi_l + y_l) = 0, \quad l = 1, \dots, d,$$

with periodic boundary conditions. We multiply the cell problem as formulated earlier by an arbitrary function $\phi \in C^1_{per}(\mathbb{T}^d)$. Integrating over the unit cell, using Remark 7.13, and Equations (12.5.1) and (12.5.2), we obtain (12.5.3). □

Using this lemma we give an alternative representation formula for the homogenized coefficients. The lemma shows that $\overline{A}$ is symmetric, whenever $A(y)$ is.

Lemma 12.4. *The effective matrix* $\overline{A}$ *has components given by*

$$\overline{a}_{ij} = a_1(\chi_j + y_j, \chi_i + y_i), \quad i, j = 1, \dots, d. \tag{12.5.4}$$

In particular, symmetry of $A(y)$ *implies symmetry of* $\overline{A}$.

Proof. Notice first that the previous lemma implies that, since $\chi_i(y) \in C^1_{per}(\mathbb{T}^d)$,

$$a_1(\chi_i, \chi_j + y_j) = 0, \quad \forall i, j, = 1, \dots, d. \tag{12.5.5}$$

We now use formula (12.3.1), together with (12.5.5), to obtain

$$\begin{aligned}
\overline{a}_{ij} &= e_i \cdot \overline{A} e_j \\
&= \int_{\mathcal{Y}} \Big(e_i \cdot Ae_j + e_i \cdot A\nabla_y \chi^T e_j \Big) dy \\
&= \int_{\mathcal{Y}} \Big(\nabla_y y_i \cdot A\nabla_y y_j + \nabla_y y_i \cdot A\nabla_y \chi_j \Big) dy \\
&= \int_{\mathcal{Y}} \Big\langle \nabla_y y_i, A\Big(\nabla_y (y_j + \chi_j)\Big) \Big\rangle \, dy \\
&= a_1(y_i, \chi_j + y_j) \\
&= a_1(y_i + \chi_i, \chi_j + y_j).
\end{aligned}$$

This proves (12.5.4). Assume now that $A(y) = A(y)^T$. This implies that the bilinear form $a_1(\cdot,\cdot)$ is symmetric. Thus

$$\begin{aligned}\overline{a}_{ij} &= a_1(y_i + \chi_i, \chi_j + y_j) \\ &= a_1(y_j + \chi_j, \chi_i + y_i) \\ &= \overline{a}_{ji},\end{aligned}$$

which shows that the homogenized matrix is symmetric. □

We now show that the homogenized matrix $\overline{A}$ is positive definite. This implies that the homogenized equation is a well-posed elliptic PDE.

Theorem 12.5. *The matrix of homogenized coefficients $\overline{A}$ is positive definite.*

Proof. Let $\xi \in \mathbb{R}^d$ be an arbitrary vector. We need to show that there exists a constant $\overline{\alpha} > 0$ such that

$$\langle \xi, \overline{A}\xi \rangle \geqslant \overline{\alpha}|\xi|^2, \quad \forall \xi \in \mathbb{R}^d.$$

We use the representation formula (12.5.4) to deduce that:

$$\langle \xi, \overline{A}\xi \rangle = a_1(w, w),$$

with $w = \xi \cdot (\chi + y)$. We now use the uniform positive definiteness of $A(y)$ to obtain

$$a_1(w, w) \geqslant \alpha \int_{\mathbb{T}^d} |\nabla_y w|^2 \, dy \geqslant 0.$$

Thus $\overline{A}$ is nonnegative.

To show that it is actually positive definite we argue as follows. Let us assume that

$$\langle \xi, \overline{A}\xi \rangle = 0$$

for some ξ. Then, since $\alpha > 0$, $\nabla_y w = 0$ and $w = c$, a constant vector; consequently

$$\xi \cdot y = c - \xi \cdot \chi.$$

The right-hand side of this equation is 1-periodic and continuous in y and consequently the left-hand side should also be. The only way this can happen is if $\xi = 0$. This completes the proof of the lemma. □

The preceding theorem shows that uniform ellipticity is a property that is preserved under the homogenization procedure. In particular, this implies that the homogenized equation is well posed, since it is a uniformly elliptic PDE with constant coefficients.

Remark 12.6. Note that homogenization does not preserve isotropy. In particular, even if the diffusion matrix A has only diagonal nonzero elements, the homogenized diffusion matrix $\overline{A}$ will in general have nonzero off-diagonal elements. To see this, let us assume that $a_{ij} = 0,\ i \neq j$. Then the off-diagonal elements of the homogenized diffusion matrix are given by the formula (no summation convention here)

$$\overline{a}_{ij} = \int_{\mathbb{T}^d} a_{ii} \frac{\partial \chi_j}{\partial y_i}\, dy, \; i \neq j.$$

This expression is not necessarily equal to zero and leads to the surprising result that an isotropic composite material can behave, in the limit as the microstructure becomes finer and finer, like an anisotropic homogeneous material. □

12.6 Applications

We present two useful illustrative examples, for which explicit solutions may be found. Essentially, the one-dimensional case is the only general setting in which the cell problem can be solved analytically and an explicit formula for the effective diffusivity can be obtained. In higher dimensions, explicit formulae for the effective diffusivities can be obtained only when the specific structure of the problem under investigation enables us to reduce the calculation of the homogenized coefficients to consideration of one-dimensional problems. Such a reduction is possible in the case of layered materials, the second example that we consider.

12.6.1 The One-Dimensional Case

Let $d = 1$ and take $\Omega = [0, L]$. Then the Dirichlet problem (12.2.1a) reduces to a two-point boundary value problem:

$$-\frac{d}{dx}\left(a\left(\frac{x}{\varepsilon}\right)\frac{du^\varepsilon}{dx}\right) = f \quad \text{for } x \in (0, L), \tag{12.6.1a}$$

$$u^\varepsilon(0) = u^\varepsilon(L) = 0. \tag{12.6.1b}$$

We assume that $a(y)$ is smooth, periodic with period 1. We also assume that there exist constants $0 < \alpha \leqslant \beta < \infty$ such that

$$\alpha \leqslant a(y) \leqslant \beta, \quad \forall y \in [0, 1]. \tag{12.6.2}$$

We also assume that f is smooth.

The cell problem becomes a boundary value problem for an ordinary differential equation with periodic boundary conditions:

$$-\frac{d}{dy}\left(a(y)\frac{d\chi}{dy}\right) = \frac{da(y)}{dy}, \quad \text{for } y \in (0, 1), \tag{12.6.3a}$$

$$\chi \text{ is 1-periodic}, \qquad \int_0^1 \chi(y)\, dy = 0. \tag{12.6.3b}$$

Since $d = 1$ we only have one effective coefficient given by the one-dimensional version of (12.3.1), namely

$$\overline{a} = \int_0^1 \left(a(y) + a(y)\frac{d\chi(y)}{dy} \right) dy$$

$$= \left\langle a(y) \left(1 + \frac{d\chi(y)}{dy} \right) \right\rangle. \qquad (12.6.4)$$

Here, and in the remainder of this chapter, we employ the notation

$$\langle f(y) \rangle := \int_{\mathbb{T}^d} f(y)\, dy,$$

for the average over $\mathbb{T}^d$.

Equation (12.6.3a) can be solved exactly. Integration from 0 to y gives

$$a(y)\frac{d\chi}{dy} = -a(y) + c_1. \qquad (12.6.5)$$

The constant c_1 is undetermined at this point. The inequality (12.6.2) allows us to divide (12.6.5) by $a(y)$ since it implies that a is strictly positive. We then integrate once again from 0 to y to deduce:

$$\chi(y) = -y + c_1 \int_0^y \frac{1}{a(y)}\, dy + c_2.$$

In order to determine the constant c_1 we use the fact that $\chi(y)$ is a periodic function. Thus $\chi(0) = \chi(1)$, and we deduce that

$$c_1 = \frac{1}{\int_0^1 \frac{1}{a(y)}\, dy} = \langle a(y)^{-1} \rangle^{-1}.$$

Thus, from (12.6.5),

$$1 + \frac{d\chi}{dy} = \frac{1}{\langle a(y)^{-1} \rangle a(y)}.$$

(Notice that c_2 is not required for the calculation of $\overline{a}$.) We substitute this expression into Equation (12.6.4) to obtain

$$\overline{a} = \langle a(y)^{-1} \rangle^{-1}. \qquad (12.6.6)$$

This is the formula that gives the homogenized coefficient in one dimension. It shows clearly that, even in this simple one-dimensional setting, the homogenized coefficient is not found by simply averaging the unhomogenized coefficients over a period of the microstructure. Rather, the homogenized coefficient is the inverse of the average of the inverse of the unhomogenized coefficient – the *harmonic average*. It is quite easy to show that the homogenized coefficient, which is given by the harmonic average (12.6.6), is bounded from above by the average of $a(y)$; see Exercise 12.

12.6.2 Layered Materials

We consider problem (12.2.1), with assumptions (12.2.2) satisfied, in two dimensions. We assume that the domain $\Omega \subset \mathbb{R}^2$ represents a *layered material*: the properties of the material change only in one direction. Hence, the coefficients $A(y)$ are functions of one variable: for $y = (y_1, y_2)^T$ we have

$$a_{ij} = a_{ij}(y_1), \quad i, j = 1, 2. \tag{12.6.7}$$

The fact that the coefficients are functions of y_1 implies the right-hand side of the cell problem (12.3.2) is a function of y_1 alone. As a consequence, the solution of the cell problem is also a function of y_1 alone and takes the form

$$\chi_\ell = \chi_\ell(y_1), \quad \ell = 1, 2. \tag{12.6.8}$$

Upon substituting this into (12.3.2) we conclude that the cell problem becomes

$$-\frac{d}{dy_1}\left(a_{11}(y_1)\frac{d\chi_\ell(y_1)}{dy_1}\right) = \frac{da_{1\ell}(y_1)}{dy_1}, \quad \ell = 1, 2 \tag{12.6.9}$$

with periodic boundary conditions. Similarly, the formula for the homogenized coefficients (12.3.1) becomes:

$$\overline{a}_{ij} = \int_0^1 \left(a_{ij}(y_1) + a_{i1}(y_1)\frac{d\chi_j(y_1)}{dy_1}\right) dy_1, \quad i, j = 1, 2. \tag{12.6.10}$$

Let us now solve Equations (12.6.9). These are ordinary differential equations, and we can solve them in exactly the same way that we solved the one-dimensional problems in the preceding subsection. To this end, we integrate from 0 to y and divide through by $a_{11}(y_1)$ to obtain

$$\frac{d\chi_\ell}{dy_1} = -\frac{a_{1\ell}}{a_{11}} + c_1\frac{1}{a_{11}}, \quad \ell = 1,\ 2 \tag{12.6.11}$$

where the constant c_1 is to be determined. We have to consider the cases $\ell = 1$ and $\ell = 2$ separately. We start with $\ell = 1$. In this case, the equation simplifies to

$$\frac{d\chi_1}{dy_1} = -1 + c_1\frac{1}{a_{11}},$$

which is precisely the equation that we considered in Section 12.6.1. Thus, we have:

$$\frac{d\chi_1}{dy_1} = -1 + \frac{1}{\langle a_{11}(y)^{-1}\rangle a_{11}(y)}. \tag{12.6.12}$$

Now we consider Equation (12.6.11) for the case $\ell = 2$:

$$\frac{d\chi_2}{dy_1} = -\frac{a_{12}}{a_{11}} + c_1\frac{1}{a_{11}}.$$

We integrate the equation once again and then determine the coefficient c_1 by requiring $\chi_2(y_1)$ to be periodic. The final result is

$$\frac{d\chi_2(y_1)}{dy_1} = -\frac{a_{12}(y_1)}{a_{11}(y_1)} + \frac{\langle a_{12}(y_1)/a_{11}(y_1)\rangle}{\langle a_{11}^{-1}(y_1)\rangle}\frac{1}{a_{11}(y_1)}. \tag{12.6.13}$$

Now we can compute the homogenized coefficients. We start with $\overline{a}_{11}$. The calculation is the same as in the one-dimensional case:

$$\overline{a}_{11} = \langle a_{11}(y_1)^{-1}\rangle^{-1}. \tag{12.6.14}$$

We proceed with the calculation of $\overline{a}_{12}$. We substitute (12.6.13) into (12.6.10) with $i = 1,\ j = 2$ to deduce:

$$\begin{aligned}
\overline{a}_{12} &= \int_0^1 \left(a_{12}(y_1) + a_{11}(y_1)\frac{d\chi_2(y_1)}{dy_1}\right) dy \\
&= \int_0^1 \left(a_{12}(y_1) + a_{11}(y_1)\left(-\frac{a_{12}(y_1)}{a_{11}(y_1)} + \frac{\langle a_{12}(y_1)/a_{11}(y_1)\rangle}{\langle a_{11}^{-1}(y_1)\rangle}\frac{1}{a_{11}(y_1)}\right)\right) dy \\
&= \int_0^1 \left(a_{12}(y_1) - a_{12}(y_1) + \frac{\langle a_{12}(y_1)/a_{11}(y_1)\rangle}{\langle a_{11}^{-1}(y_1)\rangle}\right) dy \\
&= \frac{\langle a_{12}(y_1)/a_{11}(y_1)\rangle}{\langle a_{11}^{-1}(y_1)\rangle}.
\end{aligned}$$

Hence

$$\overline{a}_{12} = \left\langle \frac{a_{12}(y_1)}{a_{11}(y_1)} \right\rangle \langle a_{11}^{-1}(y_1)\rangle^{-1}. \tag{12.6.15}$$

Similarly,

$$\overline{a}_{21} = \left\langle \frac{a_{21}(y_1)}{a_{11}(y_1)} \right\rangle \langle a_{11}^{-1}(y_1)\rangle^{-1}. \tag{12.6.16}$$

Finally we consider $\overline{a}_{22}$:

$$\begin{aligned}
\overline{a}_{22} &= \int_0^1 \left(a_{22}(y_1) + a_{21}(y_1)\frac{d\chi_2(y_1)}{dy_1}\right) dy \\
&= \int_0^1 \left(a_{22}(y_1) + a_{21}(y_1)\left(-\frac{a_{12}(y_1)}{a_{11}(y_1)} + \frac{\langle a_{12}(y_1)/a_{11}(y_1)\rangle}{\langle a_{11}^{-1}(y_1)\rangle}\frac{1}{a_{11}(y_1)}\right)\right) dy \\
&= \int_0^1 \left(a_{12}(y_1) - \frac{a_{12}(y_1)a_{21}(y_1)}{a_{11}(y_1)} + \frac{a_{21}(y_1)}{a_{11}(y_1)}\frac{\langle a_{12}(y_1)/a_{11}(y_1)\rangle}{\langle a_{11}^{-1}(y_1)\rangle}\right) dy \\
&= \left\langle \frac{a_{21}(y_1)}{a_{11}(y_1)} \right\rangle \left\langle \frac{a_{12}(y_1)}{a_{11}(y_1)} \right\rangle \langle a_{11}^{-1}(y_1)\rangle^{-1} + \left\langle a_{22}(y_1) - \frac{a_{12}(y_1)a_{21}(y_1)}{a_{11}(y_1)} \right\rangle.
\end{aligned}$$

Consequently:

$$\overline{a}_{22} = \left\langle \frac{a_{21}(y_1)}{a_{11}(y_1)} \right\rangle \left\langle \frac{a_{12}(y_1)}{a_{11}(y_1)} \right\rangle \langle a_{11}^{-1}(y_1)\rangle^{-1} + \left\langle a_{22}(y_1) - \frac{a_{12}(y_1)a_{21}(y_1)}{a_{11}(y_1)} \right\rangle. \tag{12.6.17}$$

It is evident from formulae (12.6.14), (12.6.15), (12.6.16), and (12.6.17) that the homogenized coefficients depend on the original ones in a very complicated, highly nonlinear way.

12.7 Discussion and Bibliography

The method of multiple scales was developed by various researchers in the 1970s with significant contributions from Keller, Babuska, Sanchez-Palenzia, Bensoussan, Lions, Papanicolaou, and others; see [158, 159, 26, 25, 24, 23, 91] and the references therein. A first systematic exposition of the method of multiple scales is contained in [33], where references to the earlier literature can be found. See also the book [279]. Rigorous convergence results for elliptic PDEs with rapidly oscillating coefficients were proved before the development of the method of multiple scales; see [73, 296] and the text [153]. However, the power of the method of multiple scales is its wide applicability to a variety of differing settings. In contrast, rigorous results tend to apply on a case-by-case basis, and their proofs differ substantially between different PDEs and between Markov chains, ODEs, and SDEs. (See Part III of this book.) In most cases, however, an appropriate Poisson equation (the cell problem) plays a prominent role in the analysis.

The one-dimensional problem (see Section 12.6.1) was studied in [296], without using the method of multiple scales. In the one-dimensional case, it is possible to derive the homogenized equation using the method of multiple scales even in the nonperiodic setting; see [143, 66, ch. 5]. The homogenized equation is a second-order uniformly elliptic PDE in the case of nonperiodic fast oscillatory coefficients. However, this result is most naturally obtained via the theory of H-and Γ-convergence, rather than multiple-scale expansions; see [296, 308]. In the general setting of nonperiodic, deterministic, homogenization the homogenized coefficients cannot be expressed in terms of solutions to appropriate Poisson equations and there are no explicit formulae for them. In this case, the best one can hope for is to obtain bounds on the homogenized coefficients.

The homogenized equation for layered materials (see Section 12.6.2) was derived rigorously by Murat and Tartar without any appeal to the method of multiple scales; see [232] and the references to the original papers therein. The two-dimensional case that we treated in Subsection 12.6.2 can be easily extended to the d-dimensional one, $d \geqslant 2$, i.e., to the case where $a_{ij}(y) = a_{ij}(y_1)$, $i, j = 1, \ldots, d$; see [232].

The elliptic boundary value problem (12.2.1) is a Dirichlet problem. However, an inspection of the analysis presented in Section 12.4 reveals that the boundary conditions did not play any role in the derivation of the homogenized equation. In particular, the two-scale expansion (12.4.1) that we used to derive the homogenized equation did not contain any information concerning the boundary conditions of the problem under investigation. Indeed, the boundary conditions become somewhat irrelevant in the homogenization procedure. Exactly the same calculations enable us to obtain the homogenized equation for Neumann or mixed boundary conditions. This is not surprising since the derivation of the homogenized equation is based on the analysis of local problems of the form (12.2.3). This local problem cannot really see the boundary – this is the key property of scale separation.

However, the boundary conditions become very important when trying to prove the homogenization theorem. The fact that the two-scale expansion (12.4.1) does not satisfy the boundary conditions of our PDE exactly but, rather, only up to $\mathcal{O}(\varepsilon)$, introduces boundary layers [143, ch. 3]. [1] Boundary layers affect the convergence rate at which $u^{\varepsilon}(x)$ converges to $u(x)$ as $\varepsilon \to 0$. We can solve this problem by modifying the two-scale expansion (12.4.1), adding additional terms that take care of the boundary layer and vanish exponentially fast as we move away from the boundary so that they do not affect the solution in the interior. We refer to [27] for details.

The discussion in Remark 12.2 is further elaborated in [33] and in [66]. Different expressions for the effective diffusion tensor can be useful for the proof of various properties of the effective diffusion tensor.

From the point of view of continuum mechanics, the method of homogenization enables us to obtain macroscopic *constitutive laws* for composite materials. Macroscopic constitutive laws have been derived using homogenization theory for various types of composite materials. See, e.g., [46, 108]. An alternative approach is presented in [230, 133]. The theory of composite materials is presented in the excellent monograph [229].

In the Dirichlet problem that we analyzed in Section 12.4 we assumed that the matrix $A^{\varepsilon}(x)$ depends only on the microscale, i.e.,

$$A^{\varepsilon}(x) = A\left(\frac{x}{\varepsilon}\right),$$

with $A(y)$ being a 1-periodic matrix-valued function. However, the method of multiple scales is also applicable to the case where the coefficients depend explicitly on the macroscale as well as the microscale:

$$A^{\varepsilon}(x) = A\left(x, \frac{x}{\varepsilon}\right),$$

with $A(x, y)$ being 1-periodic in y and smooth in x. When the coefficients have this form they are called *locally periodic* or *nonuniformly periodic*. Analysis similar to the one presented in Section 12.4 enables us to obtain the homogenized equation for the Dirichlet problem

$$-\nabla \cdot (A^{\varepsilon} \nabla u^{\varepsilon}) = f \quad \text{for } x \in \Omega, \tag{12.7.1a}$$

$$u^{\varepsilon} = 0 \quad \text{for } x \in \partial\Omega, \tag{12.7.1b}$$

where $A^{\varepsilon}(x) = A(x, x/\varepsilon)$. Now the homogenized coefficients $\overline{A}$ are functions of x:

$$-\nabla \cdot \left(\overline{A} \nabla u\right) = f \quad \text{for } x \in \Omega \tag{12.7.2a}$$

[1] The presence of boundary and initial layers is a common feature in all problems of singular perturbations. See the bibliographical discussions in other chapters from Part II and [143] and [161], for further details.

$$u = 0 \quad \text{for } x \in \partial\Omega, \tag{12.7.2b}$$

and the cell problem is parameterized by x since $A = A(x, y)$:

$$-\nabla_y \cdot \left(\nabla_y \chi A^T\right) = \nabla_y \cdot A^T, \quad y \in \mathbb{T}^d. \tag{12.7.3}$$

The homogenized coefficients are given by the formula:

$$\overline{A}(x) = \int_{\mathbb{T}^d} \left(A(x, y) + A(x, y)\nabla_x \chi(x, y)^T\right) dy. \tag{12.7.4}$$

We emphasize the fact that the "macroscopic variable" x enters in the preceding two equations as a parameter. Consequently, to compute the effective coefficients we need to solve the cell problem (12.7.3) and evaluate the integrals in (12.7.4) at all points $x \in \Omega$.

The method of multiple scales can also be applied to semilinear elliptic PDEs with rapidly oscillating coefficients – equations of the form

$$-\nabla \cdot \left(A^\varepsilon \nabla u^\varepsilon\right) = f(u^\varepsilon) \quad \text{for } x \in \Omega, \tag{12.7.5a}$$

$$u^\varepsilon = 0 \quad \text{for } x \in \partial\Omega. \tag{12.7.5b}$$

The homogenized equation takes the form

$$-\nabla \cdot \left(\overline{A}\nabla u\right) = f(u) \quad \text{for } x \in \Omega, \tag{12.7.6a}$$

$$u = 0 \quad \text{for } x \in \partial\Omega, \tag{12.7.6b}$$

with $\overline{A}$ as in (12.3.1).

In Section (12.2) we obtained the first two terms in the two-scale expansion for the Dirichlet problem (12.2.1). The second term is proportional – up to an unknown function of x - to the gradient of the first term in the expansion, which solves the homogenized equation, i.e.,

$$u_1\left(x, \frac{x}{\varepsilon}\right) = \chi\left(\frac{x}{\varepsilon}\right) \cdot \nabla_x u(x) + \widehat{u}_1(x), \tag{12.7.7}$$

where $\chi(y)$ solves the cell problem. We can also solve higher-order equations and obtain higher-order terms in the two-scale expansion. For example, we can solve Equation (12.4.6) and compute the third term in the expansion $u_2(x, y)$:

$$u_2(x, y) = \Theta(y) : \nabla_x \nabla_x u(x) + \widehat{u}_2(x) \tag{12.7.8}$$

where the *second-order corrector field* $\Theta(y)$ is a matrix-valued function that satisfies the boundary value problem

$$\mathcal{A}_0 \Theta = B. \tag{12.7.9}$$

Here $B(y)$ is given by

$$B(y) := -\overline{A} + A(y) + A(y)\nabla_y\chi(y)^T + \nabla_y\chi(y)A(y) + \chi(y) \otimes \left(\nabla_y \cdot A(y)^T\right).$$

All higher-order equations are of the form

$$\mathcal{A}_0 u_{k+2} = -\mathcal{A}_1 u_{k+1} - \mathcal{A}_0 u_k, \quad k = 1, 2, \dots .$$

It turns out that $u_k(x)$ is proportional to the kth order derivatives of $u(x)$; see [27].

The method of multiple scales can be extended to situations where there are k length scales in the problem, i.e., when the matrix $A^\varepsilon(x)$ has the form

$$A^\varepsilon(x) = A\left(\frac{x}{\varepsilon}, \frac{x}{\varepsilon^2}, \dots, \frac{x}{\varepsilon^k}\right),$$

and A is 1-periodic in all of its arguments. This is known as *reiterated homogenization* [33, sec. 1.8]. A rigorous analysis of reiterated homogenization in a quite general setting is presented in [8]. Reiterated homogenization has recently found applications in the problem of advection and diffusion of passive tracers in fluids. See, for example, [253, 219, 220] for details. When there are infinitely many scales in the problem, without a clear separation, the homogenization result breaks down, in the sense that the homogenized coefficient can be 0; see [16].

In general it is not possible to compute the homogenized coefficients analytically; indeed, their calculation requires the solution of the cell problem and the calculation of the integrals in (12.3.1). In most cases this can be done only numerically. It is possible, however, to obtain bounds on the magnitude of the effective coefficients. Various tools for obtaining bounds have been developed; for example, it is possible to obtain a variational characterization of the homogenized coefficients. We refer to [229, 311, 107] for various results in this direction. Many of these techniques apply to the nonperiodic setting.

The method developed in this chapter readily extends to intial/boundary value problems such as the following parabolic PDE:

$$\frac{\partial u^\varepsilon}{\partial t} - \nabla \cdot (A^\varepsilon \nabla u^\varepsilon) = f^\varepsilon \ \text{ in } \Omega \times (0,T), \tag{12.7.10a}$$

$$u^\varepsilon = 0 \ \text{ on } \partial\Omega \times (0,T) \tag{12.7.10b}$$

$$u^\varepsilon = u_{in}(x) \ \text{ in } \overline{\Omega} \times \{0\} \tag{12.7.10c}$$

under various assumptions concerning the ε dependence in A^ε and f^ε. A time-dependent situation of interest arises when the coefficients of the evolution PDE oscillate in time as well as space, i.e., $A^\varepsilon = A\left(x/\varepsilon, t/\varepsilon^k\right)$, $k > 0$ with the matrix-valued function $A(y,\tau)$ being 1-periodic in both y and τ. This means that we have to introduce two fast variables: $y = x/\varepsilon$ and $\tau = t/\varepsilon^k$. More information on homogenization for evolution equations with space-time-dependent coefficients can be found in [33, ch. 3]. We study homogenization for parabolic PDEs using the method of multiple scales in Chapters 11, 13, and 14.

One can also study the problem of homogenization for hyperbolic (wave) equations:

$$\frac{\partial^2 u^\varepsilon}{\partial t^2} - \nabla \cdot (A^\varepsilon \nabla u^\varepsilon) = f \text{ in } \Omega \times (0,T), \tag{12.7.11a}$$

$$u^\varepsilon = 0 \text{ on } \partial\Omega \times (0,T), \tag{12.7.11b}$$

$$u^\varepsilon = u_{in} \text{ in } \overline{\Omega} \times \{0\}, \tag{12.7.11c}$$

$$\frac{\partial u^\varepsilon}{\partial t} = v_{in}(x) \text{ in } \overline{\Omega} \times \{0\}. \tag{12.7.11d}$$

The method of multiple scales can be used to obtain a homogenized equation, which is a wave equation with constant coefficients and the same initial and boundary conditions. However, there is a fundamental difference between this and the parabolic case: for parabolic problems the dissipation drives the solution to lie near the null space of the leading-order operator $\mathcal{L}_0$, no matter how the initial data are chosen. For the wave equation, this does not happen and it is necessary to chose initial data close to the desired subspace. We will not study homogenization for wave equations in this book. We refer the interested reader to [66, ch. 12; 33, ch. 2; 160, 47]. Related problems arise for the Schrödinger equation with multiple scales; see [316]. Homogenization result for the Schrödinger equation and their connection to effective mass theorems are presented in [10].

The numerical evaluation of homogenized coefficients, in the periodic setting, can be performed efficiently using a spectral method. On the other hand, the numerical solution of the original boundary value problem (12.2.1) when ε is small is a very hard problem. Special methods, which in one way or another are based on homogenization, have been developed over the last few years. We refer to [145, 76, 2, 82, 9, 52, 61, 89, 90, 92, 231] and the references therein on this topic. The development and analysis of finite element methods for elliptic PDEs with a multiscale structure, and related problems arising in geophysical applications, are discussed in [60, 88, 145, 146]. Numerical methods for elliptic PDEs subject to stochastic forcing, or with stochastic coefficients, are described in [3, 141, 216, 215, 286, 287].

12.8 Exercises

1. Consider the problem of homogenization for (12.2.1) when the coefficients matrix $A(y)$ has a different period in each direction

$$A(y + \lambda_k e_k) = A(y), \quad k = 1, \dots,$$

   with $\lambda_k > 0$, $k = 1, \dots d$. Write down the formulae for the homogenized coefficients.
2. Consider the two-scale expansion (12.4.1) for problem (12.2.1). In this chapter we calculated the first three terms in the two-scale expansion: u_0 solves the homogenized equation, u_1 is given by (12.7.7), and u_2 by (12.7.8). Verify the expression for u_2 and the form of the higher-order cell problem (12.7.9).

3. Consider the Dirichlet problem (12.2.1) for a d-dimensional layered material, i.e.,
$$a_{ij}(y) = a_{ij}(y_1), \quad \text{1-periodic in } y_1, \quad i,j = 1,\dots,d.$$
We solved this problem in Subsection 12.6.2 in the case $d = 2$. Now solve the corresponding cell problem and obtain formulae for the homogenized coefficients for $d \geqslant 3$, arbitrary.
4. Consider the problem of homogenization for second-order uniformly elliptic PDEs in one dimension, i.e., the problem studied in Section 12.6.1.
   a. Calculate $\overline{a}$ for the case
   $$a(y) = \begin{cases} a_1 & : \quad y \in [0, \frac{1}{2}], \\ a_2 & : \quad y \in (\frac{1}{2}, 1], \end{cases}$$
   where a_1, a_2 are positive constants.
   b. Now calculate $\overline{a}$ for the case
   $$a(y) = \frac{1}{2 + \sin(2\pi y)}.$$
5. Consider the Dirichlet problem (12.2.1) for a d-dimensional isotropic material, i.e.,
$$a_{ij}(y) = a(y)\delta_{ij}, \quad \text{1-periodic}, \quad i,j = 1,\dots,d,$$
where δ_{ij} stands for Kronecker's delta.
   a. Use the specific structure of $A(y)$ to simplify the cell problem as much as you can.
   b. Let $d = 2$ and assume that $a(y)$ is of the form
   $$a(y) = Y_1(y_1)Y_2(y_2).$$
   Solve the two components of the cell problem and obtain formulae for the homogenized coefficients. (Hint: use separation of variables.)
6. Consider the boundary value problem (12.7.1). Assume that $A^\varepsilon = A(x, x/\varepsilon)$ where $A(x,y)$ is smooth, 1-periodic in y, and uniformly elliptic and that, furthermore, f is smooth. Use the method of multiple scales to obtain generalizations of the homogenized Equation (12.7.2), the cell problem (12.7.3), and the formula for the homogenized coefficients (12.7.4). Verify that the results of Section 12.5 still hold.
7. Consider the Dirichlet problem
$$-\nabla \cdot \left(A\left(\frac{x}{\varepsilon}, \frac{x}{\varepsilon^2}\right) \nabla u^\varepsilon \right) = f \ \text{ for } x \in \Omega \tag{12.8.1a}$$
$$u^\varepsilon(x) = 0, \ \text{ for } x \in \partial\Omega \tag{12.8.1b}$$
where the coefficients $A(y,z)$ are periodic in both y and z with period 1. Use the 3-scale expansion
$$u^\varepsilon(x) = u_0\left(x, \frac{x}{\varepsilon}, \frac{x}{\varepsilon^2}\right) + \varepsilon u_1\left(x, \frac{x}{\varepsilon}, \frac{x}{\varepsilon^2}\right) + \varepsilon^2 u_2\left(x, \frac{x}{\varepsilon}, \frac{x}{\varepsilon^2}\right) + \dots$$

to derive an effective homogenized equation, together with the formula for the homogenized coefficients and two-cell problems.

8. Repeat the previous exercise by homogenizing first with respect to $z = y/\varepsilon$ and then with respect to y:
   a. Homogenize the equation

$$-\nabla \cdot \left(A\left(y, \frac{y}{\varepsilon}\right) \nabla u^{\varepsilon} \right) = f, \quad \text{for } x \in \Omega \tag{12.8.2a}$$

$$u^{\varepsilon}(x) = 0, \quad \text{for } x \in \partial\Omega \tag{12.8.2b}$$

   by treating y as a parameter.
   b. Homogenize the equation

$$-\nabla \cdot \left(\overline{A}\left(\frac{x}{\varepsilon}\right) \nabla \overline{u}^{\varepsilon} \right) = f, \quad \text{for } x \in \Omega \tag{12.8.3a}$$

$$u^{\varepsilon}(x) = 0, \quad \text{for } x \in \partial\Omega, \tag{12.8.3b}$$

   where $\overline{A}(y)$ is given by the expression derived in the preceding section of the question.
9. Derive the homogenized equation, together with the cell problem and the formula for the homogenized coefficients, by applying the method of multiple scales to the heat Equation (12.7.10), with $A^{\varepsilon} = A(x/\varepsilon)$.
10. Consider the initial boundary value problem (12.7.10) with $A^{\varepsilon} = A(x/\varepsilon, t/\varepsilon^k)$. Explain why it is natural for the period of oscillations in time to be characterized by $k = 2$. Carry out homogenization for the cases $k = 1, 2, 3$.[2]
11. Use the method of multiple scales to derive the homogenized equation from (12.7.11).
12. Prove that the homogenized coefficient $\overline{a}$ for Equation (12.6.1) under (12.6.2) has the same upper and lower bounds as $a(y)$:

$$\alpha \leqslant \overline{a} \leqslant \beta.$$

   Moreover, show that it is bounded from above by the average of $a(y)$:

$$\overline{a} \leqslant \langle a(y) \rangle.$$

13. Show that Equation (12.7.5) can be homogenized to obtain the effective Equation (12.7.6).
14. Let $A : \mathbb{T}^d \to \mathbb{R}^{d \times d}$ be smooth and periodic and consider the eigenvalue problem

$$-\nabla \cdot \left(A^{\varepsilon} \nabla u^{\varepsilon} \right) = \lambda^{\varepsilon} u^{\varepsilon} \quad \text{for } x \in \Omega$$
$$u^{\varepsilon} = 0, \quad x \in \partial\Omega,$$

[2] See [33, ch. 3] and [253] for further details on the derivation of the homogenized equations using the method of multiple scales.

where $A^\varepsilon(x) = A(x/\varepsilon)$. Use a multiscale expansion to find an approximation to the eigenvalue problem in which $\varepsilon \to 0$ is eliminated.

15. a. Consider the eigenvalue problem

$$-\Delta u^\varepsilon + \frac{1}{\varepsilon} V^\varepsilon u^\varepsilon = \lambda^\varepsilon u^\varepsilon, \quad x \in \Omega$$
$$u^\varepsilon = 0, \quad x \in \partial\Omega.$$

Assume that $V : \mathbb{T}^d \to \mathbb{R}$ is smooth and periodic, that

$$\int_{\mathbb{T}^d} V(y)dy = 0$$

and that $V^\varepsilon(x) = V(x/\varepsilon)$. Use a multiscale expansion to find an approximation to the eigenvalue problem in which $\varepsilon \to 0$ is eliminated.

b. Are the resulting eigenvalues smaller or larger than the eigenvalues that arise when $V \equiv 0$?

13

Homogenization for Parabolic PDEs

13.1 Introduction

In this chapter we use multiscale techniques to investigate the long-time behavior of solutions to parabolic PDEs. The techniques employed are almost identical to those used in the study of homogenization for SDEs in Chapter 11. This connection will be made more explicit at the end of the chapter.

In Section 13.2 we present the full equations that we will analyze. Section 13.3 contains the simplified equations that are derived by use of the method of multiple scales in Section 13.4. Section 13.5 is devoted to various properties of the simplified equations. In Section 13.6 we study two applications of the general theory, to gradient flows (Section 13.6.1) and to divergence-free flows (Section 13.6.2). The connection between homogenization for parabolic PDEs and asymptotic problems for SDEs is made in Section 13.7. Extensions and bibliographical remarks appear in Section 13.8.

13.2 Full Equations

We study the following initial-value (Cauchy) problem

$$\frac{\partial u}{\partial t} = b \cdot \nabla u + D\Delta u \quad \text{for } (x,t) \in \mathbb{R}^d \times \mathbb{R}^+, \tag{13.2.1a}$$

$$u = u_{in} \quad \text{for } (x,t) \in \mathbb{R}^d \times \{0\}, \tag{13.2.1b}$$

with $D > 0$. In our analysis we will assume that the vector $b(x)$ is smooth and periodic in space with period 1 in all spatial directions. Furthermore, we assume that the initial conditions are slowly varying, so that

$$u_{in}(x) = g^{\varepsilon}(x) := g(\varepsilon x), \tag{13.2.2}$$

with $0 < \varepsilon \ll 1$. Since the initial data are slowly varying and so is the solution, it is natural to look at large length and time scales to see the effective behavior of the

PDE (13.2.1). If the vector field b averages to zero in an appropriate sense then, as we will show in this chapter, the effective behavior of u is that of a pure diffusion.

To see this effect we redefine the variables x, t through the rescaling

$$x = \varepsilon^{-1}x, \quad t \to \varepsilon^{-2}t \tag{13.2.3}$$

and relabel u to u^ε to emphasize this rescaling. This particular scaling of space and time, known as **diffusive scaling**, is appropriate whenever the advective effects, created by b, are expected to average out; it is then appropriate to scale time on an even longer scale than space and to seek purely diffusive effects. We will be precise about the condition that b averages out at the end of this section.

The rescaled field $u^\varepsilon(x,t)$ satisfies the equation

$$\frac{\partial u^\varepsilon}{\partial t} = \frac{1}{\varepsilon} b^\varepsilon \cdot \nabla u^\varepsilon + D\Delta u^\varepsilon \quad \text{for } (x,t) \in \mathbb{R}^d \times \mathbb{R}^+, \tag{13.2.4a}$$

$$u^\varepsilon = g \quad \text{for } (x,t) \in \mathbb{R}^d \times \{0\}. \tag{13.2.4b}$$

Here $b^\varepsilon(x) = b(x/\varepsilon)$. This equation will be the object of our study in this chapter.

Let us define the operator

$$\mathcal{L}_0 = b(y) \cdot \nabla_y + D\Delta_y \tag{13.2.5}$$

with periodic boundary conditions on $[0,1]^d$ and its L^2-adjoint $\mathcal{L}_0^*$, also with periodic boundary conditions. We refer to D as the **molecular diffusivity**. Note that $\mathcal{L}_0$ is the generator of the Markov process $y(t)$, which is the solution of the SDE

$$\frac{dy}{dt} = b(y) + \sqrt{2D}\frac{dW}{dt}$$

on the unit torus $\mathbb{T}^d$. Hence it is natural to define the **invariant distribution** $\rho(y)$ to be the stationary solution of the adjoint equation:

$$\mathcal{L}_0^*\rho = 0. \tag{13.2.6}$$

By Theorem 6.16 there is a unique solution to this equation, up to normalization, and the normalization may be chosen so that the solution is positive. In the sequel we will normalize the solution to (13.2.6) according to

$$\int_{\mathbb{T}^d} \rho(y)\, dy = 1.$$

Notice that this choice turns the measure $\mu(dy) = \rho(y)\, dy$ into a probability measure on $\mathbb{T}^d$.

In order to derive the homogenized equation for (13.2.4) we need to study equations of the form

$$-\mathcal{L}_0 v = h \tag{13.2.7}$$

with periodic boundary conditions and with h being a smooth periodic function of y. It is straightforward to check that the assumptions of Theorem 7.9 are satisfied and hence the operator $\mathcal{L}_0$ satisfies the Fredholm alternative. This implies, in particular, that $\mathcal{L}_0$ has a one-dimensional null space, comprising constants in y. It also implies that $\mathcal{L}_0^*$ has a one-dimensional null space, as stated earlier, and spanned by ρ. Furthermore, Equation (13.2.7) has a solution if and only if the right-hand side of the equation is centered with respect to the invariant distribution:

$$\int_{\mathbb{T}^d} h(y)\rho(y)\,dy = 0.$$

In this case the solution of (13.2.7) is unique up to constants. In the case where $h = b$, the vector field arising in the PDE (13.2.1), the condition is

$$\int_{\mathbb{T}^d} b(y)\rho(y)\,dy = 0. \tag{13.2.8}$$

We call this the **centering condition**. We fix the free constant in the solution (13.2.7) by requiring that the solution of (13.2.7) satisfies

$$\int_{\mathbb{T}^d} v(y)\rho(y)\,dy = 0. \tag{13.2.9}$$

When the centering condition is not satisfied it is necessary to rescale the original problem in a different fashion, to see effective *advective* behavior. In particular (13.2.3) is replaced by the **advective scaling**

$$x \to \varepsilon^{-1}x, \quad t \to \varepsilon^{-1}t. \tag{13.2.10}$$

Then averaging is used to find the effective equation, which is now of transport type; see Chapter 14.

13.3 Simplified Equations

Assume that the vector field $b(y)$ satisfies the centering condition (13.2.8). Define the vector field $\chi(y)$ to be the solution of the *cell problem*

$$-\mathcal{L}_0\chi = b, \ \chi \text{ is 1-periodic}, \ \int_{\mathbb{T}^d} \chi(y)\rho(y)dy = 0. \tag{13.3.1}$$

The *effective diffusion tensor* (or *effective diffusivity*) is defined as

$$\mathcal{K} = DI + 2D\int_{\mathbb{T}^d} \nabla_y\chi(y)^T\rho(y)\,dy + \int_{\mathbb{T}^d} \Big(b(y)\otimes\chi(y)\Big)\rho(y)\,dy. \tag{13.3.2}$$

Result 13.1. *Assume that* (13.2.8) *holds. For* $0 < \varepsilon \ll 1$ *and times* t *of* $\mathcal{O}(1)$ *the solution* u^ε *of* (13.2.4) *is approximated by* u, *the solution of the homogenized equation*

$$\frac{\partial u}{\partial t} = \mathcal{K} : \nabla_x\nabla_x u \quad \textit{for} \ (x,t) \in \mathbb{R}^d \times \mathbb{R}^+, \tag{13.3.3a}$$

$$u = g \quad \textit{for} \ (x,t) \in \mathbb{R}^d \times \{0\}. \tag{13.3.3b}$$

Remark 13.2. Since the Hessian $\nabla_x \nabla_x u$ is symmetric, it follows from property (2.2.2) applied to (13.3.2) that the following expression for $\mathcal{K}$ is equally valid:

$$\mathcal{K} = DI + D \int_{\mathbb{T}^d} \left(\nabla_y \chi(y) + \nabla_y \chi(y)^T \right) \rho(y) \, dy \\ + \frac{1}{2} \int_{\mathbb{T}^d} \Big(b(y) \otimes \chi(y) + \chi(y) \otimes b(y) \Big) \rho(y) \, dy. \qquad (13.3.4)$$

Many variants on this idea are possible. □

13.4 Derivation

Our goal now is to use the method of multiple scales to analyze the behavior of $u^\varepsilon(x,t)$, the solution of (13.2.4), in the limit as $\varepsilon \to 0$. In particular, we want to derive Result 13.1.

We introduce the auxiliary variable $y = x/\varepsilon$. [1] Let $\phi = \phi(x, x/\varepsilon)$ be scalar-valued. The chain rule gives

$$\nabla \phi = \nabla_x \phi + \frac{1}{\varepsilon} \nabla_y \phi \quad \text{and} \quad \Delta \phi = \Delta_x \phi + \frac{2}{\varepsilon} \nabla_x \cdot \nabla_y \phi + \frac{1}{\varepsilon^2} \Delta_y \phi.$$

The partial differential operator that appears on the right-hand side of Equation (13.2.4) becomes

$$\mathcal{L} = \frac{1}{\varepsilon^2} \mathcal{L}_0 + \frac{1}{\varepsilon} \mathcal{L}_1 + \mathcal{L}_2,$$

where

$$\begin{aligned} \mathcal{L}_0 &= b(y) \cdot \nabla_y + D \Delta_y, \\ \mathcal{L}_1 &= b(y) \cdot \nabla_x + 2D \nabla_x \cdot \nabla_y, \\ \mathcal{L}_2 &= D \Delta_x. \end{aligned}$$

In terms of x and y, Equation (13.2.4a) becomes

$$\frac{\partial u^\varepsilon}{\partial t} = \left(\frac{1}{\varepsilon^2} \mathcal{L}_0 + \frac{1}{\varepsilon} \mathcal{L}_1 + \mathcal{L}_2 \right) u^\varepsilon.$$

We seek a solution in the form of a multiple-scales expansion

$$u^\varepsilon(x,t) = u_0(x,y,t) + \varepsilon u_1(x,y,t) + \varepsilon^2 u_2(x,y,t) + \dots \qquad (13.4.1)$$

where $u_j(x,y,t)$, $j = 1, 2 \dots$, are periodic in y with period 1. We substitute (13.4.1) and equate terms of equal powers in ε. We obtain the following sequence of equations:

---

[1] As in the elliptic case, this is where the assumption of scale separation is exploited; **we treat x and y as independent variables**. Justifying this assumption as $\varepsilon \to 0$ is one of the main issues in the rigorous theory of homogenization; see Chapter 20.

$$\mathcal{O}(1/\varepsilon^2) \qquad -\mathcal{L}_0 u_0 = 0, \tag{13.4.2a}$$

$$\mathcal{O}(1/\varepsilon) \qquad -\mathcal{L}_0 u_1 = \mathcal{L}_1 u_0, \tag{13.4.2b}$$

$$\mathcal{O}(1) \qquad -\mathcal{L}_0 u_2 = \mathcal{L}_1 u_1 + \mathcal{L}_2 u_0 - \frac{\partial u_0}{\partial t}. \tag{13.4.2c}$$

Note that $\mathcal{L}_0$, which is a differential operator in y only, is equipped with periodic boundary conditions.

Since $\mathcal{L}_0$ has a one-dimensional null space, Equation (13.4.2a) implies that the first term in the expansion is independent of y, so that $u_0 = u(x,t)$ only. Notice that

$$\mathcal{L}_1 u_0 = b(y) \cdot \nabla_x u(x,t).$$

The centering condition (13.2.8) ensures that (13.4.2b) has a solution, by the Fredholm alternative. Since $\mathcal{L}_0$ is a differential operator in y only, we may use separation of variables to write the solution as

$$u_1(x,y,t) = \chi(y) \cdot \nabla_x u(x,t).$$

Then $\chi(y)$ solves the cell problem (13.3.1) . Our assumptions imply that there exists a unique, smooth solution to the cell problem.

Now we proceed with the analysis of the $\mathcal{O}(1)$ Equation (13.4.2c). The solvability condition (13.2.8) reads

$$\int_{\mathbb{T}^d} \left(\frac{\partial u_0}{\partial t} - \mathcal{L}_2 u_0 - \mathcal{L}_1 u_1 \right) \rho\, dy = 0.$$

The fact that $u_0 = u(x,t)$ is independent of y enables us to rewrite the preceding equation in the form

$$\frac{\partial u}{\partial t} = D\Delta u + \int_{\mathbb{T}^d} \left(\mathcal{L}_1 u_1 \right) \rho\, dy. \tag{13.4.3}$$

Now we have

$$\begin{aligned}
\mathcal{L}_1 u_1 &= \left(b \cdot \nabla_x (\chi \cdot \nabla_x u) + 2D \nabla_x \cdot \nabla_y (\chi \cdot \nabla_x u) \right) \\
&= \left(b \otimes \chi + 2D \nabla_y \chi^T \right) : \nabla_x \nabla_x u.
\end{aligned}$$

In view of the preceding calculation, Equation (13.4.3) becomes

$$\frac{\partial u}{\partial t} = \mathcal{K} : \nabla_x \nabla_x u,$$

which is the homogenized Equation (13.3.3a). The effective diffusivity $\mathcal{K}$ is given by formula (13.3.2).

13.5 Properties of the Simplified Equations

In this section we show that the effective diffusivity is positive definite. This implies that the homogenized equation is well posed. To prove this, we need to calculate the Dirichlet form associated with the operator $\mathcal{L}_0$. The following is a direct consequence of Theorem 6.12 in the case of additive noise. Recall that ρ is the invariant distribution, a nonnegative $L^1(\mathbb{T}^d)$ function in the null space of $\mathcal{L}_0^*$.

Lemma 13.3. *Let $f(y) \in C^2_{per}(\mathbb{T}^d)$. Then*

$$\int_{\mathbb{T}^d} (-\mathcal{L}_0 f(y)) f(y) \rho(y)\, dy = D \int_{\mathbb{T}^d} |\nabla_y f(y)|^2 \rho(y)\, dy. \tag{13.5.1}$$

Remark 13.4. Let $L^2_\rho(\mathbb{T}^d)$ be the L^2-space weighted by the invariant distribution $\rho(y)$ and denote the inner product and corresponding norm by $(\cdot,\cdot)_{L^2_\rho}$ and $\|\cdot\|_{L^2_\rho}$, respectively. Then, by Equation (6.3.13), the result of Lemma 13.3 can be expressed in the form

$$(-\mathcal{L}_0 f, f)_{L^2_\rho} = D \|\nabla_y f\|^2_{L^2_\rho}. \quad \square$$

The main result of this section is that the effective diffusivity is a positive-definite matrix. In particular, we have the following.

Theorem 13.5. *Let $\xi \in \mathbb{R}^d$ be an arbitrary vector and let $\chi_\xi(y) := \chi(y) \cdot \xi$. Then*

$$\langle \xi, \mathcal{K}\xi \rangle = D \int_{\mathbb{T}^d} |\xi + \nabla_y \chi_\xi(y)|^2 \rho(y)\, dy.$$

Furthermore,

$$\alpha |\xi|^2 \leqslant \langle \xi, \mathcal{K}\xi \rangle \quad \forall \xi \in \mathbb{R}^d \tag{13.5.2}$$

with

$$\alpha = D \left(\int_{\mathbb{T}^d} \rho^{-1}(y)\, dy \right)^{-1}. \tag{13.5.3}$$

Proof. Note that $-\mathcal{L}_0 \chi_\xi = \xi \cdot b$. We use the definition of $\mathcal{K}$ and Lemma 13.3 to calculate

$$\begin{aligned}
\langle \xi, \mathcal{K}\xi \rangle &= D|\xi|^2 + 2D \int_{\mathbb{T}^d} \xi \cdot \nabla_y \chi_\xi(y) \rho(y)\, dy + \int_{\mathbb{T}^d} (\xi \cdot b) \chi_\xi(y) \rho(y)\, dy \\
&= D|\xi|^2 + 2D \int_{\mathbb{T}^d} \xi \cdot \nabla_y \chi_\xi(y) \rho(y)\, dy + D \int_{\mathbb{T}^d} |\nabla_y \chi_\xi(y)|^2 \rho(y)\, dy \\
&= D \int_{\mathbb{T}^d} |\xi + \nabla_y \chi_\xi(y)|^2 \rho(y)\, dy.
\end{aligned}$$

The fact that the effective diffusivity is nonnegative definite follows immediately from the preceding equation. To show that $\mathcal{K}$ is positive definite we use the fact that the integral of derivatives of periodic functions over $\mathbb{T}^d$ is 0, together with the Cauchy-Schwarz inequality and the fact that $\rho(y)$ is everywhere positive, to calculate:

$$\begin{aligned}
D|\xi|^2 &= D \left| \int_{\mathbb{T}^d} (\xi + \nabla_y \chi_\xi)\, dy \right|^2 \\
&= D \left| \int_{\mathbb{T}^d} (\xi + \nabla_y \chi_\xi)\, \rho^{\frac{1}{2}}(y) \rho^{-\frac{1}{2}}(y)\, dy \right|^2 \\
&\leqslant D \left(\int_{\mathbb{T}^d} |\xi + \nabla_y \chi_\xi|^2 \rho(y)\, dy \right) \left(\int_{\mathbb{T}^d} \rho^{-1}(y)\, dy \right) \\
&= \langle \xi, \mathcal{K}\xi \rangle \left(\int_{\mathbb{T}^d} \rho^{-1}(y)\, dy \right),
\end{aligned}$$

from which the lower bound immediately follows. □

It is of interest to know how the effective diffusion tensor $\mathcal{K}$ compares with the original diffusion tensor DI. It turns out that $\mathcal{K}$ can be either greater or smaller than D (in the sense of matrices). This issue is discussed in detail in the next section, where we show that the effective diffusivity is smaller than D for gradient vector fields b and greater than D for divergence-free vector fields b.

13.6 Applications

In this section we consider two particular choices for the drift term b in (13.2.4a), *gradient* and *divergence-free* fields. In both cases it is possible to perform explicit calculations that yield considerable insight. In particular, we will be able to obtain a formula for the (unique) invariant distribution and, consequently, to simplify the centering condition (13.2.8). Furthermore we will be able to compare the effective diffusivity with the original diffusivity D. We will see that the effective diffusivity is smaller than D for gradient vector fields b, and that it is greater than D for divergence-free vector fields b. We also study two particular cases of gradient and divergence-free flows for which we can derive closed formulae for the effective diffusivity.

There are at least two reasons why it is interesting to consider gradient and divergence-free flows. On the one hand, parabolic PDEs of the form (13.2.1) with b being either the gradient of a scalar field or divergence-free appear frequently in applications: when $b = -\nabla V$ then Equation (13.2.1) describes Brownian motion in a periodic potential. On the other hand, when b is divergence-free, Equation (13.2.1) becomes the advection diffusion equation, which describes mixing processes in incompressible fluids. According to the Hodge decomposition theorem, every smooth vector field on $\mathbb{T}^d$ can be decomposed into the sum of a gradient and a divergence-free field:

$$b(y) = -\nabla V(y) + v(y), \quad \nabla \cdot v(y) = 0,$$

with

$$(-\nabla V(y), v(y))_{L^2(\mathbb{T}^d)} = 0.$$

Hence, by studying gradient and divergence-free flows we study the two extreme cases of this decomposition.

13.6.1 Gradient Vector Fields

We consider the case where the vector field $b(y)$ in Equation (13.2.4a) is the gradient of a smooth, scalar periodic function,

$$b(y) = -\nabla_y V(y). \tag{13.6.1}$$

The function V is called the *potential*. In this case it is straightforward to derive a formula for the solution ρ of the stationary adjoint Equation (13.2.6) with periodic boundary conditions.

Lemma 13.6. *Assume that the vector field b is a gradient given by* (13.6.1). *Let $\mathcal{L}_0^*$ denote the adjoint of $\mathcal{L}_0$ defined in* (13.2.5). *Then the equation*

$$\mathcal{L}_0^* \rho = 0, \quad \int_{\mathbb{T}^d} \rho(y) dy = 1, \tag{13.6.2}$$

subject to periodic boundary conditions on $\mathbb{T}^d$, has a unique solution given by

$$\rho(y) = \frac{1}{Z} e^{-V(y)/D}, \quad Z = \int_{\mathbb{T}^d} e^{-V(y)/D} \, dy. \tag{13.6.3}$$

Proof. Equation (13.6.2), in view of Equation (13.6.1), becomes

$$\nabla_y \cdot \big(\nabla_y V(y) \rho(y) + D \nabla_y \rho(y)\big) = 0. \tag{13.6.4}$$

We immediately check that $\rho(y)$ given by (13.6.3) satisfies

$$\nabla_y V(y) \rho(y) + D \nabla_y \rho(y) = 0,$$

and hence it satisfies (13.6.4). Furthermore, by construction we have that

$$\int_{\mathbb{T}^d} \frac{1}{Z} e^{-V(y)/D} \, dy = 1,$$

and hence $\rho(y)$ is correctly normalized. Thus we have constructed a solution of Equation (13.6.2). Uniqueness follows by the ergodicity of the stochastic process with generator $\mathcal{L}_0$ (see Theorem 6.16). □

Remark 13.7. The positive function ρ defined in (13.6.3) is called the **Gibbs distribution** and the probability measure $\rho(y)dy$ the **Gibbs measure**. The normalization constant Z is called the **partition function.** □

In the case of gradient flows the centering condition (13.2.8) is satisfied identically for any potential.

Lemma 13.8. *Consider the operator $\mathcal{L}_0$ given by* (13.2.5) *with periodic boundary conditions and assume that $b(y) = -\nabla_y V(y)$ with $V \in C^1_{per}(\mathbb{T}^d)$. Then the centering condition* (13.2.8) *is always satisfied.*

Proof. We use the divergence theorem to calculate

$$\begin{aligned}
\int_{\mathbb{T}^d} b(y) \rho(y) \, dy &= \frac{1}{Z} \int_{\mathbb{T}^d} -\nabla_y V(y) e^{-V(y)/D} \, dy \\
&= \frac{D}{Z} \int_{\mathbb{T}^d} \nabla_y e^{-V(y)/D} \, dy \\
&= 0. \quad \square
\end{aligned}$$

In the case of gradient flows, the operator $\mathcal{L}_0$ defined in (13.2.5) equipped with periodic boundary conditions becomes symmetric in the appropriate function space. We have the following.

Lemma 13.9. *Assume that condition* (13.6.1) *is satisfied and let* ρ *denote the Gibbs distribution* (13.6.3). *Then the operator* $\mathcal{L}_0$ *given in* (13.2.5) *satisfies*

$$\int_{\mathbb{T}^d} f(y)\Big(\mathcal{L}_0 h(y)\Big)\rho(y)\,dy = \int_{\mathbb{T}^d} h(y)\Big(\mathcal{L}_0 f(y)\Big)\rho(y)\,dy \tag{13.6.5}$$

for all $f,\ h \in C^2_{per}(\mathbb{T}^d)$.

Proof. Using the divergence theorem we have

$$\begin{aligned}
\int_{\mathbb{T}^d} f\mathcal{L}_0 h\rho\,dy &= \frac{1}{Z}\int_{\mathbb{T}^d} f\big(-\nabla_y V\cdot\nabla_y h\big)e^{-V/D}\,dy + \frac{D}{Z}\int_{\mathbb{T}^d} f\Delta_y h e^{-V/D}\,dy \\
&= \frac{D}{Z}\int_{\mathbb{T}^d} f\nabla_y h\cdot\nabla_y\left(e^{-V/D}\right)dy - \frac{D}{Z}\int_{\mathbb{T}^d}\left(\nabla_y f\cdot\nabla_y h\right)e^{-V/D}\,dy \\
&\quad -\frac{D}{Z}\int_{\mathbb{T}^d} f\nabla_y h\cdot\nabla_y\left(e^{-V/D}\right)dy \\
&= -D\int_{\mathbb{T}^d}\Big(\nabla_y f\cdot\nabla_y h\Big)\rho\,dy.
\end{aligned}$$

The expression in the last line is symmetric in $f,\ h$, and hence (13.6.5) follows. □

Remark 13.10. The symmetry of $\mathcal{L}_0$ arises quite naturally from the identity (6.3.11) used in proving Theorem 6.12. Furthermore, the calculation used in the proof of Lemma 13.9 gives us the following useful formula

$$\int_{\mathbb{T}^d} f(-\mathcal{L}_0 h)\rho\,dy = D\int_{\mathbb{T}^d}\Big(\nabla_y f\cdot\nabla_y h\Big)\rho\,dy \tag{13.6.6}$$

for all $f,\ h \in C^2_{per}(\mathbb{T}^d)$. The Dirichlet form Lemma 13.3 follows from this on setting $f = h$. Now let $\phi,\ \psi \in C^2_{per}(\mathbb{T}^d;\mathbb{R}^d)$. In view of (13.6.6) we also have

$$\int_{\mathbb{T}^d}\big(\phi\otimes(-\mathcal{L}_0\psi)\big)\rho\,dy = D\int_{\mathbb{T}^d}\Big(\nabla_y\phi\otimes\nabla_y\psi\Big)\rho\,dy. \qquad \square \tag{13.6.7}$$

Remark 13.11. Using the notation introduced in Remark 13.4 we can express the result of Lemma 13.9 by saying that $\mathcal{L}_0$ is symmetric as an operator from L^2_ρ to L^2_ρ. Furthermore, identity (13.6.6) can be written in the form

$$(f, -\mathcal{L}_0 h)_{L^2_\rho} = D\,(\nabla_y f, \nabla_y h)_{L^2_\rho}\,.$$

Ergodic Markov processes whose generator is a symmetric operator in L^2_ρ are called **reversible**. Thus we have shown that SDEs with additive noise and with a drift that is a gradient field are reversible. □

Now we are ready to prove various properties of the effective diffusivity. For this we will need the following integration-by-parts formula, which follows from the divergence theorem and the periodicity of χ and ρ :

$$\int_{\mathbb{T}^d} (\nabla_y \chi) \rho \, dy = \int_{\mathbb{T}^d} \left(\nabla_y (\chi \rho) - \chi \otimes \nabla_y \rho \right) dy = - \int_{\mathbb{T}^d} (\chi \otimes \nabla_y \rho) \, dy. \quad (13.6.8)$$

Theorem 13.12. *Assume that* $b(y)$ *is a gradient so that* (13.6.1) *holds and let* $\rho(y)$ *denote the Gibbs distribution* (13.6.3). *Then the effective diffusivity* (13.3.2) *satisfies the upper and lower bounds*

$$\frac{D}{Z\widehat{Z}} \leqslant \langle \xi, \mathcal{K} \xi \rangle \leqslant D|\xi|^2 \quad \forall \xi \in \mathbb{R}^d, \qquad (13.6.9)$$

where

$$\widehat{Z} = \int_{\mathbb{T}^d} e^{V(y)/D} \, dy.$$

In particular, diffusion is always depleted when compared to molecular diffusivity. Furthermore, the effective diffusivity is symmetric. [2]

Proof. The lower bound follows from the general lower bound (13.5.2), Equation (13.5.3), and the formula for the Gibbs measure. To establish the upper bound, we use (13.6.8) and (13.6.7) to obtain

$$\begin{aligned}
\mathcal{K} &= DI + 2D \int_{\mathbb{T}^d} (\nabla \chi)^T \rho \, dy + \int_{\mathbb{T}^d} -\nabla_y V \otimes \chi \rho \, dy \\
&= DI - 2D \int_{\mathbb{T}^d} \nabla_y \rho \otimes \chi \, dy + \int_{\mathbb{T}^d} -\nabla_y V \otimes \chi \rho \, dy \\
&= DI - 2 \int_{\mathbb{T}^d} -\nabla_y V \otimes \chi \rho \, dy + \int_{\mathbb{T}^d} -\nabla_y V \otimes \chi \rho \, dy \\
&= DI - \int_{\mathbb{T}^d} -\nabla_y V \otimes \chi \rho \, dy \\
&= DI - \int_{\mathbb{T}^d} (-\mathcal{L}_0 \chi) \otimes \chi \rho \, dy \\
&= DI - D \int_{\mathbb{T}^d} (\nabla_y \chi \otimes \nabla_y \chi) \rho \, dy. \qquad (13.6.10)
\end{aligned}$$

Hence, for $\chi_\xi = \chi \cdot \xi$,

$$\begin{aligned}
\langle \xi, \mathcal{K} \xi \rangle &= D|\xi|^2 - D \int_{\mathbb{T}^d} |\nabla_y \chi_\xi|^2 \rho \, dy \\
&\leqslant D|\xi|^2.
\end{aligned}$$

This proves depletion. The symmetry of $\mathcal{K}$ follows from (13.6.10). □

[2] Notice that the Cauchy-Schwarz inequality shows that $Z\widehat{Z} \geqslant 1$.

The One-Dimensional Case

The one-dimensional case is always in gradient form: $b(y) = -\partial_y V(y)$. Furthermore in one dimension we can solve the cell problem (13.3.1) in closed form and calculate the effective diffusion coefficient explicitly – up to quadratures. We start with the following calculation concerning the structure of the diffusion coefficient.

$$\begin{aligned}
\mathcal{K} &= D + 2D\int_0^1 \partial_y \chi \rho \, dy + \int_0^1 -\partial_y V \chi \rho \, dy \\
&= D + 2D\int_0^1 \partial_y \chi \rho \, dy + D\int_0^1 \chi \partial_y \rho \, dy \\
&= D + 2D\int_0^1 \partial_y \chi \rho \, dy - D\int_0^1 \partial_y \chi \rho \, dy \\
&= D\int_0^1 (1 + \partial_y \chi)\rho \, dy. \qquad (13.6.11)
\end{aligned}$$

The cell problem (13.3.1) in one dimension is

$$D\partial_{yy}\chi - \partial_y V \partial_y \chi = \partial_y V. \qquad (13.6.12)$$

We multiply Equation (13.6.12) by $e^{-V(y)/D}$ to obtain

$$\partial_y \left(\partial_y \chi e^{-V(y)/D}\right) = -\partial_y \left(e^{-V(y)/D}\right).$$

We integrate this equation from 0 to y and multiply by $e^{V(y)/D}$ to obtain

$$\partial_y \chi(y) = -1 + c_1 e^{V(y)/D}.$$

Another integration yields

$$\chi(y) = -y + c_1 \int_0^y e^{V(y)/D} \, dy + c_2.$$

The periodic boundary conditions imply that $\chi(0) = \chi(1)$, from which we conclude that

$$-1 + c_1 \int_0^1 e^{V(y)/D} \, dy = 0.$$

Hence

$$c_1 = \frac{1}{\widehat{Z}}, \quad \widehat{Z} = \int_0^1 e^{V(y)/D} \, dy.$$

We deduce that

$$\partial_y \chi = -1 + \frac{1}{\widehat{Z}} e^{V(y)/D}.$$

We substitute this expression into (13.6.11) to obtain

$$\begin{aligned}\mathcal{K} &= \frac{D}{Z}\int_0^1 (1+\partial_y\chi(y))\, e^{-V(y)/D}\, dy \\ &= \frac{D}{Z\widehat{Z}}\int_0^1 e^{V(y)/D}e^{-V(y)/D}\, dy \\ &= \frac{D}{Z\widehat{Z}},\end{aligned} \tag{13.6.13}$$

with

$$Z = \int_0^1 e^{-V(y)/D}\, dy, \quad \widehat{Z} = \int_0^1 e^{V(y)/D}\, dy. \tag{13.6.14}$$

Notice that in the one-dimensional case the formula for the effective diffusivity is precisely the lower bound in (13.6.9). This shows that the lower bound is sharp.

Example 13.13. Consider the potential

$$V(y) = \begin{cases} a_1 & : \quad y \in [0, \frac{1}{2}], \\ a_2 & : \quad y \in (\frac{1}{2}, 1], \end{cases} \tag{13.6.15}$$

where a_1, a_2 are positive constants.[3]

It is straightforward to calculate the integrals in (13.6.14) to obtain the formula

$$\mathcal{K} = \frac{D}{\cosh^2\left(\frac{a_1-a_2}{D}\right)}. \tag{13.6.16}$$

In Figure 13.1 we plot the effective diffusivity given by (13.6.16) as a function of the molecular diffusivity D, on a log scale. We observe that $\mathcal{K}$ decays exponentially fast in the limit as $D \to 0$. □

13.6.2 Divergence-Free Fields

In this section we consider the problem of homogenization for (13.2.4a) in the case where the vector field $b(y)$ is divergence-free (or incompressible):

$$\nabla \cdot b(y) = 0. \tag{13.6.17}$$

The incompressibility of $b(y)$ simplifies the analysis considerably because the advection operator

$$\widehat{\mathcal{L}}_0 = b(y) \cdot \nabla_y,$$

with periodic boundary conditions, is antisymmetric in $L^2(\mathbb{T}^d)$.

[3] Of course, this potential is not even continuous, let alone smooth, and the theory as developed in this chapter does not apply. It is possible, however, to consider a regularized version of this discontinuous potential, and then homogenization theory applies.

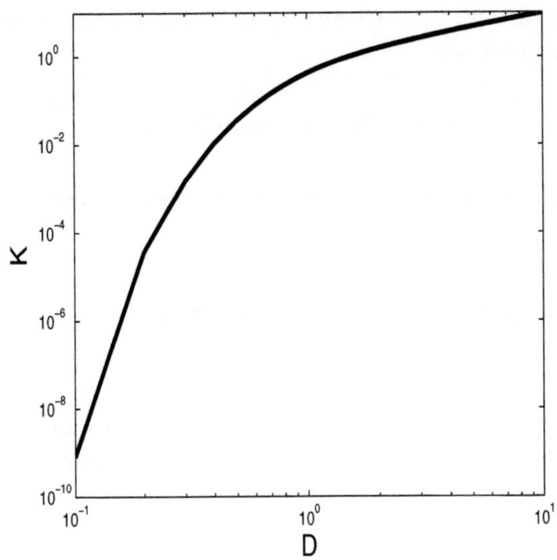

Fig. 13.1. Log-log plot of the effective diffusivity versus molecular diffusivity for the potential (13.6.15).

Lemma 13.14. *Let* $b(y) \in C^1_{per}(\mathbb{T}^d;\mathbb{R}^d)$ *satisfy* (13.6.17). *Then for all* $f(y),\ h(y) \in C^1_{per}(\mathbb{T}^d)$ *we have*

$$\int_{\mathbb{T}^d} f(y)\,(b(y)\cdot\nabla_y h(y))\,dy = -\int_{\mathbb{T}^d} h(y)\,(b(y)\cdot\nabla_y f(y))\,dy.$$

In particular,

$$\int_{\mathbb{T}^d} f(y)\,(b(y)\cdot\nabla_y f(y))\,dy = 0. \tag{13.6.18}$$

Proof. We use the incompressibility of $b(y)$, together with the periodicity of $f(y)$, $h(y)$, and $b(y)$ to calculate

$$\begin{aligned}\int_{\mathbb{T}^d} f(y)\,(b(y)\cdot\nabla_y h(y))\,dy &= \int_{\mathbb{T}^d} f(y)\nabla_y\cdot\big(b(y)h(y)\big)\,dy\\ &= -\int_{\mathbb{T}^d} \nabla_y f(y)\cdot(b(y)h(y))\,dy\\ &= -\int_{\mathbb{T}^d} h(y)\big(b(y)\cdot\nabla_y f(y)\big)\,dy.\end{aligned}$$

Equation (13.6.18) follows from this calculation on setting $f = h$. □

Using the previous lemma it is easy to prove that the unique invariant measure of the fast process is the Lebesgue measure.

Lemma 13.15. *Let $\mathcal{L}_0$ denote the operator defined in* (13.2.5) *with periodic boundary conditions and with $b(y)$ satisfying* (13.6.17)*. Let $\mathcal{L}_0^*$ denote the L^2-adjoint of $\mathcal{L}_0$. Then the adjoint equation*

$$\mathcal{L}_0^* \rho = 0, \qquad \int_{\mathbb{T}^d} \rho(y) dy = 1, \tag{13.6.19}$$

with periodic boundary conditions on $\mathbb{T}^d$ has a unique classical solution given by

$$\rho(y) = 1. \tag{13.6.20}$$

Proof. Lemma 13.14 implies that the L^2-adjoint of $\mathcal{L}_0$ is

$$\mathcal{L}_0^* = -b(y) \cdot \nabla_y + D\Delta_y, \tag{13.6.21}$$

with periodic boundary conditions. Let $\rho(y)$ be a solution of Equation (13.6.19). We multiply the equation by $\rho(y)$, integrate over $\mathbb{T}^d$, and use Lemma 13.14 to obtain

$$\int_{\mathbb{T}^d} |\nabla_y \rho(y)|^2 \, dy = 0, \tag{13.6.22}$$

from which we deduce that $\rho(y)$ is a constant. Hence, the unique normalized solution of (13.6.19) is given by (13.6.20). □

Remark 13.16. The solution $\rho(y) = 1$ can be seen to be in the null space of (13.6.21) by inspection. Uniqueness can then be proved by appealing to ergodicity of the process with generator $\mathcal{L}_0$ (see Theorem 6.16), or by use of the maximum principle. □

Remark 13.17. An immediate corollary of Proposition 13.15 is that for divergence-free fields the solvability condition (13.2.8) becomes

$$\int_{\mathbb{T}^d} b(y) \, dy = 0.$$

Thus, it is straightforward to check whether a given periodic divergence-free field satisfies the solvability condition – the field must average to zero over the unit torus. □

Now let $\chi(y)$ be the solution of the cell problem (13.3.1) with $b(y)$ satisfying (13.6.17). The periodicity of $\chi(y)$, together with (13.6.20), implies that the second term on the right-hand side of Equation (13.3.2) vanishes and the formula for the effective diffusivity reduces to

$$\mathcal{K} = DI + \int_{\mathbb{T}^d} b(y) \otimes \chi(y) \, dy. \tag{13.6.23}$$

The effective diffusivity as given in (13.3.2) is symmetric for gradient flows. This is not true for divergence-free flows. However, only the symmetric part of $\mathcal{K}$ enters into the homogenized equation by Remark 13.2. For this reason we redefine the effective diffusivity to be the symmetric part of $\mathcal{K}$:

$$\mathcal{K} := DI + \frac{1}{2}\int_{\mathbb{T}^d} \big(b(y)\otimes\chi(y) + \chi(y)\otimes b(y)\big)\,dy. \tag{13.6.24}$$

Our goal now is to show that the homogenization procedure enhances diffusion, i.e., that the effective diffusivity is always greater than the molecular diffusivity D. For this we need an alternative representation formula for $\mathcal{K}$.

Theorem 13.18. *The effective diffusivity $\mathcal{K}$ given by the expression* (13.6.24) *can be written in the form*

$$\mathcal{K} = DI + D\int_{\mathbb{T}^d} \nabla_y\chi(y)\otimes\nabla_y\chi(y)\,dy. \tag{13.6.25}$$

Proof. We take the outer product of the cell problem (13.3.1) with $\chi(y)$ to the left and integrate over the unit cell to obtain

$$-D\int_{\mathbb{T}^d}\chi(y)\otimes\Delta_y\chi(y)\,dy - \int_{\mathbb{T}^d}\chi(y)\otimes\big(\nabla_y\chi(y)b(y)\big)\,dy = \int_{\mathbb{T}^d}\chi(y)\otimes b(y)\,dy.$$

We apply the divergence theorem to the *two integrals* on the left-hand side of the equation, using periodicity and the fact that b is divergence-free, to obtain

$$D\int_{\mathbb{T}^d}\nabla_y\chi(y)\otimes\nabla_y\chi(y)\,dy + \int_{\mathbb{T}^d}\big(\nabla\chi(y)b(y)\big)\otimes\chi(y)\,dy = \int_{\mathbb{T}^d}\chi(y)\otimes b(y)\,dy. \tag{13.6.26}$$

Alternatively we may take the outer product with χ in (13.3.1) to the right and use the divergence theorem only on the first integral, to obtain

$$D\int_{\mathbb{T}^d}\nabla_y\chi(y)\otimes\nabla_y\chi(y)\,dy - \int_{\mathbb{T}^d}\big(\nabla\chi(y)b(y)\big)\otimes\chi(y)\,dy = \int_{\mathbb{T}^d}b(y)\otimes\chi(y)\,dy. \tag{13.6.27}$$

We add Equations (13.6.26) and (13.6.27) to obtain:

$$\frac{1}{2}\int_{\mathbb{T}^d}\big(b(y)\otimes\chi(y) + \chi(y)\otimes b(y)\big)\,dy = D\int_{\mathbb{T}^d}\nabla_y\chi(y)\otimes\nabla_y\chi(y)\,dy.$$

Equation (13.6.25) follows, on substituting this expression into Equation (13.6.24). □

We can now obtain upper and lower bounds for the effective diffusivity.

Theorem 13.19. *Assume that $b(y)$ is divergence-free. Then the effective diffusivity satisfies the upper and lower bounds*

$$D|\xi|^2 \leqslant \langle\xi,\mathcal{K}\xi\rangle \leqslant \left(D + \frac{C}{D}\right)|\xi|^2, \tag{13.6.28}$$

where $C = C(b,\Omega) > 0$ is explicitly computable. [4] *The lower bound becomes an equality for all ξ only when $b(y)\equiv 0$.*

[4] Indeed $C = (C_p\|b\|_{L_2})^2$, where C_p is the Poincaré constant from inequality (2.4.7).

Proof. The lower bound follows from the general bound (13.5.2), Equation (13.5.3), and the fact that for divergence-free flows $\rho(y) = 1$. Furthermore, Equation (13.6.25) implies that

$$\langle \xi, \mathcal{K}\xi \rangle := D|\xi|^2 + D \int_{\mathbb{T}^d} |\nabla_y \chi_\xi(y)|^2 \, dy, \tag{13.6.29}$$

where $\chi_\xi = \chi \cdot \xi$. Clearly the equality $\langle \xi, \mathcal{K}\xi \rangle = D|\xi|^2$ for all ξ implies that $\chi_\xi = 0$ for all ξ implying that $\chi(y) \equiv 0$. By (13.3.1) this implies that $b \equiv 0$.

For the upper bound we take the inner product of the cell problem with an arbitrary vector $\xi \in \mathbb{R}^d$ to obtain

$$-\mathcal{L}_0 \chi_\xi = b \cdot \xi.$$

We multiply this equation with χ_ξ, integrate over $\mathbb{T}^d$, and use Lemma 13.14 and the Poincaré inequality to calculate

$$\begin{aligned} D\|\nabla_y \chi^\xi\|_{L^2}^2 &= (-\mathcal{L}_0 \chi_\xi, \chi_\xi) = (b \cdot \xi, \chi_\xi) \\ &\leqslant \|b \cdot \xi\|_{L^2} \|\chi_\xi\|_{L^2} \\ &\leqslant C_p \|b\|_{L^2} \|\nabla_y \chi_\xi\|_{L^2} |\xi|, \end{aligned}$$

where C_p is the Poincaré constant on $\mathbb{T}^d$. From this estimate we deduce that

$$\|\nabla_y \chi_\xi\|_{L^2} \leqslant \frac{\sqrt{C}}{D} |\xi|$$

with $C = \left(C_p \|b\|_{L^2}\right)^2$. The result follows from (13.6.29). □

Shear Flow in 2D

In this section we study an example of a divergence-free flow for which the cell problem can be solved in closed form, that of a *shear flow*. The structure of a shear velocity field is such that the cell problem becomes an ordinary differential equation.

Let $y = (y_1, y_2)^T$. We consider the problem of homogenization for (13.2.4a) in two dimensions for the following velocity field:

$$b(y) = (0, b_2(y_1))^T, \tag{13.6.30}$$

where $b_2(y_1)$ is a smooth, 1-periodic function with mean zero. Notice that the velocity field (13.6.30) is incompressible:

$$\nabla \cdot b(y) = \frac{\partial b_1}{\partial y_1} + \frac{\partial b_2}{\partial y_2} = \frac{\partial b_2(y_1)}{\partial y_2} = 0.$$

The two components of the cell problem satisfy

$$-D\Delta_y \chi_1(y) - b_2(y_1) \frac{\partial \chi_1(y)}{\partial y_2} = 0, \tag{13.6.31a}$$

$$-D\Delta_y \chi_2(y) - b_2(y_1)\frac{\partial \chi_2(y)}{\partial y_2} = b_2(y_1), \tag{13.6.31b}$$

as well as periodicity and the normalization condition that χ integrates to zero over the unit cell $\mathcal{Y}$.

If we multiply the first Equation (13.6.31a) by $\chi_1(y)$ and integrate by parts over $\mathbb{T}^d$, then we deduce that

$$\int_{\mathbb{T}^d} |\nabla_y \chi_1(y)|^2 \, dy = 0.$$

Hence $\chi_1(y) = 0$, since we impose the normalization $\langle \chi(y) \rangle = 0$ with $\langle \cdot \rangle := \int_{\mathbb{T}^d} \cdot dy$. On the other hand, since the right-hand side of (13.6.31b) depends only on y_1, it is reasonable to assume that the solution $\chi_2(y)$ is independent of y_2; we seek a solution of this form and then, provided that we can find such a solution, uniqueness of solutions to the cell problem implies that it is the only solution. Equation (13.6.31b) becomes:

$$-D\frac{d^2\chi_2(y_1)}{dy_1^2} = b_2(y_1). \tag{13.6.32}$$

If ψ is a periodic solution to

$$-\frac{d^2\psi(y_1)}{dy_1^2} = b_2(y_1), \tag{13.6.33}$$

then ψ is independent of D and $\chi_2 = \psi/D$.

By (13.6.24) the effective diffusivity $\mathcal{K}$ is the following 2×2 matrix:

$$\begin{aligned}\mathcal{K} &= \begin{pmatrix} D + \int_{\mathbb{T}^2} (b_1\chi_1)\, dy & \frac{1}{2}\int_{\mathbb{T}^2} (b_2\chi_1 + b_1\chi_2)\, dy \\ \frac{1}{2}\int_{\mathbb{T}^2} (b_2\chi_1 + b_1\chi_2)\, dy & D + \int_{\mathbb{T}^2} (b_2\chi_2)\, dy \end{pmatrix} \\ &= \begin{pmatrix} D & 0 \\ 0 & \mathcal{K}_{22} \end{pmatrix},\end{aligned}$$

where we have used the fact that $b_1 = \chi_1 = 0$. Using the fact that b_2, χ_2 depend only on y_1 we obtain

$$\begin{aligned}\mathcal{K}_{22} &:= D + \int_0^1 b_2\chi_2 dy_1 \\ &= D + \int_0^1 -D\frac{d^2\chi_2}{dy_1^2}\chi_2 dy_1 \\ &= D + D\int_0^1 \left|\frac{d\chi_2}{dy_1}\right|^2 dy_1 \\ &= D + \frac{1}{D}\int_0^1 \left|\frac{d\psi}{dy_1}\right|^2 dy_1.\end{aligned}$$

Notice the remarkable fact that, since ψ is independent of D, the formula shows that the effective diffusion coefficient scales as D^{-1} as the original molecular diffusion coefficient D tends to zero. This demonstrates that the upper bound in

Theorem 13.19 is sharp. The intuition behind this scaling is that, for small D, the equation is approximately a transport equation in the direction x_2. The direction of transport is slowly modulated, leading to overall diffusive behavior, but on long time scales the predominant effect is transport. This enhances the diffusivity.

It is possible to express ψ as an integral operator acting on b_2 and to show that

$$\mathcal{K}_{22} = D + \frac{1}{D}\|b_2\|^2_{H^{-1}_{per}(0,1)}. \tag{13.6.34}$$

See Exercise 10.

Example 13.20. Consider the case

$$b_2(y_1) = \sin(2\pi y_1). \tag{13.6.35}$$

We use formula (13.6.34) and Exercise 10 to obtain

$$\mathcal{K}_{22} = D + \frac{1}{8\pi^2 D}. \tag{13.6.36}$$

In Figure 13.2 we plot the effective diffusivity given by (13.6.36) as a function of the molecular diffusivity D, on a log scale. We observe that $\mathcal{K}$ diverges like $1/D$ in the limit as $D \to 0$. □

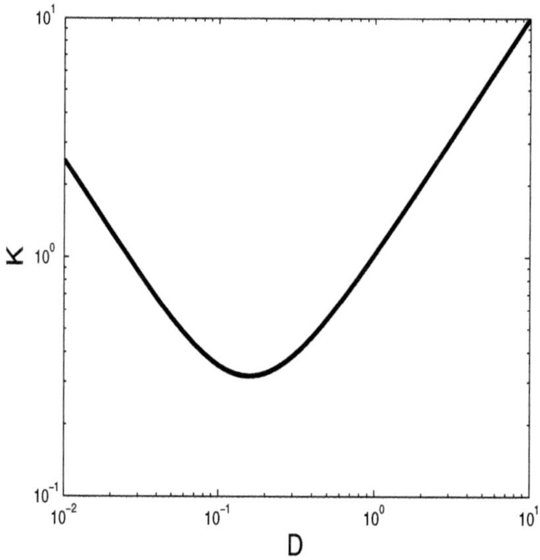

Fig. 13.2. Log-log plot of the effective diffusivity versus molecular diffusivity for the sine shear flow (13.6.35).

13.7 The Connection to SDEs

Equation (13.2.1) is the backward Kolmogorov equation associated with the SDE

$$\frac{dx}{dt} = b(x) + \sqrt{2D}\,\frac{dW}{dt}, \tag{13.7.1}$$

where W denotes standard Brownian motion on $\mathbb{R}^d$. Unsurprisingly, then, the homogenization results derived in this chapter have implications for the behavior of solutions to this SDE. To see this we first apply the rescaling used to derive (13.2.4) from (13.2.1) to the SDE (13.7.1). That is, we relabel according to

$$x \to x/\varepsilon, \quad t \to t/\varepsilon^2,$$

giving the SDE

$$\frac{dx}{dt} = \frac{1}{\varepsilon} b\Big(\frac{x}{\varepsilon}\Big) + \sqrt{2D}\frac{dW}{dt}. \tag{13.7.2}$$

(Recall Remark 6.3 regarding the behavior of white noise under time rescaling.)

If we introduce the variable $y = x/\varepsilon$, then we can write this SDE in the form

$$\frac{dx}{dt} = \frac{1}{\varepsilon} b(y) + \sqrt{2D}\frac{dW}{dt},$$
$$\frac{dy}{dt} = \frac{1}{\varepsilon^2} b(y) + \frac{1}{\varepsilon}\sqrt{2D}\frac{dW}{dt}.$$

Here we view x as being an element of $\mathbb{R}^d$ while y is on the torus $\mathbb{T}^d$. This is very similar to the form (11.2.1), which we analyzed in Chapter 11. The only difference is that the noises appearing in the x and y equations are *correlated* (in fact $U = V = W$). This has the effect of changing the operator $\mathcal{L}_1$ in that chapter, so that the results derived there do not apply directly. They can, however, be readily extended to the study of correlated noise; see Chapter 11, Exercises 5 and 1. Notice that the centering condition (13.2.8) is precisely the condition (11.2.5) since ρ is the stationary solution of the same Fokker–Planck equation.

The calculations in this chapter show how the backward Kolmogorov equation for the coupled SDE in (x, y) can be approximated by a diffusion equation in the x variable alone. Indeed, the diffusion equation is the backward Kolmogorov equation for pure Brownian motion. Interpreted in terms of the SDE we obtain the following result.

Result 13.21. *Assume that the centering condition* (13.2.8) *holds. For* $\varepsilon \ll 1$ *and* $t = \mathcal{O}(1)$, *x solving the SDE* (13.7.2) *can be approximated by X solving*

$$\frac{dX}{dt} = \sqrt{(\mathcal{K} + \mathcal{K}^T)}\frac{dW}{dt}$$

where the matrix $\mathcal{K}$ is given by (13.3.2).

If the centering condition is not satisfied then the appropriate rescaling of (13.7.1) is an advective one, leading to Equations (14.6.1) considered in the next chapter.

13.8 Discussion and Bibliography

The problem of homogenization for second-order parabolic PDEs and its connection to the study of the long-time asymptotics of solutions of SDEs is studied in [33, ch. 3]. References to the earlier literature can be found there. See also [238]. SDEs of the form (13.7.1), whose drift is the gradient of a periodic scalar function, describe Brownian motion in periodic potentials. This a very important problem in many applications, for example, in solid-state physics and biology; see [271, ch. 11; 267] and the references therein. Multiscale techniques were applied to this problem in [257]. Periodic homogenization for gradient flows is also discussed in [238, 256, 323, 118]. Formula (13.6.13) for the effective diffusivity of a Brownian particle moving in a one-dimensional periodic potential was derived in [191] without any appeal to homogenization theory; see also [138, sec. VII]. Brownian motion in a two-scale periodic potential in one dimension is studied in [342]. The multidimensional problem is analyzed in [258].

On the other hand, the SDE (13.7.1) with divergence-free drift occurs naturally in the modeling of diffusion processes in fluids. Homogenization for periodic, incompressible flows is a part of the theory of *turbulent diffusion* [200, 99]; see also [221, 100, 101]. In this context an interesting question concerns the dependence of the effective diffusivity on the molecular diffusion D. It turns out that the small D-asymptotics of the effective diffusivity depend sensitively on the streamline topology of the fluid velocity field $b(y)$; see [63, 294, 295, 62, 140, 20, 22, 21]. Interesting experimental results concerning the dependence of the effective diffusivity on D or, rather, on the *Peclet number* Pe are reported in [293, 292]; rescaling enables these results to be interpreted in terms of molecular diffusivity. Homogenization for compressible flows with applications to atmospheric transport phenomena is studied in [223].

It is possible to derive a homogenized equation even when the centering condition (13.2.8) is not satisfied. In this case it is necessary to use a frame co-moving with the *mean flow*

$$\overline{b} = \int_{\mathbb{T}^d} b(y)\rho(y)\,dy. \tag{13.8.1}$$

Then it is possible to derive a homogenized equation of the form (13.3.3) for the rescaled field

$$u^{\varepsilon}(x,t) = u\left(\frac{x}{\varepsilon} - \frac{\overline{b}t}{\varepsilon^2}, \frac{t}{\varepsilon^2}\right).$$

The effective diffusivity is given by the formula

$$\mathcal{K} = DI + 2D\int_{\mathbb{T}^d} \nabla_y \chi(y)^T \rho(y)\,dy + \int_{\mathbb{T}^d} \left(b(y) - \overline{b}\right) \otimes \chi(y)\rho(y)\,dy. \tag{13.8.2}$$

The cell problem (13.3.1) is also modified:

$$-\mathcal{L}_0\chi = b - \overline{b}. \tag{13.8.3}$$

See Exercise 5 in Chapter 14.

The mean flow $\overline{b}$ can have a dramatic effect in the small D-asymptotics of the effective diffusivity for periodic divergence-free flows; in particular, the scaling of $\mathcal{K}$ with D for $D \ll 1$ depends on whether the mean flow is a rational or irrational vector. See [201, 222, 35, 295, 175]. A similar discontinuous dependence of the effective diffusivity on the wavelengths of the inhomogeneities was observed for gradient flows in [126].

It is proved in Section 13.6.1 that for gradient flows the diffusion is always depleted. In fact, much sharper results can be obtained: the effective diffusivity is "exponentially" smaller than D, for D sufficiently small. That is, there exist positive constants c_1 and c_2 such that

$$\langle \xi, \mathcal{K}\xi \rangle = c_1 e^{-c_2/D}, \quad D \ll 1.$$

See [54] and the references therein. On the other hand, the effective diffusion coefficient can become arbitrarily large, when compared to the molecular diffusivity, when a constant external force is added to the gradient drift, see [268, 282].

The fact that the effective diffusivity along the direction of the shear is inversely proportional to the molecular diffusivity, formula (13.6.34), was discovered in [313], without any appeal to homogenization theory. This phenomenon is often referred to as *Taylor dispersion*; see also [11]. A similar result for time-dependent periodic shear flows was obtained in [340] through a direct calculation with the advection–diffusion equation.

To derive the expression (13.6.34) for the effective diffusion coefficient (from Exercise 10) it is necessary to use formal calculations with Fourier series. Of course, we have to prove that we can differentiate the Fourier series and that the Fourier series that we get for the second derivative of $\chi(y)$ makes sense. For various properties of Fourier series we refer the reader to [132, ch. 3].

We showed that the effective diffusion tensor is symmetric for gradient flows. The effective diffusivity, however, is not necessarily symmetric for general vector fields. Despite the fact that the antisymmetric part of the effective diffusivity does not affect the homogenized equation, it is of physical significance: it gives rise to a component of the flux that is perpendicular to the concentration gradient, [174]. Whether the effective diffusivity is symmetric or not depends on the symmetry properties of the underlying vector field $b(y)$.[5] This issue is studied for divergence-free flows in [174, 253]; in those references the dependence of the antisymmetric part of the effective diffusivity on the Peclet number is also studied.

In addition to the Eulerian definition of the effective diffusivity giving rise to the effective diffusion tensor $\mathcal{K}$ we can also define a Lagrangian effective diffusivity through the long-time average of the variance of the underlying stochastic process $x(t)$:

$$D^L_{eff} := \lim_{t\to\infty} \frac{\langle ((x(t) - \langle x(t)\rangle) \otimes (x(t) - \langle x(t)\rangle)\rangle}{2t}. \tag{13.8.4}$$

[5] For example, in the case of gradient flows the effective diffusivity is symmetric because of the reversibility (which, of course, is a symmetry property) of gradient flows.

Notice that D^L_{eff} is a symmetric tensor. It is straightforward to show that the the Lagrangian effective diffusivity (13.8.4) agrees with the *symmetric part* of the Eulerian effective diffusivity.

The method of multiple scales can also be used to study the problem of homogenization for parabolic PDEs with time-dependent coefficients that are periodic in both x and t; see, e.g., [118, 228, 323, 257, 42].

Monte Carlo methods for advection-diffusion and for transport PDEs are presented in [187]. Numerical methods for advection-diffusion equations with a multiscale structure are developed in [1].

13.9 Exercises

1. Derive a formula for $u_2(x, x/\varepsilon, t)$, the third term in the expansion (13.4.1).
2. Consider the problem of homogenization for

$$\frac{\partial u^\varepsilon}{\partial t} = -\frac{1}{\varepsilon}\nabla V\left(\frac{x}{\varepsilon}\right)\cdot\nabla u^\varepsilon + D\Delta u^\varepsilon$$

in one dimension with the (1-periodic) potential

$$V(y) = \begin{cases} y & : \quad y \in [0, \frac{1}{2}], \\ 1-y & : \quad y \in (\frac{1}{2}, 1]. \end{cases}$$

Calculate the effective diffusivity $\mathcal{K}$. Use Laplace's method to study the small D-asymptotics of $\mathcal{K}$.

3. Carry out the program from the previous exercise for the potential $V(y) = \sin(2\pi y)$. (Hint: Use Bessel functions.)
4. Calculate the effective diffusivity (13.3.2) for the 2-dimensional vector field $b(y) = (b_1(y_1), b_2(y_1))$.
5. Consider the problem of homogenization for the reaction–advection–diffusion equation

$$\frac{\partial u^\varepsilon}{\partial t} = \frac{1}{\varepsilon}b\left(\frac{x}{\varepsilon}\right)\cdot\nabla u^\varepsilon + \Delta u^\varepsilon + \frac{1}{\varepsilon}c\left(\frac{x}{\varepsilon}\right)u^\varepsilon, \qquad (13.9.1)$$

where the vector field $b(y)$ and the scalar function $c(y)$ are smooth and periodic. Use the method of multiple scales to homogenize the preceding PDE. In particular:

a) Derive the solvability condition.
b) Obtain the conditions that $b(y)$ and $c(y)$ should satisfy so that you can derive the homogenized equation.
c) Derive the homogenized equation, the cell problem(s), and the formula for the homogenized coefficients.
d) Suppose that the reaction term is nonlinear: the zeroth-order term in Equation (13.9.1) is replaced by

$$c\left(\frac{x}{\varepsilon}, u^\varepsilon\right),$$

where the function $c(y, u)$ is 1-periodic in y for every u. Can you homogenize Equation (13.9.1) in this case?

6. Consider the problem of homogenization for the PDE

$$\frac{\partial u^\varepsilon}{\partial t} = \left(b_1(x) + \frac{1}{\varepsilon} b_2\left(\frac{x}{\varepsilon}\right)\right) \cdot \nabla u^\varepsilon + \Delta u^\varepsilon, \tag{13.9.2}$$

where the vector field $b_2(y)$ is smooth and periodic and $b_1(x)$ is periodic. Use the method of multiple scales to homogenize the preceding PDE. In particular:
   a) Derive the solvability condition.
   b) Obtain the conditions that $b_2(y)$ should satisfy so that you can derive the homogenized equation.
   c) Show that the homogenized equation is

$$\frac{\partial u}{\partial t} = b \cdot \nabla u + \mathcal{K} : \nabla\nabla u \tag{13.9.3}$$

   and derive the cell problem(s) and the formulae for the homogenized coefficients b and $\mathcal{K}$.
7. Consider the problem of homogenization for the PDE (13.9.2) in the case where

$$b_1(x) = -\nabla V(x) \quad \text{and} \quad b_2(y) = -\nabla p(y),$$

where $p(y)$ is periodic.
   a) Show that in this case there exists a symmetric matrix $\widehat{\mathcal{K}}$ such that

$$\mathcal{K} = D\widehat{\mathcal{K}}, \quad B = -\widehat{\mathcal{K}}\nabla V.$$

   b) Let

$$\mathcal{L} := b \cdot \nabla + \mathcal{K} : \nabla\nabla u.$$

      1. Derive a formula for $\mathcal{L}^*$, the L^2-adjoint of $\mathcal{L}$.
      2. Show that the function

$$\rho(y) := \frac{1}{Z} e^{-V(y)/D}, \quad Z = \int_{\mathbb{T}^d} e^{-V(y)/D} \, dy$$

      solves the homogeneous adjoint equation

$$\mathcal{L}^* \rho = 0.$$

8. Consider the problem of homogenization for the following PDE

$$\frac{\partial u^\varepsilon}{\partial t} = b^\varepsilon \cdot \nabla u^\varepsilon + A^\varepsilon : \nabla_x \nabla_x u^\varepsilon$$

where $A^\varepsilon = A(x/\varepsilon)$, $b^\varepsilon = b(x/\varepsilon)$, the vector field $b(y)$ and the matrix $A(y)$ are smooth and periodic, and $A(y)$ is positive definite. Use the method of multiple scales to derive the homogenized equation. In particular:
   a) Derive the solvability condition.
   b) Obtain conditions on $b(y)$ that ensure the existence of a homogenized equation.

c) Derive the homogenized equation, the cell problem, and the formula for the homogenized coefficients.
d) Prove that the homogenized matrix is positive definite.

9. Consider the problem of homogenization for the following PDE

$$\frac{\partial u^\varepsilon}{\partial t} = \frac{1}{\varepsilon} b\left(\frac{x}{\varepsilon}, \frac{t}{\varepsilon^2}\right) \cdot \nabla u^\varepsilon + D \Delta u^\varepsilon,$$

where the vector field $b(y, \tau)$ is smooth, divergence-free, and 1-periodic in both y and τ. Use the method of multiple scales to derive the homogenized equation. In particular:
a) Derive the solvability condition.
b) Obtain conditions on $b(y, \tau)$ that ensure the existence of a homogenized equation.
c) Derive the homogenized equation, the cell problem, and the formula for the homogenized coefficients.
d) Prove that the homogenized matrix is positive definite.

10. The H^{-1}_{per} norm of a real-valued, periodic function with period 1 can be expressed in terms of Fourier series (see the discussion in Section 2.7) as follows:

$$\|f\|^2_{H^{-1}_{per}(0,1)} = \frac{1}{2\pi^2} \sum_{k=1}^{\infty} \frac{|f_k|^2}{|k|^2}.$$

Use this definition and the Fourier series representation of the solution ψ of the problem (13.6.31b), to establish formula (13.6.34) from the expression for $\mathcal{K}_{22}$ in terms of ψ.

11. Consider Exercise 9 in dimension $d = 2$ and with the velocity field

$$b(y_1, y_2, \tau) = (0, b_2(y_1, \tau)).$$

Derive a formula for the effective diffusivity $\mathcal{K}$. How does $\mathcal{K}$ depend on D?

12. Repeat the calculations of Section 13.6.2 and Exercise 11 for the 2D velocity fields

$$b(y_1, y_2) = (V, b(y_1))$$

and

$$b(y_1, y_2, \tau) = (V, b(y_1, \tau)),$$

respectively, where $V \in \mathbb{R}$. (Hint: You need to use Equations (13.8.2) and (13.8.3) and their generalizations for time-dependent velocity fields.)

13. Let $b(y)$ be a smooth, real-valued 1-periodic, mean-zero function and let $\{b_k\}_{k=-\infty}^{+\infty}$ be its Fourier coefficients. Prove that $b_0 = 0$ and that $b_{-k} = \overline{b}_k$.

14

Averaging for Linear Transport and Parabolic PDEs

14.1 Introduction

In this chapter we investigate the long-time behavior of solutions to the linear transport (or advection) equation and to the parabolic (advection–diffusion) equation from the previous chapter, when the centering condition is not satisfied. The techniques we employ are sometimes referred to as homogenization techniques in the literature. However, in terms of the classification in Section 1.3 the methods are actually averaging methods. We use this terminology.

In Sections 14.2 and 14.3 we set up the problem of interest and then state the approximation result. Section 14.4 contains the derivation of the averaged equation, when the starting point is a parabolic equation. Section 14.5 is devoted to the case where the averaging is for a pure transport equation; the issues here are more subtle (no Fredholm alternative for the leading-order linear operator), and this is why we devote a separate section to it. In Section 14.6 we make the connection to averaging for ODEs and SDEs. Section 14.7 contains bibliographical remarks.

14.2 Full Equations

We study the long-time behavior of solutions to the linear parabolic equation corresponding to advection–diffusion in a steady periodic velocity field b:

$$\frac{\partial u}{\partial t} = b \cdot \nabla u + D\Delta u \quad \text{for } (x,t) \in \mathbb{R}^d \times \mathbb{R}^+, \tag{14.2.1a}$$

$$u = u_{in} \quad \text{for } (x,t) \in \mathbb{R}^d \times \{0\}. \tag{14.2.1b}$$

This is the parabolic Equation (13.2.1) and, in the case $D = 0$, it reduces to a linear transport equation. As in Chapter 13 we study the case where

$$u_{in}(x) = g(\varepsilon x),$$

and rescale the equation in both space and time in order to understand the behavior of solutions to Equation (14.2.1) at length and time scales that are long when compared to those of the velocity field $b(x)$. In this setting, the small parameter in the problem is the ratio between the characteristic length (time) scale of the velocity field – its period – and the largest length (time) scale of the problem – the one at which we are looking for an averaged description. In contrast to the analysis of the advection–diffusion equation in the previous chapter, we rescale time and space in the same fashion, namely

$$x \to \varepsilon^{-1} x, \quad t \to \varepsilon^{-1} t. \tag{14.2.2}$$

In the parabolic case $D > 0$ this is because we do not assume that the centering condition (13.2.8) holds; thus the advective effects do not average out. The transformation (14.2.2) is also natural in the case $D = 0$ since the transport PDE (14.2.1a) is then of first order in both space and time.

The initial value problem that we wish to investigate becomes:

$$\frac{\partial u^\varepsilon}{\partial t} = b^\varepsilon \cdot \nabla u^\varepsilon + \varepsilon D \Delta u^\varepsilon \quad \text{for } (x,t) \in \mathbb{R}^d \times \mathbb{R}^+, \tag{14.2.3a}$$

$$u^\varepsilon = f \quad \text{for } x \in \mathbb{R}^d \times \{0\}. \tag{14.2.3b}$$

Here $b^\varepsilon(x) = b(x/\varepsilon)$, as in the previous chapter.

As in the previous chapter we define the operator

$$\mathcal{L}_0 = b(y) \cdot \nabla_y + D\Delta_y \tag{14.2.4}$$

with periodic boundary conditions. Note that constants in y are in the null space of this operator; furthermore, for $D > 0$ the null space is one-dimensional and comprises only constants. The L^2-adjoint of $\mathcal{L}_0$ is $\mathcal{L}_0^*$, also with periodic boundary conditions. Recall from Chapter 13 that, for $D > 0$, the **invariant distribution** $\rho(y)$ is the unique stationary solution of the adjoint equation

$$\mathcal{L}_0^* \rho = 0, \quad \int_{\mathbb{T}^d} \rho(y)\, dy = 1, \tag{14.2.5}$$

equipped with periodic boundary conditions. For $D > 0$ both operators $\mathcal{L}_0$ and $\mathcal{L}^*$ satisfy a Fredholm alternative.

14.3 Simplified Equations

In this and the following section we simply assume that the operator $\mathcal{L}_0$ has a one-dimensional null space comprising constants and that the same holds for its adjoint $\mathcal{L}_0^*$, with null space spanned by ρ. This follows from the Fredholm alternative for $D > 0$. For $D = 0$ it requires some form of ergodicity of the underlying ODE for which $\mathcal{L}_0$ is the generator. We discuss this ergodicity issue in Sections 14.5 and 14.6.

Under the stated assumptions on $\mathcal{L}_0$ we have the following result.

Result 14.1. *Let b be a smooth periodic vector field. Assume that the operator $\mathcal{L}_0$ defined in* (14.2.4) *satisfies*

$$\mathcal{N}(\mathcal{L}_0) = span(1), \quad \mathcal{N}(\mathcal{L}_0^*) = span(\rho).$$

Then, for $\varepsilon \ll 1$ and times t of $\mathcal{O}(1)$, the solution $u^\varepsilon(x,t)$ of (13.2.4) *is approximated by $u(x,t)$, the solution of the averaged equation:*

$$\frac{\partial u}{\partial t} - \overline{b} \cdot \nabla_x u = 0, \quad \overline{b} := \int_{\mathbb{T}^d} \rho(y) b(y)\, dy,$$

together with the same initial condition as for u^ε.

The calculations leading to this approximation result take the rescaled parabolic Equation (14.2.3a) as a starting point and recover a transport equation by means of averaging. Naively it might appear that the diffusion term in (14.2.3a) simply disappears from the averaging calculation, since it is multiplied by ε. *This viewpoint is wrong*: the diffusion coefficient plays an essential role. In general the form of the stationary distribution, against which b is averaged, depends crucially on $D > 0$, through ρ.[1]

Note that the centering condition (13.2.8) simply states that $\overline{b} = 0$. This is why a different scaling of space and time is used in Chapter 13 from that used here: specifically a longer time scale is used there, in order to see nonnegligible effects.

14.4 Derivation

We use the method of multiple scales as introduced in the two preceding chapters. We introduce the auxiliary variable $y = x/\varepsilon$. Let $\phi = \phi(x, x/\varepsilon)$ be scalar-valued. The chain rule gives

$$\nabla\phi = \nabla_x \phi + \frac{1}{\varepsilon}\nabla_y \phi \quad \text{and} \quad \Delta\phi = \Delta_x \phi + \frac{2}{\varepsilon}\nabla_x \cdot \nabla_y \phi + \frac{1}{\varepsilon^2}\Delta_y \phi.$$

The partial differential operator that appears on the right-hand side of Equation (14.2.3) has the form

$$\mathcal{L} = \frac{1}{\varepsilon}\mathcal{L}_0 + \mathcal{L}_1 + \varepsilon \mathcal{L}_2,$$

where

$$\begin{aligned}
\mathcal{L}_0 &= b(y) \cdot \nabla_y + D\Delta_y, \\
\mathcal{L}_1 &= b(y) \cdot \nabla_x + 2D\nabla_x \cdot \nabla_y, \\
\mathcal{L}_2 &= D\Delta_x.
\end{aligned}$$

In terms of x and y Equation (14.2.3) becomes

[1] An exception is the case of divergence-free flows: the invariant measure ρ is the Lebesgue measure on the unit torus for all $D > 0$; see Lemma 13.15.

$$\frac{\partial u^\varepsilon}{\partial t} = \left(\frac{1}{\varepsilon}\mathcal{L}_0 + \mathcal{L}_1 + \varepsilon\mathcal{L}_2\right) u^\varepsilon.$$

We look for a solution in the form of a two-scale expansion:

$$u^\varepsilon(x,t) = u_0\left(x, \frac{x}{\varepsilon}, t\right) + \varepsilon u_1\left(x, \frac{x}{\varepsilon}, t\right) + \dots . \tag{14.4.1}$$

We assume that all terms in the expansion $u_j(x,y,t)$, $j = 0, 1, \dots$ are 1-periodic in y and treat x and $y := x/\varepsilon$ as independent variables. [2] We substitute (14.4.1) into Equation (14.2.3a), use the assumed independence of x and y, and collect equal powers of ε to obtain the following set of equations:

$$\mathcal{O}(1/\varepsilon) \quad -\mathcal{L}_0 u_0 = 0, \tag{14.4.2a}$$

$$\mathcal{O}(1) \quad -\mathcal{L}_0 u_1 = \mathcal{L}_1 u_0 - \frac{\partial u_0}{\partial t}, \tag{14.4.2b}$$

where $u_j(x,y)$ is 1-periodic in y.

We can now complete the averaging procedure. From the first equation in (14.4.2) and our assumptions on $\mathcal{L}_0$, we deduce that the first term in the expansion is independent of the oscillations that are expressed through the auxiliary variable y:

$$u_0 = u(x,t).$$

We use this to compute:

$$\mathcal{L}_1 u_0 = \frac{\partial u(x,t)}{\partial t} - b(y) \cdot \nabla_x u(x,t).$$

Since ρ is in the null space of $\mathcal{L}_0^*$, the second equation in (14.4.2) implies that

$$0 = \frac{\partial u(x,t)}{\partial t} - \left(\int_{\mathbb{T}^d} \rho(y) b(y)\, dy\right) \cdot \nabla_x u(x,t). \tag{14.4.3}$$

We have thus obtained the desired averaged equation:

$$\frac{\partial u(x,t)}{\partial t} - \overline{b} \cdot \nabla_x u(x,t) = 0, \quad \overline{b} := \int_{\mathbb{T}^d} \rho(y) b(y)\, dy,$$

together with the same initial conditions as for u^ε.

14.5 Transport Equations: $D = 0$

We have indicated that, in general, the averaged transport equation depends subtly on the diffusion coefficient D through the invariant distribution ρ against which b

[2] As in the elliptic and parabolic homogenization procedures in the previous two chapters, this is where we exploit scale separation: **we treat x and y as independent variables**. Justifying this assumption as $\varepsilon \to 0$ is one of the main issues in the rigorous theory of averaging; see Chapter 21.

is averaged. Existence and uniqueness of the stationary distribution ρ are automatic when $D > 0$ but require some form of ergodicity, which will depend on the properties of b, when $D = 0$. It is therefore a nontrivial question to ask when, and to what extent, the preceding averaging calculations extend to the case $D = 0$. The calculations rely on the null spaces of $\mathcal{L}_0$ and $\mathcal{L}_0^*$ being one-dimensional, something ensured by the Fredholm alternative in the case $D > 0$. We discuss the analogs of these results in the case $D = 0$.

Let

$$\mathcal{L}_0 = b(y) \cdot \nabla_y \tag{14.5.1}$$

with domain $C^1_{per}(\mathbb{T}^d)$. We can extend this operator to $\mathcal{D}(\mathcal{L}_0) \subset L^\infty_{per}(\mathbb{T}^d)$ as in (4.3.8). We assume for the moment that there are no nontrivial functions in the null space $\mathcal{N}$ of $\mathcal{L}_0$:

$$\mathcal{N}(\mathcal{L}_0) = \{\text{constants in } y\} \tag{14.5.2}$$

viewing the operator as acting on $\mathcal{D}(\mathcal{L}_0)$. From Chapter 4 we know that this is essentially an ergodicity assumption on the ODE with vector field b; see Theorem 4.13(iii). In relation to this, the idea that $\mathcal{L}_0^*$ is nonempty with domain viewed as being $C^1_{per}(\mathbb{T}^d)$ implies the existence of an invariant measure that is absolutely continuous with respect to the Lebesgue measure; see Theorem 4.12(iii). Thus ergodicity with respect to absolutely continuous invariant measure μ provides us with the necessary tools to carry out the formal perturbation expansions of this chapter in the case $D = 0$. In particular, in the ergodic case, (14.4.2a) implies that u_0 is independent of y and also that a necessary condition for a solution u_1 of (14.4.2b) to exist is Equation (14.4.3).

Note that if b is divergence-free (the velocity field is incompressible), then $\mathcal{L}$ is skew-symmetric (Lemma 13.14) and so we deduce from (14.5.2) that

$$\mathcal{N}(\mathcal{L}_0^*) = \{\text{constants in } y\}. \tag{14.5.3}$$

However, in the general ergodic case, ρ will not be a constant function.

14.5.1 The One-Dimensional Case

Consider the rescaled transport Equation (14.2.3a) in one dimension:

$$\frac{\partial u^\varepsilon}{\partial t} - b^\varepsilon \frac{\partial u^\varepsilon}{\partial x} = 0 \quad \text{for } (x,t) \in \mathbb{R} \times \mathbb{R}^+, \tag{14.5.4a}$$

$$u = g \quad \text{for } (x,t) \in \mathbb{R} \times \{0\}, \tag{14.5.4b}$$

where $g = g(x)$ is independent of the oscillations.[3] We assume that $b(y)$ is a strictly positive, smooth, 1-periodic function. The stationary Liouville equation

$$\mathcal{L}_0^* \rho = 0, \quad \rho > 0, \text{ 1-periodic}, \tag{14.5.5}$$

together with the normalization condition

[3] This is not necessary; see Exercise 3 from Chapter 21.

$$\int_0^1 \rho(y)\, dy = 1,$$

has unique normalized solution the probability density

$$\rho(y) = \frac{C}{b(y)}, \quad C = \langle b(y)^{-1} \rangle^{-1}; \tag{14.5.6}$$

here we have used the notation $\langle \cdot \rangle$ to denote averaging over $[0, 1]$, as in Chapter 12. Positivity of b is key to this existence and uniqueness result, and also to the ergodicity of the underlying flow. These issues are discussed in Example 4.14.

We obtain the averaged equation

$$\frac{\partial u}{\partial t} - \bar{b}\frac{\partial u}{\partial x} = 0, \tag{14.5.7}$$

with the same initial conditions as in (14.5.4b) and with

$$\bar{b} = \langle b(y)^{-1} \rangle^{-1}.$$

Notice that, in contrast to the ergodic divergence-free case presented in the next subsection, it is the harmonic average of the velocity field that appears in the averaged Equation (14.5.7) rather than the standard average. (Note that the harmonic average also arises in the one-dimensional elliptic case; see Subsection 12.6.1.)

14.5.2 Divergence-Free Velocity Fields

If b is divergence-free (the velocity field is incompressible), then $\mathcal{L}$ given by (14.5.1) is skew-symmetric (Lemma 13.14) and so we deduce that, if (14.5.2) holds, then

$$\mathcal{N}(\mathcal{L}_0^*) = \{\text{constants in } y\}. \tag{14.5.8}$$

(See Example 4.15.) Unfortunately, even for divergence-free fields, the ergodic hypothesis leading to (14.5.2) is often not satisfied. Consider an equation in the form (14.4.2a):

$$\mathcal{L}_0 u := b(y) \cdot \nabla_y u = 0 \tag{14.5.9}$$

with periodic boundary conditions. Although $u \equiv 1$ solves this equation, it is rare that this solution is unique: the null space of the operator $\mathcal{L}_0$ contains, in general, nontrivial functions of y. As an example, consider the smooth, 1-periodic, divergence-free field

$$b(y) = (\sin(2\pi y_2), \sin(2\pi y_1)).$$

It is easy to check that the function

$$u(y) = \cos(2\pi y_1) - \cos(2\pi y_2)$$

solves Equation (14.5.9). Consequently, the null space of $\mathcal{L}_0$ depends on the velocity field $b(y)$, and it does not consist, in general, merely of constants in y. This implies that we cannot carry out the averaging procedure using the method of multiple scales.

It is natural to ask whether there is a way of deciding whether a given divergence-free velocity field on $\mathbb{T}^d$ is ergodic or not. This is indeed possible in two dimensions. A result along these lines is the following.

Theorem 14.2. *Let $b(y) : \mathbb{T}^2 \to \mathbb{R}^2$ be a smooth divergence-free velocity field satisfying*

$$b_1(y) \neq 0 \quad \forall y \in \mathbb{T}^2$$

so that it has no stagnation points. Let $\overline{b}_i$, $i = 1, 2$, denote the average of the ith component of the velocity field over $\mathbb{T}^2$ and define the rotation number as

$$\gamma = \frac{\overline{b}_1}{\overline{b}_2}.$$

Then there exists a smooth change of variables $y \mapsto z$ under which the ODEs

$$\frac{dy_1}{dt} = b_1(y), \quad \frac{dy_2}{dt} = b_2(y) \tag{14.5.10}$$

transform into

$$\frac{dz_1}{dt} = g(z), \quad \frac{dz_2}{dt} = \gamma g(z) \tag{14.5.11}$$

where $g(z)$ is a nonvanishing smooth scalar function. Assume furthermore that γ is irrational. Then the null space of the generator $\mathcal{L}_0$ is one-dimensional in $D(\mathcal{L}_0)$.

Proof. The first part of the theorem can be proved by constructing explicitly the transformation that maps (14.5.10) into (14.5.11):[4]

$$z_1 = \frac{1}{\overline{b}_2} \int_0^{y_1} b_2(\xi, 0)\, d\xi, \quad z_2 = \frac{1}{\overline{b}_1} \int_0^{y_2} b_1(y_1, \xi)\, d\xi.$$

The second part of the theorem can be proved using Fourier analysis; see Exercise 7.
□

Thus, under the conditions of this theorem, Theorem 4.13 holds and the formal perturbation expansions of this chapter may be applied.

14.6 The Connection to ODEs and SDEs

We consider first the case where $D = 0$. Recall from Chapter 4 that the solution of (14.2.3) is given by

$$u(x,t) = g(\varphi^t(x)),$$

where $\varphi^t(x)$ solves the ODE

$$\frac{d}{dt}\varphi^t(x) = b\Big(\frac{\varphi^t(x)}{\varepsilon}\Big),$$
$$\varphi^t(x) = x.$$

Result 14.1 shows that, when the ergodicity assumption holds so that $\mathcal{L}_0$ has one-dimensional null space, this equation is well approximated by

[4] Under the additional assumption that $b_2 \neq 0$, which can be removed later.

$$\overline{\varphi}^t(x) = \overline{b}t + x,$$

the solution of

$$\frac{d}{dt}\overline{\varphi}^t(x) = \overline{b},$$
$$\overline{\varphi}^t(x) = x.$$

Here

$$\overline{b} = \int_{\mathbb{T}^d} \rho(y) b(y) dy.$$

In dimension $d = 1$ we have

$$\overline{b} = \langle b(y)^{-1} \rangle$$

by (14.5.6).

Another way to see this result is as follows. Let $x = \varphi^t(x_0)$ and $y = x/\varepsilon$. Then

$$\frac{dx}{dt} = b(y),$$
$$\frac{dy}{dt} = \frac{1}{\varepsilon} b(y).$$

Under the ergodic hypothesis the fast process y has invariant measure ρ on the torus $\mathbb{T}^d$. Thus the averaging Result 10.1 gives that x is well approximated by the solution of the equation

$$\frac{dX}{dt} = \overline{b}.$$

This is precisely the approximation just derived.

Example 14.3. In the one-dimensional case it is possible to derive the averaged Equation (14.5.7) using the method of characteristics. To see this, consider the equation

$$\frac{dx}{dt} = b\Big(\frac{x}{\varepsilon}\Big)$$

in one dimension and under the same assumptions as before. If we set $y = x/\varepsilon$ then it is straightforward to show that

$$\frac{dy}{dt} = \frac{1}{\varepsilon} b(y),$$

so that, if we define T by

$$T = \int_0^1 \frac{1}{b(z)} dz = \frac{1}{\overline{b}},$$

then

$$y(n\varepsilon T) = \frac{x(0)}{\varepsilon} + n.$$

Hence

$$x(n\varepsilon T) = x(0) + n\varepsilon.$$

It follows from continuity that $x(t)$ converges to $X(t)$ where

$$X(t) = x(0) + \frac{t}{T}.$$

This limiting function $X(t)$ satisfies the homogenized equation

$$\frac{dX}{dt} = \frac{1}{T} = \overline{b}. \quad \square$$

If $D > 0$, then Equation (14.2.3) is the backward Kolmogorov equation for the SDE

$$\frac{dx}{dt} = b\Big(\frac{x}{\varepsilon}\Big) + \sqrt{2\varepsilon D}\frac{dW}{dt}.$$

Another way to interpret the averaging result is thus as follows. Let $y = x/\varepsilon$ to obtain

$$\frac{dx}{dt} = b(y) + \sqrt{2\varepsilon D}\frac{dW}{dt},$$

$$\frac{dy}{dt} = \frac{1}{\varepsilon}b(y) + \sqrt{\frac{2D}{\varepsilon}}\frac{dW}{dt}. \tag{14.6.1}$$

Under the ergodic hypothesis the fast process y has invariant measure ρ on the torus $\mathbb{T}^d$. Thus a generalization of the averaging Result 10.1 gives that x is well approximated by the ODE

$$\frac{dx}{dt} = \overline{b}.$$

14.7 Discussion and Bibliography

The perturbation expansion used here is analogous to that used in the method of averaging, for Markov chains, ODE, and SDE, in Chapters 9 and 10. The problem of averaging for linear transport equations has been studied by many authors. See for example [80, 147, 312, 51]. Averaging for SDEs is studied in detail in [111].

When $D = 0$ the method of multiple scales enables us to obtain the averaged linear transport Equation (14.2.3a) only in the case where the velocity field is ergodic. The method of multiple scales breaks down when the velocity field is not ergodic, since in this case we do not have a solvability condition that would enable us to average. In fact, when the velocity field is not ergodic, the $\varepsilon \to 0$ limit becomes much more complicated and the limiting process cannot be expressed through a simple PDE. In order to study the problem for general velocity fields, not necessarily ergodicones, it is possible to use the method of two-scale convergence. This will be done in Chapter 21.

Theorem 14.2 is proved in [312], where the result is proved for two-dimensional flows that have a smooth invariant density, not only divergence-free flows (for which the invariant density is simply 1). It is not the sharpest result that one can prove, but the assumption $b_1 \neq 0$ leads to a particularly simple proof. The proof of the analogous theorem under the assumptions that there are no stagnation points can be found in [290]. A similar theorem holds for velocity fields with an invariant measure other than the Lebesgue measure on $\mathbb{T}^2$; see [312].

The example studied in Section 14.5.1 can be found in [81, 312]. Monte Carlo methods for advection–diffusion and for transport PDEs are presented in [187].

14.8 Exercises

1. How does the dynamics of the ODE studied in Section 14.5.1 change if b is allowed to change sign?
2. Consider the equation
$$\frac{dx}{dt} = a\left(\frac{x}{\varepsilon}\right)b\left(\frac{t}{\varepsilon^\alpha}\right)$$
in one dimension and under the assumption that a (resp. b) is smooth, 1-periodic, and $\inf_x a > 0$ (resp. $\inf_y b > 0$). Find the averaged equations.
3. Study the problem of averaging for (14.2.3) with a smooth periodic (shear) velocity field $b : \mathbb{T}^2 \mapsto \mathbb{R}^2$ of the form
$$b(y) = (0, b_2(y_1))^T.$$
4. Study the problem of averaging for (14.2.3) with a velocity field $b : \mathbb{T}^2 \mapsto \mathbb{R}^2$ of the form
$$b(y) = \widehat{b}(y)(0, \gamma)^T,$$
where $\widehat{b}(y)$ is a smooth, 1-periodic scalar function, and $\gamma \in \mathbb{R}$.
5. Consider Equation (13.2.4) in the case where the centering condition (13.2.8) does not hold. Show that it is possible to derive a homogenized equation of the form (13.3.3) for the rescaled field
$$u^\varepsilon(x,t) = u\left(\frac{x}{\varepsilon} - \frac{\overline{b}t}{\varepsilon^2}, \frac{t}{\varepsilon^2}\right)$$
where u solves (14.2.1) and $\overline{b}$ is given in Result 14.1. Show that the cell problem becomes
$$-\mathcal{L}_0\chi = b - \overline{b}. \tag{14.8.1}$$
and that the effective diffusivity is given by the formula
$$\mathcal{K} = DI + 2D\int_{\mathbb{T}^d} \nabla_y\chi(y)^T\rho(y)\,dy + \int_{\mathbb{T}^d}\Big(\big(b(y) - \overline{b}\big)\otimes\chi(y)\rho(y)\Big)\,dy. \tag{14.8.2}$$
6. Study the problem of homogenization for the ODE
$$\frac{dx}{dt} = -\nabla V\left(\frac{x}{\varepsilon}\right) + F$$
where $V(y)$ is a smooth periodic function and F is a constant vector.
7. Complete the details in the proof of Theorem 14.2.

Part III

Theory

15

Invariant Manifolds for ODEs: The Convergence Theorem

15.1 Introduction

In this chapter we describe a rigorous theory substantiating the perturbation expansions for invariant manifolds in Chapter 8. The approximation theorem is stated in Section 15.2. We prove this straightforward estimate, which is valid on any finite time interval, in Section 15.3, using basic techniques from the theory of ODEs. In the Discussion and Bibliography section, Section 15.4, we point to more sophisticated techniques that establish results on arbitrary time intervals.

15.2 The Theorem

We study the equations

$$\frac{dx}{dt} = f(x, y), \tag{15.2.1a}$$

$$\frac{dy}{dt} = \frac{1}{\varepsilon} g(x, y), \tag{15.2.1b}$$

for $\varepsilon \ll 1$ and $x \in \mathbb{R}^l$, $y \in \mathbb{R}^{d-l}$. We assume that the dynamics for y with x frozen has a unique exponentially attracting fixed point, uniformly in x. Specifically we assume that there exists $\eta : \mathbb{R}^l \to \mathbb{R}^{d-l}$ and $\alpha > 0$ such that, for all $x \in \mathbb{R}^l$ and all $y_1, y_2 \in \mathbb{R}^{d-l}$,

$$g(x, \eta(x)) = 0, \tag{15.2.2a}$$

$$\langle g(x, y_1) - g(x, y_2), y_1 - y_2 \rangle \leqslant -\alpha |y_1 - y_2|^2. \tag{15.2.2b}$$

We will refer to (15.2.2b) as the *contractivity condition.*

The dynamics with x frozen at ξ satisfies

$$\frac{d}{dt}\varphi_\xi^t(y) = g(\xi, \varphi_\xi^t(y)), \qquad \varphi_\xi^0(y) = y. \tag{15.2.3}$$

We assume that the vector field $(f^T, g^T)^T$ is Lipschitz on $\mathbb{R}^d$ so that both (15.2.1) and (15.2.3) have globally defined solutions.

Our assumptions on η and g imply the following exponential convergence of $\varphi_\xi^t(y)$ to its globally attracting fixed point $\eta(\xi)$.

Lemma 15.1. *Under assumption* (15.2.2) *we have that, for all* $y \in \mathbb{R}^{d-l}$,

$$|\varphi_\xi^t(y) - \eta(\xi)| \leqslant e^{-\alpha t}|y - \eta(\xi)|.$$

Proof. Since $\eta(\xi)$ is time-independent we have

$$\frac{d}{dt}\varphi_\xi^t(y) = g(\xi, \varphi_\xi^t(y)),$$
$$\frac{d}{dt}\eta(\xi) = g(\xi, \eta(\xi)) = 0.$$

Hence

$$\frac{1}{2}\frac{d}{dt}|\varphi_\xi^t(y) - \eta(\xi)|^2 = \langle g(\xi, \varphi_\xi^t(y)) - g(\xi, \eta(\xi)), \varphi_\xi^t(y) - \eta(\xi)\rangle$$
$$\leqslant -\alpha|\varphi_\xi^t(y) - \eta(\xi)|^2.$$

The result follows from the differential form of the Gronwall Lemma 4.4. □

In essence we wish to prove a result like this for $y(t)$ instead of $\varphi_\xi^t(y)$ when x is no longer frozen at ξ but rather evolves on its own time scale, which is slow compared to that of y. We make the following standing assumptions. These simplify the analysis and make the ideas of the proof of the basic result clearer; however, they can all be weakened in various different ways. The assumptions are the existence of a constant $C > 0$ such that:[1]

$$|f(x,y)| \leqslant C \quad \forall (x,y) \in \mathbb{R}^d, \tag{15.2.4a}$$
$$|\nabla_x f(x,y)| \leqslant C \quad \forall (x,y) \in \mathbb{R}^d, \tag{15.2.4b}$$
$$|\nabla_y f(x,y)| \leqslant C \quad \forall (x,y) \in \mathbb{R}^d, \tag{15.2.4c}$$
$$|\eta(x)| \leqslant C \quad \forall x \in \mathbb{R}^l, \tag{15.2.4d}$$
$$|\nabla\eta(x)| \leqslant C \quad \forall x \in \mathbb{R}^l. \tag{15.2.4e}$$

With these assumptions we prove that x is close to X solving

$$\frac{dX}{dt} = f(X, \eta(X)), \quad X(0) = x(0). \tag{15.2.5}$$

Theorem 15.2. *Assume that assumptions* (15.2.2) *and* (15.2.4) *hold. Then there are constants* $K, c > 0$ *such that* $x(t)$ *solving* (15.2.1) *and* $X(t)$ *solving* (15.2.5) *satisfy*

$$|x(t) - X(t)|^2 \leqslant ce^{Kt}\big(\varepsilon|y(0) - \eta(x(0))|^2 + \varepsilon^2\big).$$

---

[1] In the first and fourth items in this list the norms are standard vector norms; in the second, third, and fifth they are matrix (operator) norms. All are Euclidean. In fact the properties assumed of η follow from suitable assumptions on g, because of the uniformity of the contraction assumption (15.2.2b) in x.

Note that the error is of size $\sqrt{\varepsilon}$ for times of $\mathcal{O}(1)$. However, this can be reduced to size ε if the initial deviation $y(0)-\eta(x(0))$ is of size $\sqrt{\varepsilon}$. Note, furthermore, that the distance between $x(t)$ and $X(t)$ grows exponentially fast in time. Hence, for given ε, Theorem 15.2 is useful only on intervals of order $\ln \varepsilon^{-1}$. Stronger results are possible and are discussed in Section 15.4.

15.3 The Proof

Define $z(t)$ by $y(t)=\eta(x(t))+z(t)$. Then

$$\begin{aligned}\frac{dz}{dt} &= \frac{dy}{dt} - \nabla\eta(x)\frac{dx}{dt}\\ &= \frac{1}{\varepsilon}g(x,\eta(x)+z) - \nabla\eta(x)f(x,\eta(x)+z)\\ &= \frac{1}{\varepsilon}\Big(g(x,\eta(x)+z) - g(x,\eta(x))\Big) - \nabla\eta(x)f(x,\eta(x)+z),\end{aligned}$$

using assumption (15.2.2a). Now assumption (15.2.2b) implies that

$$\langle g(x,\eta(x)+z) - g(x,\eta(x)), z\rangle \leqslant -\alpha|z|^2.$$

Furthermore, assumptions (15.2.4) imply that

$$\langle z, \nabla\eta(x)f(x,\eta(x)+z)\rangle \leqslant C^2|z|.$$

Hence, by using the Cauchy-Schwarz inequality with $\delta^2=\varepsilon/\alpha$,

$$\begin{aligned}\frac{1}{2}\frac{d}{dt}|z|^2 &= \langle z, \frac{dz}{dt}\rangle\\ &= \frac{1}{\varepsilon}\langle z, g(x,\eta(x)+z) - g(x,\eta(x))\rangle - \langle z, \nabla\eta(x)f(x,\eta(x)+z)\rangle\\ &\leqslant -\frac{\alpha}{\varepsilon}|z|^2 + C^2|z|\\ &\leqslant -\frac{\alpha}{\varepsilon}|z|^2 + \frac{\delta^2}{2}C^4 + \frac{1}{2}\frac{|z|^2}{\delta^2}\\ &\leqslant -\frac{\alpha}{2\varepsilon}|z|^2 + \frac{\varepsilon}{2\alpha}C^4.\end{aligned}$$

By Gronwall's lemma,

$$|z(t)|^2 \leqslant e^{-\frac{\alpha}{\varepsilon}t}|z(0)|^2 + \left(1 - e^{-\frac{\alpha}{\varepsilon}t}\right)\frac{\varepsilon^2C^4}{\alpha^2}. \tag{15.3.1}$$

Now,

$$\frac{dX}{dt} = f(X,\eta(X))$$

and

$$\frac{dx}{dt} = f(x, \eta(x) + z).$$

Subtracting gives

$$\begin{aligned}\frac{d}{dt}(x - X) &= f(x, \eta(x) + z) - f(X, \eta(X)) \\ &= f(x, \eta(x) + z) - f(x, \eta(X)) + f(x, \eta(X)) - f(X, \eta(X)).\end{aligned}$$

We use now the Cauchy-Schwarz inequality, together with the Lipschitz continuity of $f(\cdot, \cdot)$ and $\eta(\cdot)$ to estimate:

$$\begin{aligned}\frac{1}{2}\frac{d}{dt}|x - X|^2 &= \langle x - X, \frac{d}{dt}(x - X)\rangle \\ &= \langle x - X, f(x, \eta(x) + z) - f(x, \eta(X))\rangle \\ &\quad + \langle x - X, f(x, \eta(X)) - f(X, \eta(X))\rangle \\ &\leqslant C|\eta(x) - \eta(X) + z||x - X| + C|x - X|^2 \\ &\leqslant (C^2 + C)|x - X|^2 + C|z||x - X|.\end{aligned}$$

It follows that

$$\frac{d}{dt}|x - X|^2 \leqslant (3C^2 + 2C)|x - X|^2 + |z|^2.$$

Letting $K = 3C^2 + 2C$ and using the bound (15.3.1) for $|z(t)|^2$, Gronwall's lemma, and the fact that $x(0) = X(0)$ gives the desired result. □

15.4 Discussion and Bibliography

Theorem 15.2 shows that $x(t)$ from the full Equations (15.2.1) remains close to $X(t)$ solving the reduced Equations (15.2.5) over time scales that are of the order $\ln(\varepsilon^{-1})$. On longer time scales the individual solutions can diverge, because of the exponential separation of trajectories which may be present in any dynamical system. Notice, however, that estimate (15.3.1) shows that

$$\limsup_{t \to \infty} |y(t) - \eta(x(t))| \leqslant \frac{C^4}{\alpha^2}\varepsilon^2,$$

suggesting that $y(t)$ is approximately slaved to $x(t)$, via $y = \eta(x)$, for arbitrary time intervals. There are results concerning the approximation of $x(t)$ over arbitrarily long times, using this slaving idea. Specifically, these long-time approximation results are built on making rigorous the construction of an invariant manifold as described in Chapter 8. The idea is as follows. Consider the equations

$$\begin{aligned}\frac{dx}{dt} &= f(x, \eta(x) + z), \\ \frac{dz}{dt} &= \frac{1}{\varepsilon}g(x, \eta(x) + z) - \nabla\eta(x) f(x, \eta(x) + z).\end{aligned}$$

Notice that $y = \eta(x) + z$. Using the fact that $\nabla_y g(x, \eta(x))$ is negative definite it is possible to prove the existence of an invariant manifold for z with the form

$$z = \varepsilon\eta_1(x; \varepsilon)$$

with η_1 bounded uniformly in $\varepsilon \to 0$. To be precise, the equations for x and z started with initial conditions $z(0) = \varepsilon\eta_1(x(0); \varepsilon)$ will satisfy $z(t) = \varepsilon\eta_1(x(t); \varepsilon)$ for all positive times. Furthermore the manifold is attracting so that

$$|z(t) - \varepsilon\eta_1(x(t); \varepsilon)| \to 0$$

as $t \to \infty$. Thus we have an attractive invariant manifold for y with the form

$$y = \eta(x) + \varepsilon\eta_1(x; \varepsilon).$$

The existence and uniqueness of invariant manifolds can be proved by a variety of techniques, predominantly the Lyapunov-Perron approach ([136, 314]) and the Hadamard graph transform ([330]). Important work in this area is due to Fenichel [104, 105], who set up a rather general construction of normally hyperbolic invariant manifolds. The book [57] has a clear introduction to the Lyapunov-Perron approach to proving existence of invariant manifolds. The graph transform approach, for both maps and flows, is overviewed in [305].

15.5 Exercises

1. Show that, under the assumptions on g stated at the beginning of the chapter, $\varphi_\xi^t : \mathbb{R}^{d-l} \to \mathbb{R}^{d-l}$ (15.2.3) is a contraction mapping for any $t > 0$. What is its fixed point?
2. Prove a result similar to Theorem 15.2 but removing the assumption that η and f are globally bounded; use instead linear growth assumptions on η and f.
3. Consider the equations

$$\begin{aligned} \frac{dx}{dt} &= Ax + \varepsilon f_0(x, y), \\ \frac{dy}{dt} &= -\tfrac{1}{\varepsilon} By + g_0(x, y), \end{aligned} \tag{15.5.1}$$

for $0 < \varepsilon \ll 1$ and $x \in \mathbb{R}^l$, $y \in \mathbb{R}^{d-l}$. Assume that B is symmetric positive definite. Let $z(t, x_0; \eta)$ solve the equation

$$\frac{dz}{dt} = Az + \varepsilon f_0(z, \eta(z)), \quad z(0, x_0; \eta) = x_0.$$

Given $\eta : \mathbb{R}^l \to \mathbb{R}^{d-l}$, define $T\eta : \mathbb{R}^l \to \mathbb{R}^{d-l}$ by

$$(T\eta)(x_0) = \int_{-\infty}^0 e^{-Bs/\varepsilon} g_0(z(s, x_0; \eta), \eta(z(s, x_0; \eta)))ds.$$

Show that if η is a fixed point of T then $y = \eta(x)$ is an invariant manifold for the Equations (15.5.1). (This is known as the Lyapunov-Perron approach to the construction of invariant manifolds.)

4. Assume that f_0, g_0 and all derivatives are uniformly bounded. Prove that T from the previous question has a fixed point. (Hint: Apply a contraction mapping argument in a space of Lipschitz graphs η that are sufficiently small and have sufficiently small Lipschitz constant.)

16

Averaging for Markov Chains: The Convergence Theorem

16.1 Introduction

In this chapter we prove a result concerning averaging for Markov chains. The techniques presented lead to a weak-convergence-type result showing that expectations under the original chain and under the averaged chain are close. The technique is to work with the backward equation for the two Markov chains. The fundamental estimate (5.2.2) plays a central role. This estimate is analogous to the maximum principle for parabolic PDEs. In Chapter 20 we use techniques similar to those in this chapter, based on the maximum principle, to prove a homogenization result for parabolic PDEs. The main theorem is stated in Section 16.2 and is proved in Section 16.3. The chapter concludes with bibliographical notes in Section 16.4.

16.2 The Theorem

The setup is as in Chapter 9. To make the proofs transparent we concentrate on the finite state space case. Let $\mathcal{I}_x$, $\mathcal{I}_y \subseteq \{1, 2, \cdots\}$ be finite sets. Consider a continuous-time Markov chain

$$z(t) = \begin{pmatrix} x(t) \\ y(t) \end{pmatrix}$$

on $\mathcal{I}_x \times \mathcal{I}_y$. We assume that this Markov chain is parameterized by ε and that the backward equation has the form

$$\frac{dv}{dt} = \frac{1}{\varepsilon} Q_0 v + Q_1 v \tag{16.2.1}$$

where Q_0, Q_1 are given by (9.2.4). Let $X(t)$ be a Markov chain on $\mathcal{I}_x$ with backward equation

$$\frac{dv_0}{dt} = \overline{Q}_1 v_0, \tag{16.2.2}$$

and with $\bar{Q}_1$ given by (9.3.1). We are interested in approximating $x(t)$ by $X(t)$. Note that the formula for the approximate process implied by the Kolmogorov equation is exactly that derived in Chapter 9 by means of formal asymptotics.

Note that $x(t)$ is not itself Markovian; only the pair $(x(t), y(t))$ is. Thus we are approximating a non-Markovian stochastic process by a Markovian one. To be precise, we prove that, at any fixed time, the statistics of $x(t)$ are close to those of $X(t)$. That is, we prove weak convergence of $x(t)$ to $X(t)$ at any fixed time t.

Theorem 16.1. *For any* $t > 0$, $x(t) \Rightarrow X(t), \quad$ *as* $\quad \varepsilon \to 0.$

16.3 The Proof

Let v_0 be defined as in (16.2.2). We then have

$$v_0 \in \mathcal{N}(Q_0), \quad \frac{dv_0}{dt} - Q_1 v_0 \perp \mathcal{N}(Q_0^T),$$

by construction. Hence there exists v_1 so that

$$\begin{aligned} Q_0 v_0 &= 0, \\ Q_0 v_1 &= \frac{dv_0}{dt} - Q_1 v_0. \end{aligned}$$

We can make v_1 unique by insisting that it is orthogonal to the null space of Q_0^*, although this particular choice is not necessary. We simply ask that a solution is chosen that is bounded, with bounded derivative in time.

For any such v_1 and for v_0 given by (16.2.2), define

$$r = v - v_0 - \varepsilon v_1.$$

Substituting $v = v_0 + \varepsilon v_1 + r$ into (16.2.1) and using the properties of v_0, v, we obtain

$$\frac{dv_0}{dt} + \varepsilon \frac{dv_1}{dt} + \frac{dr}{dt} = \frac{1}{\varepsilon} Q_0 v_0 + Q_0 v_1 + \frac{1}{\varepsilon} Q_0 r + Q_1 v_0 + \varepsilon Q_1 v_1 + Q_1 r.$$

Hence

$$\begin{aligned} \frac{dr}{dt} &= \Big(\frac{1}{\varepsilon} Q_0 + Q_1\Big) r + \varepsilon q, \\ q &= Q_1 v_1 - \frac{dv_1}{dt}. \end{aligned}$$

Now $Q := \frac{1}{\varepsilon} Q_0 + Q_1$ is the generator of a Markov chain. Hence using $|\cdot|_\infty$ to denote the supremum norm on vectors over the finite set $\mathcal{I}_x \times \mathcal{I}_y$, as well as the induced operator norm, we have

$$|e^{Qt}|_\infty = 1. \tag{16.3.1}$$

This follows from (5.2.2) because e^{Qt} is a stochastic matrix. It is a noteworthy estimate because, although Q depends on ε in a singular fashion, this estimate does not.

By the variation-of-constants formula we have

$$r(t) = e^{Qt}r(0) + \varepsilon \int_0^t e^{Q(t-s)}q(s)ds, \tag{16.3.2}$$

viewing $r(t), q(t)$ as vectors on $\mathcal{I}_x \times \mathcal{I}_y$, for each t. We assume that $v(i,k,0) = \phi(i), v_0(i,0) = \phi(i)$ and then

$$v(i,k,t) = \mathbb{E}\Big(\phi(x(t))|x(0) = i, y(0) = k\Big), \tag{16.3.3a}$$

$$v_0(i,t) = \mathbb{E}\Big(\phi(X(t))|X(0) = i\Big). \tag{16.3.3b}$$

Weak convergence of $x(t)$ to $X(t)$, for fixed t, is proved if

$$v(i,k,t) \to v_0(i,t), \quad \text{as} \quad \varepsilon \to 0,$$

for any $\phi : \mathcal{I}_x \to \mathbb{R}$. Equations (16.3.3) imply that $r(0) = -\varepsilon v_1(0)$. Using (16.3.1) we have, from (16.3.2),

$$\begin{aligned}
|r(t)|_\infty &\leqslant \varepsilon |e^{Qt}|_\infty |v_1(0)|_\infty + \varepsilon \int_0^t |e^{Q(t-s)}|_\infty |q(s)|_\infty ds \\
&\leqslant \varepsilon |v_1(0)|_\infty + \varepsilon \int_0^t |q(s)|_\infty ds \\
&\leqslant \varepsilon \left(|v_1(0)|_\infty + t \sup_{0 \leqslant s \leqslant t} |q(s)|_\infty \right).
\end{aligned}$$

Hence, for any fixed $t > 0$, $r(t) \to 0$ as $\varepsilon \to 0$, and it follows that $v \to v_0$ as $\varepsilon \to 0$.
□

Remark 16.2. The proof actually gives a convergence rate because it shows that, for $0 \leqslant t \leqslant T$,

$$|v(t) - v_0(t)|_\infty \leqslant C(T)\varepsilon.$$

16.4 Discussion and Bibliography

The result we prove only shows that the random variable $x(t)$ converges to $X(t)$ for each fixed t. It is also possible to prove the more interesting result that the weak convergence result actually occurs on path space, in the Skorokhod topology (defined in [120]); see [291]. It is also possible to derive diffusion limits of Markov chains; see [182, 183, 184].

16.5 Exercises

1. Consider the two-state continuous-time Markov chain y with generator

$$L = \frac{1}{\varepsilon}\begin{pmatrix} -a & a \\ b & -b \end{pmatrix}$$

and state space $\mathcal{I} = \{-1, +1\}$. Consider the ODE on $\mathbb{T}^d$ given by

$$\frac{dx}{dt} = f(x, y)$$

where $f : \mathbb{T}^d \times \mathcal{I} \to \mathbb{R}^d$.

a. Write down the generator for this process.

b. Using multiscale analysis, show that the averaged SDE is

$$\frac{dX}{dt} = F(X)$$

where

$$F(x) = \lambda f(x, +1) + (1 - \lambda) f(x, -1)$$

and $\lambda \in (0, 1)$ should be specified.

2. Prove the assertions made in the preceding exercise: show that $x(t) \Rightarrow X(t)$ for each fixed $t \geqslant 0$.
3. Conjecture what the behavior of x is in the case where $F(x) = 0$ and large times are considered.
4. Justify the two-state Markov chain approximation derived in Chapter 9, Exercise 6.

17

Averaging for SDEs: The Convergence Theorem

17.1 Introduction

The goal of this chapter is to develop a rigorous theory based on the averaging principle for SDEs that we developed in Chapter 10. We introduce the use of the Itô formula, applied to the solution of a carefully chosen elliptic PDE (the **Poisson equation**), in order to estimate integrals containing rapidly varying components (see Remark 6.17). The existence of this estimation technique makes the treatment of the SDE case considerably easier than the pure ODE case. Pointers to the literature on the pure ODE case are given in the Discussion and Bibliography section. We also consider only the case where the averaged equation is an ODE. This allows the proof of straightforward *strong* convergence results. If the averaged equation is itself an SDE then *weak* convergence results are more natural. These weak convergence methods are illustrated, in the context of homogenization for SDEs, in the next chapter.

Section 17.2 contains the theorem statement, and Section 17.3 the proof. Generalizations of the results presented in this chapter and bibliographical comments are given in Section 17.4.

17.2 The Theorem

To allow for a simplified, but prototypical, theorem and statement, we study the following problem on the torus $\mathbb{T}^d$:

$$\frac{dx}{dt} = f(x,y), \quad x(0) = x_0, \tag{17.2.1a}$$

$$\frac{dy}{dt} = \frac{1}{\varepsilon} g(x,y) + \frac{1}{\sqrt{\varepsilon}} \beta(x,y) \frac{dV}{dt}, \quad y(0) = y_0. \tag{17.2.1b}$$

Here V is a standard Brownian motion on $\mathbb{R}^{d-l}$, $f : \mathbb{T}^l \times \mathbb{T}^{d-l} \mapsto \mathbb{R}^l$, $g : \mathbb{T}^l \times \mathbb{T}^{d-l} \mapsto \mathbb{R}^{d-l}$, $\beta : \mathbb{T}^l \times \mathbb{T}^{d-l} \mapsto \mathbb{R}^{(d-l)\times(d-l)}$ are smooth periodic functions. Let $B(x,y) = \beta(x,y)\beta(x,y)^T$. Assume that, writing $z = (x^T, y^T)^T$,

$$\exists \bar{\beta} > 0 : \langle \xi, B(x,y)\xi \rangle \geqslant \bar{\beta}|\xi|^2 \quad \forall \xi \in \mathbb{R}^{d-l}, \quad z \in \mathbb{T}^d. \tag{17.2.2}$$

We will also assume that the initial conditions are deterministic. This is not necessary for the averaging result but simplifies the presentation.

Recall that, under assumption (17.2.2), the process found by freezing $x = \xi$ in (17.2.1b), is ergodic (Theorem 6.16). Thus we expect that an effective equation for the evolution of x can be found by averaging f over the invariant measure of this ergodic process. We now make this idea precise.

The process φ_ξ^t given by (10.5.1) is ergodic and has a smooth invariant density $\rho^\infty(y;\xi)$. This invariant density spans the null space of $\mathcal{L}_0^*$, found as the adjoint of $\mathcal{L}_0$ given by

$$\mathcal{L}_0 = g(x,y) \cdot \nabla_y + \frac{1}{2} B(x,y) : \nabla_y \nabla_y, \tag{17.2.3}$$

evaluated at $x = \xi$; both $\mathcal{L}_0$ and $\mathcal{L}_0^*$ have periodic boundary conditions. The averaged equations are then

$$\frac{dX}{dt} = F(X), \tag{17.2.4a}$$

$$F(\xi) = \int_{\mathbb{T}^{d-l}} f(\xi,y)\rho^\infty(y;\xi)\, dy. \tag{17.2.4b}$$

Note that $F : \mathbb{T}^l \mapsto \mathbb{R}^l$ is periodic by construction. The resulting formulae are exactly those given in Chapter 10, specialized to the particular drift and diffusion coefficients in (17.2.1).

Theorem 17.1. *Let $p > 1$ and let $X(0) = x_0$. Then the function $x(t)$ solving (17.2.1) converges to $X(t)$ solving (17.2.4) in $L^p(\Omega, C([0,T],\mathbb{T}^l))$: for any $T > 0$, there is $C = C(T)$ such that*

$$\mathbb{E}\left(\sup_{0 \leqslant t \leqslant T} |x(t) - X(t)|^p \right) \leqslant C\varepsilon^{p/2}.$$

17.3 The Proof

Recall that $\mathcal{L}_0$ is the generator for $\varphi_x^t(y)$, with x viewed as a fixed parameter, given by (17.2.3). Thus $\mathcal{L}_0$ is a differential operator in y only; x appears as a parameter. Now let $\phi(x,y)$ solve the elliptic boundary value problem

$$\begin{aligned} \mathcal{L}_0 \phi(x,y) &= f(x,y) - F(x), \\ \int_{\mathbb{T}^{d-l}} \phi(x,y)\rho^\infty(y;x)dy &= 0, \\ \phi(x,\cdot) \quad & \text{is periodic on } \mathbb{T}^{d-l}. \end{aligned}$$

This is known as a Poisson equation. By construction

$$\int_{\mathbb{T}^{d-l}} \big(f(x,y) - F(x)\big)\rho^{\infty}(y;x)\,dy = 0$$

and ρ^{∞} spans $\mathcal{N}(\mathcal{L}_0^*)$. Hence, by the Fredholm alternative, or by Theorem 7.8, ϕ has a unique solution.

Lemma 17.2. *The functions $f, \phi, \nabla_x\phi, \nabla_y\phi$, and β are smooth and bounded.*

Proof. The properties of f, β follow from the fact that they are defined on the torus and have derivatives of all orders by assumption. Since the invariant density ρ^{∞} is the solution of an elliptic eigenvalue problem on the torus, it is also smooth and periodic. Hence F is smooth and periodic. Consequently $f - F$ is smooth and periodic. Hence ϕ and all its derivatives are smooth and periodic. □

Proof of Theorem 17.1 Notice that the generator for (17.2.1) is

$$\mathcal{L} = \frac{1}{\varepsilon}\mathcal{L}_0 + \mathcal{L}_1.$$

Here $\mathcal{L}_0$ is given by (17.2.3) and

$$\mathcal{L}_1 = f(x,y)\cdot\nabla_x.$$

Now we apply the Itô formula (Lemma 6.5) to $\phi(x(t), y(t))$ to obtain the following informal expression, with precise interpretation found by integrating in time:

$$\frac{d\phi}{dt}(x,y) = \frac{1}{\varepsilon}(\mathcal{L}_0\phi)(x,y) + f(x,y)\cdot\nabla_x\phi(x,y) + \frac{1}{\sqrt{\varepsilon}}\nabla_y\phi(x,y)\beta(x,y)\frac{dV}{dt}.$$

Since $\mathcal{L}_0\phi = f - F$, we obtain

$$\begin{aligned}\frac{dx}{dt} &= F(x) + (\mathcal{L}_0\phi)(x,y)\\ &= F(x) + \varepsilon\frac{d\phi}{dt} - \varepsilon f(x,y)\cdot\nabla_x\phi(x,y)\\ &\qquad -\sqrt{\varepsilon}\nabla_y\phi(x,y)\beta(x,y)\frac{dV}{dt}\end{aligned} \tag{17.3.1}$$

(again a formal expression made rigorous by time integration). The functions f, ϕ, and $\nabla_x\phi$ are all smooth and bounded by Lemma 17.2. Hence there is a constant $C > 0$ so that

$$\theta(t) := \big(\phi(x(t), y(t)) - \phi(x(0), y(0))\big) - \int_0^t f(x(s), y(s))\cdot\nabla_x\phi(x(s), y(s))\,ds$$

satisfies, with probability one,

$$\sup_{0\leqslant t\leqslant T}|\theta(t)| \leqslant C.$$

Now consider the martingale term

$$M(t) := -\int_0^t \nabla_y\phi(x(s),y(s))\beta(x(s),y(s))\,dV(s).$$

Since $\nabla_y\phi, \beta$ are smooth and bounded, by Lemma 17.2, the Itô isometry gives

$$\begin{aligned}\mathbb{E}|\langle M\rangle_t|^2 &\leqslant C\int_0^t \mathbb{E}|\nabla_y\phi(x(s),y(s))\beta(x(s),y(s))|_{\mathrm{F}}^2 ds\\ &\leqslant Ct.\end{aligned}$$

Similarly (see (3.4.8)), for $p \geqslant 1$,

$$\mathbb{E}|\langle M\rangle_t|^{p/2} \leqslant C. \tag{17.3.2}$$

Now the rigorous interpretation of (17.3.1) is

$$x(t) = x(0) + \int_0^t F(x(s))ds + \varepsilon\theta(s) + \sqrt{\varepsilon}M(t).$$

Also, from (17.2.4),

$$X(t) = X(0) + \int_0^t F(X(s))ds.$$

Let $e(t) = x(t) - X(t)$ so that, using $e(0) = 0$,

$$e(t) = \int_0^t \big(F(x(s)) - F(X(s))\big)ds + \varepsilon\theta(t) + \sqrt{\varepsilon}M(t).$$

Since F is Lipschitz on $\mathbb{T}^l$ we obtain, for $t \in [0,T]$,

$$|e(t)| \leqslant \int_0^t L|e(s)|ds + \varepsilon C + \sqrt{\varepsilon}|M(t)|.$$

Hence, by (17.3.2) and the Burkholder-Davis-Gundy inequality, Theorem 3.22, we obtain[1]

$$\begin{aligned}\mathbb{E}\left(\sup_{0\leqslant t\leqslant T}|e(t)|^p\right) &\leqslant C\Big(\varepsilon^p + \varepsilon^{p/2}\mathbb{E}\left(\sup_{0\leqslant t\leqslant T}|M(t)|^p\right) + L^pT^{p-1}\int_0^T \mathbb{E}|e(s)|^p ds\Big)\\ &\leqslant C\Big(\varepsilon^p + \varepsilon^{p/2}\mathbb{E}\big(|\langle M\rangle_T|\big)^{p/2} + L^pT^{p-1}\int_0^T \mathbb{E}|e(s)|^p ds\Big)\\ &\leqslant C\Big(\varepsilon^{p/2} + \int_0^T \mathbb{E}\sup_{0\leqslant\tau\leqslant s}|e(\tau)|^p ds\Big).\end{aligned}$$

By the integrated version of the Gronwall inequality in Lemma 4.4 we deduce that

$$\mathbb{E}\left(\sup_{0\leqslant t\leqslant T}|e(t)|^p\right) \leqslant C\varepsilon^{p/2}$$

and the theorem is proved. □

[1] For C a constant, independent of ε, changing from occurrence to occurrence.

17.4 Discussion and Bibliography

Our convergence result proves *strong convergence*: we compare each path of the SDE for (x, y) with the approximating ODE for X. This strong convergence result is possible primarily because the limiting approximation is in this case deterministic. When the approximation is itself stochastic, as arises for (10.7.1), then it is more natural to study weak-convergence-type results (but see Exercise 1). For results of the latter type, see [94].

A key role is played in the proof presented here by the Poisson equation for ϕ and by application of the Itô formula to ϕ. Averaging theorems for SDEs are proved in [111, ch. 7]. Notice, however, that no systematic use of an appropriate Poisson equation is made in that book.

Averaging results for ODEs are proved by somewhat different techniques, because an Itô-formula-based methodology does not apply, essentially because the Poisson equation is no longer elliptic. Much of the original motivation for the study of averaging in the context of ODEs comes from averaging of perturbed integrable Hamiltonian systems, expressed in action-angle variables; see [135] for references.

17.5 Exercises

1. Consider Equation (10.7.1) in the case where $\alpha(x, y) \equiv 1$. Write down the averaged dynamics for X and modify the techniques of this chapter to prove strong convergence of x to X.
2. Consider Equation (10.7.1) in the case where $d = 2, l = 1$, and $\alpha(x, y) \equiv y^2$. If y is a scalar OU process (6.4.4), independent of x, write down the averaged dynamics for X and use the properties of the OU process to prove weak convergence.
3. Consider the fast/slow system

$$\dot{x} = f(x)\eta^2, \quad \dot{\eta} = -\frac{1}{\varepsilon}\eta + \frac{1}{\sqrt{\varepsilon}}\dot{W},$$

where $W(t)$ is a standard 1-dimensional Brownian motion. Prove a strong convergence theorem by applying the Itô formula to the function $f(x)\eta^2$. Make a connection with the approach based on the Poisson equation that we use in this chapter.

18

Homogenization for SDEs: The Convergence Theorem

18.1 Introduction

In this chapter we develop a rigorous theory based on the homogenization principle derived in Chapter 11. We consider a simple setting where the entire problem is posed on the torus, in order to elucidate the principle ideas. Furthermore, we work in the skew-product setting where the fast process is independent of the slow process. Finally we assume that the fluctuating term in the slow process depends only on the fast process. These assumptions allow for a simplified proof, which nonetheless contains the essence of the main ideas. As in the previous chapter, the use of an appropriate Poisson equation and of the Itô formula plays a central role (see Remark 6.17). Section 18.2 contains the theorem statement and Section 18.3 its proof. Generalizations of the convergence theorem proved in this chapter, together with bibliographical remarks, are presented in Section 18.4.

18.2 The Theorem

Consider the following system of SDEs on $\mathbb{T}^d$

$$\frac{dx}{dt} = \frac{1}{\varepsilon} f_0(y) + f_1(x, y) + \alpha \frac{dU}{dt}, \tag{18.2.1a}$$

$$\frac{dy}{dt} = \frac{1}{\varepsilon^2} g(y) + \frac{1}{\varepsilon} \beta(y) \frac{dV}{dt}, \tag{18.2.1b}$$

where U (resp. V) is a standard Brownian motion on $\mathbb{R}^l$ (resp. $\mathbb{R}^{d-l}$) and $\alpha \in \mathbb{R}^{l \times l}$, a constant matrix. The two Brownian motions are assumed to be independent. The functions $f_0 : \mathbb{T}^{d-l} \mapsto \mathbb{R}^l$, $f_1 : \mathbb{T}^l \times \mathbb{T}^{d-l} \mapsto \mathbb{R}^l$, $g : \mathbb{T}^{d-l} \mapsto \mathbb{R}^{d-l}$, $\beta : \mathbb{T}^{d-l} \mapsto \mathbb{R}^{(d-l)\times(d-l)}$ are smooth and periodic. Let $B(y) = \beta(y)\beta(y)^T$. We also assume that

$$\exists \bar{\beta} > 0 : \langle \xi, B(y)\xi \rangle \geqslant \bar{\beta}|\xi|^2 \quad \forall \xi \in \mathbb{R}^{d-l}, \quad y \in \mathbb{T}^{d-l}.$$

Recall that, under this assumption, the process y in (18.2.1b) is ergodic (see Result 6.16.) Thus an effective equation for the evolution of x can be found by averaging f_1

over the invariant measure of this ergodic process and by examining the fluctuations induced by f_0. The aim of this chapter is to make these ideas rigorous.

The generator of the process y is given by

$$\mathcal{L}_0 = g(y) \cdot \nabla_y + \frac{1}{2} B(y) : \nabla_y \nabla_y,$$

equipped with periodic boundary conditions. This is simply the formula (17.2.3) for $\mathcal{L}_0$ in the case where the fast process is x-independent. Then $\varphi^t(y)$ is given by (10.5.1) with the ξ-dependence removed. The process $\varphi^t(y)$ is ergodic and has a smooth invariant density $\rho^\infty(y)$; this function spans the null space of $\mathcal{L}_0^*$. Note that $\varphi^t(\cdot)$ and $\rho^\infty(\cdot)$ are independent of ξ here, because g, β depend only on y. We assume that

$$\int_{\mathbb{T}^{d-l}} f_0(y) \rho^\infty(y) dy = 0.$$

Under this assumption the calculations in Chapter 11 apply and we may homogenize the SDE (18.2.1) to obtain

$$\frac{dX}{dt} = F(X) + A \frac{dW}{dt}, \tag{18.2.2}$$

where

$$F(\xi) = \int_{\mathbb{T}^{d-l}} f_1(\xi, y) \rho^\infty(y) dy$$

and

$$AA^T = \alpha\alpha^T + \int_{\mathbb{T}^{d-l}} \big(f_0(y) \otimes \Phi(y) + \Phi(y) \otimes f_0(y)\big) \rho^\infty(y) dy.$$

Recall that $\Phi(y)$ solves the cell problem

$$\begin{aligned} \mathcal{L}_0 \Phi(y) &= -f_0(y), \\ \int_{\mathbb{T}^{d-l}} \Phi(y) \rho^\infty(y) &= 0, \\ \Phi(y) &\quad \text{periodic on } \mathbb{T}^{d-l}. \end{aligned} \tag{18.2.3}$$

This has a unique solution, by the Fredholm alternative, as applied to elliptic PDEs with periodic boundary conditions; see Section 7.2.3. The resulting formulae for A and F are exactly those given in Chapter 11, specialized to the particular drift and diffusion coefficients studied in this chapter.

Notice that, for $\xi \in \mathbb{R}^l$, $\phi = \Phi \cdot \xi$, the proof of Theorem 11.3 shows that AA^T is positive definite and that

$$\langle \xi, AA^T \xi^T \rangle = |\alpha^T \xi|^2 + \int_{\mathbb{T}^{d-l}} |\beta^T(y) \nabla \phi(y)|^2 \rho^\infty(y) dy.$$

Hence the SDE for X is well-defined. Remark 11.4 shows that

$$AA^T = \alpha\alpha^T + \alpha_2\alpha_2^T$$
$$\alpha_2\alpha_2^T = \int_{\mathbb{T}^{d-l}} \rho^\infty(y)\Big(\nabla\Phi(y)\beta(y) \otimes \nabla\Phi(y)\beta(y)\Big)dy. \tag{18.2.4}$$

This form for the effective diffusion matrix arises naturally in the Poisson-equation-based proof that we use in this chapter.

Theorem 18.1. *Let $x(t)$ solve (18.2.1), let $X(t)$ solve (18.2.2) with $X(0) = x(0)$. Then, for any $T > 0$, $x \Rightarrow X$ in $C([0,T],\mathbb{T}^l)$.*

18.3 The Proof

The structure of the proof is as follows. We show that the process $x(t)$ satisfies

$$x(t) = x(0) + \int_0^t F(x(s))\,ds + \alpha U(t) + M_2(t) + \eta(t). \tag{18.3.1}$$

The desired limit process satisfies

$$X(t) = x(0) + \int_0^t F(X(s))\,ds + \alpha U(t) + \alpha_2 W(t)$$

where the diffusion coefficient α_2 is defined in (18.2.4). In the equation for x, the term representing deviation from the desired limit process satisfies

$$(\alpha U, \eta, M_2) \Rightarrow (\alpha U, 0, \alpha_2 W), \quad \text{in} \quad C([0,T],\mathbb{R}^{3l}). \tag{18.3.2}$$

We then use the following lemma to deduce the desired convergence, using the fact that weak convergence of probability measures is preserved under continuous mappings.

Lemma 18.2. *Let $w \in C([0,T],\mathbb{R}^r)$,[1] let $F \in C^1(\mathbb{T}^l;\mathbb{T}^l)$, and let $D \in \mathbb{R}^{l\times r}$. There is a unique $u \in C([0,T],\mathbb{T}^l)$ satisfying the integral equation*

$$u(t) = u(0) + \int_0^t F(u(s))ds + Dw(t)$$

and the mapping $w \mapsto u$ is a continuous mapping from $C([0,T],\mathbb{R}^r)$ into $C([0,T],\mathbb{T}^l)$.

Proof. Existence and uniqueness follow by a standard contraction mapping argument, based on the iteration

$$u^{(n+1)}(t) = u(0) + \int_0^t F(u^n(s))ds + Dw(t),$$

[1] Here $w(t)$ is an arbitrary continuous path, not necessarily a Brownian path.

and using the fact that F is a globally Lipschitz function with constant L. For continuity consider the equations

$$u^{(i)}(t) = u(0) + \int_0^t F(u^{(i)}(s))ds + Dw^{(i)}(t)$$

for $i = 1, 2$. Subtracting and letting $e = u^1 - u^2, \delta = w^1 - w^2$, we get

$$e(t) = \int_0^t \Big(F(u^1(s)) - F(u^2(s))\Big)ds + D\delta(s).$$

Hence

$$|e(t)| \leqslant \int_0^t L|e(s)|ds + |D||\delta(t)|.$$

From the integrated form of the Gronwall inequality in Lemma 4.4 it follows that

$$\sup_{0\leqslant t\leqslant T} |e(s)| \leqslant C \sup_{0\leqslant t\leqslant T} |\delta(t)|.$$

This establishes continuity. □

In establishing the theorem, we will apply the Itô formula twice: once to $\Phi(y(s))$ where Φ is the solution of the cell problem (18.2.3), and once to $\chi(x(s), y(s))$, where χ solves the following Poisson equation:

$$\begin{aligned} \mathcal{L}_0\chi(x,y) &= f_1(x,y) - F(x), \\ \int_{\mathbb{T}^{d-l}} \chi(x,y)\rho^\infty(y)dy &= 0, \\ \chi(x,y) &\quad \text{periodic on } \mathbb{T}^{d-l}. \end{aligned} \tag{18.3.3}$$

This has a unique solution, by the Fredholm alternative, or by Theorem 7.8, since $f_1 - F$ averages to zero over $\mathbb{T}^{d-l}$. The proof of the following lemma is very similar to the proof of Lemma 17.2; hence we omit it.

Lemma 18.3. *The functions f_0, f_1, Φ, χ and all their derivatives are smooth and bounded.*

Proof of Theorem 18.1 Let $\chi = \chi(x(t), y(t))$, where $(x(t), y(t))$ is the solution of (18.2.1). Notice that, by the Itô formula, Lemma 6.5, we have that

$$\frac{d\chi}{dt} = \frac{1}{\varepsilon^2}\mathcal{L}_0\chi + \frac{1}{\varepsilon}\mathcal{L}_1\chi + \mathcal{L}_2\chi + \nabla_x\chi\alpha\frac{dU}{dt} + \frac{1}{\varepsilon}\nabla_y\chi\beta\frac{dV}{dt}; \tag{18.3.4}$$

the rigorous interpretation is, as usual, the integrated form. Here the operator $\mathcal{L}_0$ is as defined earlier and $\mathcal{L}_1, \mathcal{L}_2$ are given by

$$\begin{aligned} \mathcal{L}_1 &= f_0(y)\cdot\nabla_x, \\ \mathcal{L}_2 &= f_1(x,y)\cdot\nabla_x + \frac{1}{2}\alpha\alpha^T : \nabla_x\nabla_x. \end{aligned}$$

Now define

$$\theta(t) = \varepsilon^2\Big(\chi(x(t),y(t)) - \chi(x(0),y(0))\Big) - \varepsilon\int_0^t \Big(\mathcal{L}_1\chi\Big)(x(s),y(s))\,ds - \varepsilon^2\int_0^t \Big(\mathcal{L}_2\chi\Big)(x(s),y(s))\,ds \quad (18.3.5)$$

and

$$M_1(t) = \varepsilon^2\int_0^t \nabla_x\chi(x(s),y(s))\alpha\,dU(s) + \varepsilon\int_0^t \nabla_y\chi(x(s),y(s))\beta(y(s))\,dV(s). \quad (18.3.6)$$

Since χ, f_0, and f_1 and all their derivatives are bounded, we have

$$\mathbb{E}\sup_{0\leqslant t\leqslant T}|\theta(t)|^p \leqslant C\varepsilon^p. \quad (18.3.7)$$

Furthermore

$$\mathbb{E}|\langle M_1\rangle|^2 = \varepsilon^4\int_0^t |\nabla_x\chi(x(s),y(s))\alpha|_F^2\,ds + \varepsilon^2\int_0^t |\nabla_y\chi(x(s),y(s))\beta(y(s))|_F^2\,ds.$$

More generally,

$$\mathbb{E}|\langle M_1\rangle|^{p/2} \leqslant C_1\varepsilon^{2p}\int_0^t |\nabla_x\chi(x(s),y(s))\alpha|_F^{p}\,ds + C_2\varepsilon^p\int_0^t |\nabla_y\chi(x(s),y(s))\beta(y(s))|_F^{p}\,ds$$

for every $p \geqslant 1$. By the Burkholder–Davis–Gundy inequality of Theorem 3.22, and by (3.4.8), we deduce that

$$\mathbb{E}\sup_{0\leqslant t\leqslant T}|M_1(t)|^p \leqslant C\varepsilon^p. \quad (18.3.8)$$

Hence, by (18.3.4) and (18.3.3), we deduce that

$$\int_0^t \big(f_1(x(s),y(s)) - F(x(s))\big)ds = r(t) := \theta(t) - M_1(t), \quad (18.3.9)$$

where we have shown that

$$r = \mathcal{O}(\varepsilon) \quad \text{in} \quad L^p\big(\Omega, C([0,T],\mathbb{R}^l)\big). \quad (18.3.10)$$

Now apply the Itô formula to Φ solving (18.2.3) to obtain, since Φ is independent of x,

$$\frac{d\Phi}{dt} = \frac{1}{\varepsilon^2}\mathcal{L}_0\Phi + \frac{1}{\varepsilon}\nabla_y\Phi\beta\frac{dV}{dt}.$$

(As usual the rigorous interpretation is in integrated form.) Hence

$$\frac{1}{\varepsilon}\int_0^t f_0(y(s))ds = \varepsilon\big(\Phi(y(0)) - \Phi(y(s))\big) + \int_0^t \nabla_y\Phi(y(s))\beta(y(s))dV(s). \tag{18.3.11}$$

Since Φ is bounded on $\mathbb{T}^{d-l}$, we deduce that, as $\varepsilon \to 0$,

$$\varepsilon\big(\Phi(y(\cdot)) - \Phi(y(0))\big) = \mathcal{O}(\varepsilon) \quad \text{in} \quad L^p\big(\Omega, C([0,T],\mathbb{R}^l)\big). \tag{18.3.12}$$

We set

$$M_2(t) = \int_0^t (\nabla_y\Phi)(y(s))\beta(y(s))\,dV(s).$$

Note that $y(s) = \varphi^{s/\varepsilon^2}(y)$, where $y(0) = y$. Rescaling time, and using the scaling properties of Brownian motion encapsulated in Remark 6.3, we obtain the identity, in law,

$$M_2(t) = \varepsilon\int_0^{t/\varepsilon^2} (\nabla_y\Phi)(\varphi^\tau(y))\beta(\varphi^\tau(y))\,dV(\tau).$$

The Martingale central limit Theorem 3.33 implies that, as $\varepsilon \to 0$,

$$M_2 \Rightarrow \alpha_2 W, \tag{18.3.13}$$

in $C([0,T],\mathbb{R}^l)$, where $W(t)$ is standard Brownian motion and

$$\begin{aligned}\alpha_2\alpha_2^T &= \lim_{t\to\infty}\frac{1}{t}\int_0^t \nabla_y\Phi\varphi^s(y))\beta(\varphi^s(y)) \otimes \nabla_y\Phi(\varphi^s(y))\beta(\varphi^s(y))ds \\ &= \int_{\mathbb{T}^{d-l}} \rho^\infty(y)\Big(\nabla_y\Phi(y)\beta(y) \otimes \nabla_y\Phi(y)\beta(y)\Big)dy.\end{aligned}$$

Combining (18.3.9), (18.3.11) in (18.2.1a) we obtain (18.3.1) where

$$\eta(t) = r(t) + \varepsilon\big(\Phi(y(0)) - \Phi(y(t))\big).$$

We have that

$$(\alpha U, M_2) \Rightarrow (\alpha U, \alpha_2 W), \quad \text{in} \quad C([0,T],\mathbb{R}^{2l})$$

because U and M_2 are independent. By (18.3.10), (18.3.12) we have that

$$\eta \to 0, \quad \text{in} \quad L^p\big(\Omega, C([0,T],\mathbb{R}^l)\big).$$

Hence, by Theorem 3.30, (18.3.2) holds:

$$(\alpha U, \eta, M_2) \Rightarrow (\alpha U, 0, \alpha_2 W), \quad \text{in} \quad C([0,T],\mathbb{R}^{3l}).$$

The mapping $(U, \eta, M_2) \to x$ in (18.3.1) is continuous from $C([0,T],\mathbb{R}^{3l})$ into $C([0,T],\mathbb{T}^l)$, by Lemma 18.2. Hence, because weak convergence is preserved under continuous mappings (see Theorem 3.28), we deduce that $x(t) \Rightarrow X(t)$ in $C([0,T],\mathbb{T}^l)$, where X solves

$$X(t) = x(0) + \int_0^t F(X(s))ds + \alpha U(t) + \alpha_2 W(t).$$

18.4 Discussion and Bibliography

Limit theorems for singularly perturbed systems of SDEs, of the type considered in this chapter, have been studied since the early 1960s; see, for example, [300, 301, 166]. The theory was developed further in the 1970s; see [291; 33; 243; 94, ch. 12] and the references therein.

The proofs presented here have been simplified by the assumption that the limiting SDE for X has additive noise. Thus we were able to use the martingale central limit theorem in a very straightforward way. In the general case where the limiting process has general state-dependent noise, proofs of convergence are more complicated and the martingale formulation of the solution of an SDE is used; see [94, ch. 12].

The fundamental role that Poisson equations play in the proof of limit theorems for SDEs has been known for a long time [181, 240, 241, 246]. A systematic use of an appropriate Poisson equation in the proof of limit theorems (diffusion approximations) for singularly perturbed SDEs in a noncompact state space has undertaken recently by Pardoux and Veretennikov in a series of papers [249, 250, 251].

It is also possible to obtain corvergence rates, by introducing additional Poisson equations and applying the Itô formula to their solutions; this is then combined with the Dambis–Dubins–Schwarz theorem; see Section 3.6 and [134]. These error estimates play an important role in the study of the parameter estimation for SDEs with a multiscale structure; see [258]. A quite general strong approximation theorem for martingales is proved in [40].

18.5 Exercises

1. Consider the coupled pair of scalar SDEs

$$\frac{dx}{dt} = f(x,y) - \frac{1}{\varepsilon}y + \big(y^2 + a(x)\big)\frac{dU}{dt}, \tag{18.5.1a}$$

$$\frac{dy}{dt} = -\frac{1}{\varepsilon^2}y + \frac{1}{\varepsilon}\sqrt{2}\frac{dV}{dt}. \tag{18.5.1b}$$

   a. Write down the homogenized equations.
   b. Assume that f, a, and all derivatives are bounded. Using the exact solution of the OU process prove a convergence theorem related to the conjectured homogenized equation.

2. a. Consider the fast/slow system

$$\dot{x} = \frac{f(x)y}{\varepsilon}, \quad \dot{y} = -\frac{\alpha}{\varepsilon^2}y + \sqrt{\frac{2\lambda}{\varepsilon^2}}\dot{\beta}.$$

   b. Prove that the homogenized equation is

$$\dot{X} = \frac{\lambda}{\alpha^2}f(X)f'(X) + \sqrt{\frac{2\lambda}{\alpha^2}}f(X)\dot{\beta}$$

by applying the Itô formula to appropriate functions of $x(t)$ and $y(t)$. Use this method to obtain error estimates.

c. Assume that the fast process is stationary. Use the Itô formula to prove a strong (pathwise) limit theorem. (Hint: You need to use pathwise estimates on the Ornstein-Uhlenbeck process.)

3. Consider the problem studied in this chapter in one dimension. Combine the analysis presented here with the Dambis–Dubins–Schwarz theorem to prove a strong (pathwise) approximation theorem.
4. Consider the SDEs (18.2.1) for $x \in \mathbb{R}^l$, $y \in \mathbb{R}^{d-l}$. State carefully the properties that solutions of the Poisson equation should have in this case in order to be able to prove the homogenization theorem.

19

Homogenization for Elliptic PDEs: The Convergence Theorem

19.1 Introduction

In this chapter we prove two homogenization theorems for second-order uniformly elliptic PDEs with periodic coefficients and Dirichlet boundary conditions. Our method of proof is to use two-scale convergence. This technique provides an elegant way of deriving the homogenized equation, and the cell problem, which is required for its definition, via a coupled system of equations called the **two-scale system**. The method of two-scale convergence is also applicable to transport equations and we use it in that context in Chapter 21. Two theorems are stated in Section 19.2; they are proved in each of the following two sections. We finish the chapter with comments and bibliographical remarks.

19.2 The Theorems

Theorem 19.1. *Let u^ε be the weak solution of*

$$-\nabla \cdot (A^\varepsilon \nabla u^\varepsilon) = f \;\; \textit{for } x \in \Omega, \tag{19.2.1a}$$

$$u^\varepsilon = 0 \;\; \textit{for } x \in \partial\Omega \tag{19.2.1b}$$

with $f = f(x) \in L^2(\Omega)$, $\Omega \subset \mathbb{R}^d$ bounded and $A^\varepsilon = A\left(\frac{x}{\varepsilon}\right)$, $A \in M_{per}(\alpha, \beta, \mathbb{T}^d)$, $0 < \alpha \leqslant \beta < \infty$. Furthermore, let u be the weak solution of the homogenized problem

$$-\nabla \cdot \left(\overline{A} \nabla u\right) = f \;\; \textit{for } x \in \Omega \tag{19.2.2a}$$

$$u = 0 \;\; \textit{for } x \in \partial\Omega, \tag{19.2.2b}$$

with $\overline{A}$ given by

$$\overline{A} = \int_{\mathbb{T}^d} \left(A(y) + A(y)\nabla_y \chi(y)^T\right) dy, \tag{19.2.3}$$

and where the vector field $\chi(y)$ is a weak solution of the cell problem

$$-\nabla_y \cdot \left(\nabla_y \chi A^T\right) = \nabla_y \cdot A^T, \quad \chi(y) \text{ is 1-periodic.} \tag{19.2.4}$$

Then

$$u^\varepsilon \rightharpoonup u \text{ weakly in } H_0^1(\Omega)$$

and

$$u^\varepsilon \to u \text{ strongly in } L^2(\Omega).$$

In addition to the basic homogenization theorem, we will prove a result that shows that retaining extra terms in the multiscale expansion does indeed give improved approximations. The following result says that we can get strong convergence in $H^1(\Omega)$ provided that we take the first-order corrector field into account.

Theorem 19.2. *Consider $u^\varepsilon(x)$ and $u(x)$ as in Theorem 19.1. Assume that $f \in L^2(\Omega)$, that $\partial\Omega$ is sufficiently smooth so that $u \in H^2(\Omega) \cap H_0^1(\Omega)$, and that the coefficient matrix A is such that the cell problem* (19.2.4) *has solution $\chi \in C_{per}^1(\mathbb{T}^d)$. Then*

$$\lim_{\varepsilon \to 0} \left\| u^\varepsilon(x) - \left(u(x) + \varepsilon \chi\left(\frac{x}{\varepsilon}\right) \cdot \nabla u(x)\right) \right\|_{H^1(\Omega)} = 0. \tag{19.2.5}$$

Remark 19.3. The assumption concerning the smoothness of $\partial\Omega$ is met if it is a C^2 hypersurface or a convex polytope, for example. In order for $\chi \in C_{per}^1(\mathbb{T}^d)$, $A(y)$ has to be sufficiently regular; $A(y) \in C_{per}^1(\mathbb{T}^d; \mathbb{R}^{d \times d})$ is more than sufficient. □

19.3 The Proof: Strong Convergence in L^2

In this section we prove the homogenization theorem, Theorem 19.1, using the method of two-scale convergence. Before starting the proof, we make some remarks on our approach. The first step in our analysis is to use the energy estimates from Chapter 7 to deduce that u^ε and ∇u^ε have two-scale convergent subsequences (Lemma 19.4) defined via a pair of functions $\{u(x), u_1(x,y)\}$. The second step is to use a test function of the form

$$\phi^\varepsilon(x) = \phi_0(x) + \varepsilon \phi_1\left(x, \frac{x}{\varepsilon}\right), \tag{19.3.1}$$

in the definition of weak solution of (19.2.2), in order to characterize the two-scale limits. In this way we obtain a coupled system of equations for functions appearing in the first step of the analysis, $\{u,\ u_1\}$; we call these equations the two-scale system; see Lemma 19.5. The third step, in Lemma 19.6, is to prove existence and uniqueness of this system using the Lax-Milgram theorem. The final step, Lemma 19.7, is to decouple this system of equations using separation of variables, showing that it gives rise to the homogenized equations (19.2.2)–(19.2.4) identified by perturbation expansions in Chapter 12.

In the following we will use the space H defined by (2.4.6) and from it define the space

$$X = H_0^1(\Omega) \times L^2(\Omega; H).$$

This is a Hilbert space with inner product

$$(U,V)_X = (\nabla u, \nabla v)_{L^2(\Omega)} + (\nabla_y u_1, \nabla_y v_1)_{L^2(\Omega\times\mathbb{T}^d)}$$

for all $U = \{u, u_1\}$, $V = \{v, v_1\}$, and induced norm

$$\|U\|_X^2 = \|\nabla u\|_{L^2(\Omega)}^2 + \|\nabla_y u_1\|_{L^2(\Omega\times\mathbb{T}^d)}^2.$$

This turns out to be the right space in which to describe the two-scale limits, which we now introduce.

Lemma 19.4. *Let $u^\varepsilon(x)$ be the solution of* (19.2.1) *with the assumptions of Theorem 19.1. Then there exist functions $\{u(x),\ u_1(x,y)\} \in X$ such that, along a subsequence, u^ε and ∇u^ε two-scale converge to $u(x)$ and to $\nabla_x u + \nabla_y u_1$, respectively.*

Proof. We have that $\|u^\varepsilon\|_{H_0^1(\Omega)} \leqslant C$ by Theorem 7.5, which implies the existence of a subsequence converging weakly to a limit u in $H_0^1(\Omega)$. By Theorem 2.36 there exist functions $u \in H_0^1(\Omega)$, $u_1 \in L^2(\Omega; H)$ such that, possibly on a further subsequence,

$$u^\varepsilon \overset{2}{\rightharpoonup} u, \tag{19.3.2a}$$

$$\nabla u^\varepsilon \overset{2}{\rightharpoonup} \nabla_x u + \nabla_y u_1. \quad \square \tag{19.3.2b}$$

We now show that $\{u, u_1\}$ satisfy the two-scale system

$$-\nabla_y \cdot \Big(A(y)\,(\nabla_x u + \nabla_y u_1)\Big) = 0 \quad \text{in } \Omega\times\mathbb{T}^d, \tag{19.3.3a}$$

$$-\nabla_x \cdot \Big(\int_{\mathbb{T}^d} A(y)\,(\nabla_x u + \nabla_y u_1)\,dy\Big) = f \quad \text{in } \Omega, \tag{19.3.3b}$$

$$u(x) = 0 \ \text{ for } x \in \partial\Omega, \quad u_1(x,y) \text{ is periodic in } y. \tag{19.3.3c}$$

To define the weak formulation of the two-scale system we introduce the bilinear form

$$a[U,\Phi] = \int_\Omega \int_{\mathbb{T}^d} \langle A\,(\nabla_x u + \nabla_y u_1),\, \nabla_x \phi_0 + \nabla_y \phi_1\rangle\, dydx,$$

with $\Phi := \{\phi_0,\ \phi_1\} \in X$. The weak formulation of the two-scale system (19.3.3) is to seek $U \in X$ such that

$$a[U,\Phi] = (f, \phi_0) \quad \forall \Phi \in X. \tag{19.3.4}$$

To see this, we set $\phi_0 = 0$ to obtain:

$$\int_\Omega \int_{\mathbb{T}^d} \langle A\,(\nabla_x u + \nabla_y u_1),\, \nabla_y \phi_1\rangle\, dydx = 0.$$

This is precisely the weak formulation of (19.3.3a). Now setting $\phi_1 = 0$ in (19.3.4) we get

$$\int_\Omega \int_{\mathbb{T}^d} \langle A\,(\nabla_x u + \nabla_y u_1),\, \nabla_x \phi_0\rangle\, dydx = (f, \phi_0).$$

This is the weak formulation of (19.3.3b). The boundary conditions (19.3.3c) follow from the fact that $u \in H_0^1(\Omega)$ and $u_1 \in L^2(\Omega; H)$.

Lemma 19.5. *Let $u^\varepsilon(x)$ be the weak solution of* (19.2.1) *with the assumptions of Theorem 19.1. Then any limit point $\{u,\, u_1\}$ from Lemma 19.4 is a weak solution of the two-scale system* (19.3.3).

Proof. The weak formulation of (19.2.1) is to find $u^\varepsilon \in H_0^1(\Omega)$ such that

$$\int_\Omega \langle A^\varepsilon \nabla u^\varepsilon, \nabla \phi^\varepsilon \rangle \, dx = (f, \phi^\varepsilon) \quad \forall \phi^\varepsilon \in H_0^1(\Omega). \tag{19.3.5}$$

We use a test function of the form (19.3.1) for $\phi_0 \in C_0^\infty(\Omega)$, $\phi_1 \in C_0^\infty(\Omega; C_{per}^\infty(\mathbb{T}^d))$. We clearly have that $\phi^\varepsilon \in H_0^1(\Omega)$. Upon using this test function in (19.3.5) and rearranging terms we obtain:

$$I_1 + \varepsilon I_2 = (f, \phi_0 + \varepsilon \phi_1)$$

where

$$I_1 = \int_\Omega \left\langle \nabla u^\varepsilon, (A^\varepsilon)^T \left(\nabla_x \phi_0(x) + \nabla_y \phi_1 \left(x, \frac{x}{\varepsilon}\right)\right)\right\rangle dx$$

$$I_2 = \int_\Omega \left\langle \nabla u^\varepsilon, (A^\varepsilon)^T \nabla_x \phi_1 \left(x, \frac{x}{\varepsilon}\right)\right\rangle dx.$$

Now the function $A^\varepsilon(x)^T \left(\nabla_x \phi_0(x) + \nabla_y \phi_1\left(x, \frac{x}{\varepsilon}\right)\right)$ is of the form $\psi(x,y) := \psi_1(y)\psi_2(x,y)$, $y = x/\varepsilon$, with $\psi_1 \in L^\infty(\mathbb{T}^d)$ and $\psi_2 \in L^2(\Omega; C_{per}(\mathbb{T}^d))$. Hence, by Lemma 2.31, we can use $\psi(x,y)$ as a test function and pass to the two-scale limit to obtain:

$$I_1 \to \int_\Omega \int_{\mathbb{T}^d} \langle A(y)\left(\nabla_x u + \nabla_y u_1\right), \left(\nabla_x \phi_0 + \nabla_y \phi_1\right)\rangle \, dy dx \quad \text{as } \varepsilon \to 0.$$

The function $A^\varepsilon(x)^T \nabla_x \phi_1(x,y)$ is also an admissible test function. Passing to the two-scale limit similarly in I_2 we obtain $\varepsilon I_2 \to 0$. Moreover, by Theorem 2.29, $\phi_0 + \varepsilon\phi_1 \rightharpoonup \phi_0$ weakly in $L^2(\Omega)$. This implies that

$$(f, \phi_0 + \varepsilon \phi_1) \to (f, \phi_0).$$

Putting these considerations together we obtain the limiting equation

$$\int_\Omega \int_{\mathbb{T}^d} \langle A(y)\left(\nabla_x u + \nabla_y u_1\right), \nabla_x \phi_0 + \nabla_y \phi_1\rangle \, dy dx = (f, \phi_0).$$

Thus we have derived (19.3.4). In deriving this identity we assumed that the test functions ϕ_0, ϕ_1 are smooth; a density argument enables us to conclude that it holds for every $\phi_0 \in H_0^1(\Omega)$, $\phi_1 \in L^2(\Omega; H)$. □

Next we prove that the two-scale system has a unique solution.

Lemma 19.6. *Under the assumptions of Theorem 19.1 the two-scale system* (19.3.3) *has a unique weak solution $\{u, u_1\} \in X$.*

Proof. We will use the Lax-Milgram theorem. The weak formulation of the two-scale system is given by (19.3.4). We have to check that the bilinear form a is continuous and coercive. To establish both of these properties it will be helpful to note the following. Let $a = a(x)$ and $b = b(x, y)$ be smooth functions, with b periodic in y. Then

$$\int_\Omega \int_{\mathbb{T}^d} |\nabla_x a + \nabla_y b|^2 dydx = \|\{a, b\}\|_X^2.$$

This follows from the fact that

$$\int_\Omega \int_{\mathbb{T}^d} \langle \nabla_x a, \nabla_y b \rangle dydx = \int_\Omega \Big(\int_{\mathbb{T}^d} \nabla_y \cdot (b \nabla_x a) dy \Big) dx = 0,$$

using the divergence theorem, and periodicity of $b\nabla_x a$ in y (see Remark 7.13).

We start with continuity. We use the L^∞ bound on $A(y)$, together with the Cauchy-Schwarz inequality, to obtain:

$$\begin{aligned} a[U, \Phi] &= \int_\Omega \int_{\mathbb{T}^d} \langle A \left(\nabla_x u + \nabla_y u_1 \right), \nabla_x \phi_0 + \nabla_y \phi_1 \rangle \; dydx \\ &\leqslant \beta \int_\Omega \int_{\mathbb{T}^d} |\nabla_x u + \nabla_y u_1| \, |\nabla_x \phi_0 + \nabla_y \phi_1| \, dydx \\ &\leqslant \beta \|U\|_X \|\Phi\|_X. \end{aligned}$$

We proceed with coercivity. We use the divergence theorem and periodicity as earlier to obtain:

$$\begin{aligned} a[U, U] &= \int_\Omega \int_{\mathbb{T}^d} \langle A \left(\nabla_x u + \nabla_y u_1 \right), \nabla_x u + \nabla_y u_1 \rangle \; dydx \\ &\geqslant \alpha \int_\Omega \int_{\mathbb{T}^d} |\nabla_x u + \nabla_y u_1|^2 \; dydx \\ &= \alpha \|U\|_X^2 \end{aligned}$$

and consequently

$$a[U, U] \geqslant \alpha \|U\|_X^2.$$

Hence, the bilinear form $a[U, \Phi]$ is continuous and coercive and the Lax-Milgram lemma applies. This proves existence and uniqueness of solutions of the two-scale system in X. □

Now we relate the two-scale system to the form of the homogenized equation that we derived in Chapter 12, stated at the start of this chapter.

Lemma 19.7. *Consider the unique solution* $\{u, u_1\} \in X$ *of the two-scale system* (19.3.3)*. Then* u *is the unique solution of the homogenized Equation* (19.2.2) *and* $u_1(x, y)$ *is given by*

$$u_1(x, y) = \chi(y) \cdot \nabla u(x), \tag{19.3.6}$$

where $\chi(y)$ *is the solution of the cell problem* (19.2.4).

Proof. We substitute (19.3.6) into (19.3.3a) to obtain

$$-\nabla_y \cdot \left(\nabla_y \chi A^T\right) \cdot \nabla_x u = \left(\nabla_y \cdot A^T\right) \cdot \nabla_x u.$$

This equation is satisfied if $\chi \in H$ is the unique solution of the cell problem (19.2.4). Equation (19.3.3b) becomes

$$-\nabla_x \cdot \left(\int_{\mathbb{T}^d} A \left(\nabla_x u + (\nabla_y \chi)^T \nabla_x u\right) \, dy \right) = f$$

so that

$$-\nabla_x \cdot \left(\int_{\mathbb{T}^d} A \left(I + \nabla_y \chi^T\right) \, dy \right) \nabla_x u = f$$

and hence

$$-\nabla_x \cdot \left(\overline{A} \nabla_x u\right) = f.$$

This is precisely the homogenized equation with the homogenized coefficients given by (19.2.3).

The fact that the choice (19.3.6) for u_1 enables us to solve the two-scale system, provided that u_0 satisfies the homogenized equation, implies that this is the only possible set of functions $\{u, u_1\}$ that solves the two-scale system, since we have already proved uniqueness of solutions. □

We may now conclude the proof of Theorem 19.1. The first lemma shows that u^ε two-scale converges to $\{u, u_1\}$, along a subsequence, with u independent of y. Furthermore u^ε converges weakly in $H_0^1(\Omega)$ and strongly in $L^2(\Omega)$. But the limit is unique, as we proved by applying the Lax-Milgram theorem to the two-scale system. Hence the whole sequence must converge, by the subsequence principle, and the proof is complete. □

19.4 The Proof: Strong Convergence in H^1

Theorem 19.1 implies that u^ε converges to $u(x)$ strongly in $L^2(\Omega)$. Thus, in order to prove Theorem 19.2, it is enough to prove that

$$\lim_{\varepsilon \to 0} \left\| \nabla u^\varepsilon(x) - \nabla \left(u(x) + \varepsilon u_1 \left(x, \frac{x}{\varepsilon}\right)\right) \right\|_{L^2(\Omega;\mathbb{R}^d)} = 0,$$

or, equivalently,

$$\lim_{\varepsilon \to 0} \left\| \nabla u^\varepsilon(x) - \left(\nabla u(x) + \varepsilon \nabla_x u_1 \left(x, \frac{x}{\varepsilon}\right) + \nabla_y u_1 \left(x, \frac{x}{\varepsilon}\right)\right) \right\|_{L^2(\Omega;\mathbb{R}^d)} = 0.$$

Since $f \in L^2(\Omega)$ we have that $u(x) \in H^2(\Omega) \cap H_0^1(\Omega)$. Hence, and since by assumption $\chi \in C_{per}^1(\mathbb{T}^d)$, we have that $u_1(x, y) = \chi(y) \cdot \nabla u(x) \in H^1(\Omega; C_{per}^1(\mathbb{T}^d))$. This implies, by Theorem 2.28, that

$$\left\| \nabla_x u_1\left(x, \frac{x}{\varepsilon}\right) \right\|_{L^2(\Omega;\mathbb{R}^d)} \leqslant C.$$

Consequently,

$$\left\| \varepsilon \nabla_x u_1\left(x, \frac{x}{\varepsilon}\right) \right\|_{L^2(\Omega;\mathbb{R}^d)} \to 0.$$

Hence, it is enough to prove that

$$\lim_{\varepsilon\to 0} \left\| \nabla u^\varepsilon(x) - \left(\nabla u(x) + \nabla_y u_1\left(x, \frac{x}{\varepsilon}\right)\right) \right\|_{L^2(\Omega);\mathbb{R}^d} = 0.$$

The uniform ellipticity of A now implies:

$$\begin{aligned}
\alpha \left\| \nabla u^\varepsilon(x) \right. & - \left. \left(\nabla u(x) + \nabla_y u_1\left(x, \frac{x}{\varepsilon}\right)\right) \right\|^2_{L^2(\Omega;\mathbb{R}^d)} \\
&= \alpha \int_\Omega \left| \nabla u^\varepsilon(x) - \left(\nabla u(x) + \nabla_y u_1\left(x, \frac{x}{\varepsilon}\right)\right) \right|^2 dx \\
&\leqslant \int_\Omega \left\langle A\left(\frac{x}{\varepsilon}\right)\left(\nabla_x u^\varepsilon(x) - \nabla_x u(x) - \nabla_y u_1\left(x, \frac{x}{\varepsilon}\right)\right), \right. \\
&\qquad \left. \nabla_x u^\varepsilon(x) - \nabla_x u(x) - \nabla_y u_1\left(x, \frac{x}{\varepsilon}\right) \right\rangle dx \\
&\leqslant \int_\Omega \left\langle A\left(\frac{x}{\varepsilon}\right) \nabla_x u^\varepsilon(x), \nabla_x u^\varepsilon(x) \right\rangle dx \\
&\quad + \int_\Omega \left\langle A\left(\frac{x}{\varepsilon}\right)\left(\nabla_x u(x) + \nabla_y u_1\left(x, \frac{x}{\varepsilon}\right)\right), \nabla_x u(x) + \nabla_y u_1\left(x, \frac{x}{\varepsilon}\right) \right\rangle dx \\
&\quad - \int_\Omega \left\langle \nabla_x u^\varepsilon, \left(A^\varepsilon + (A^\varepsilon)^T\right)\left(\nabla_x u + \nabla_y u_1\right) \right\rangle dx \\
&=: (f, u^\varepsilon) + I_1^\varepsilon + I_2^\varepsilon.
\end{aligned}$$

By Theorem 19.1, u^ε conveges to u strongly in $L^2(\Omega)$. Since strong convergence implies weak convergence, we have that

$$(f, u^\varepsilon) \to (f, u) = a[U, U],$$

where the last equality follows from (19.3.4). Furthermore, since

$$A \in L^\infty_{per}(\mathbb{T}^d; \mathbb{R}^{d\times d}),\ u \in H^2(\Omega) \cap H^1_0(\Omega)$$

and $\chi \in C^1_{per}(\mathbb{T}^d)$, Lemma 2.31 implies that

$$A\left(\frac{x}{\varepsilon}\right)\left(\nabla_x u(x) + \nabla_y u_1\left(x, \frac{x}{\varepsilon}\right)\right)$$

and

$$\left(A^\varepsilon + (A^\varepsilon)^T\right)\left(\nabla_x u + \nabla_y u_1\right)$$

can be used as test functions in I_1^ε and I_2^ε, respectively. We pass to the two-scale limit in these two expressions to obtain that, as $\varepsilon \to 0$,

$$I_1^\varepsilon \to \int_\Omega \int_{\mathbb{T}^d} \left\langle A(y)\left(\nabla_x u(x) + \nabla_y u_1(x,y)\right), \nabla_x u(x) + \nabla_y u_1(x,y) \right\rangle dydx$$

and

$$I_2^{\varepsilon} \to -\int_{\Omega}\int_{\mathbb{T}^d} \Big\langle \Big(A(y) + (A(y))^T\Big) \left(\nabla_x u(x) + \nabla_y u_1(x,y)\right), \nabla_x u(x) + \nabla_y u_1(x,y)\Big\rangle \, dydx.$$

Thus we have proved that

$$\lim_{\varepsilon\to 0} \alpha \left\|\nabla u^{\varepsilon}(x) - \left(\nabla u(x) + \nabla_y u_1\left(x, \frac{x}{\varepsilon}\right)\right)\right\|^2_{L^2(\Omega;\mathbb{R}^d)} \leqslant a[U,U] + a[U,U] - 2a[U,U] = 0.$$

Consequently,

$$\lim_{\varepsilon\to 0} \left\|\nabla u^{\varepsilon}(x) - \left(\nabla u(x) + \nabla_y u_1\left(x, \frac{x}{\varepsilon}\right)\right)\right\|_{L^2(\Omega;\mathbb{R}^d)} = 0,$$

and the theorem is proved □

19.5 Discussion and Bibliography

The proofs of Theorems 19.1 and 19.2 are taken from [6]. The regularity assumptions in the corrector result, Theorem 19.2, are by no means optimal. See, e.g., [6, Theorem 2.6] or [66, prop. 9.12] for sharper results. A certain amount of regularity is, however, needed to be able to prove corrector results.

The proof of the homogenization theorem essentially consists of two steps: first one proves the existence of a limit, and then one tries to characterize the limit. The existence of a limit follows from a priori estimates, which are quite often not hard to obtain. On the other hand, the characterization of the limit-i.e., the rigorous justification that the limit of the sequence u^{ε} satisfies the homogenized PDE – is usually a much more delicate issue. The idea of using appropriate test functions for characterizing the limit of a sequence of functions is very common in the theory of PDEs. See [95] and the references therein. In the context of homogenization, test functions of the form (19.3.1) have been used by Kurtz in [181]. See also the perturbed test function approach of Evans in, [97, 96].

Tartar's method of oscillating test functions is based on constructing appropriate test functions using the cell problem; see [307; 311; 66, ch. 8] and the references therein. Homogenization results for elliptic PDEs can also be proved by using methods from the calculus of variations and the concept of Γ-convergence; see [214]. These techniques apply to various homogenization problems for nonlinear PDEs. It is relatively straightforward to prove a homogenization theorem for *monotone operators* using the method of two-scale convergence; see Exercise 7.

The bootstrapping technique employed in Chapter 16 for the proof of the averaging theorem for Markov chains and in Chapter 20 for the proof of the homogenization theorem for parabolic PDEs can also be used for the proof of the homogenization theorem for elliptic PDEs; see Exercise 6 in this chapter and Exercise 4 in the next. This method enables error estimates to be found; see [66, ch. 7].

It is not always possible to decouple the two-scale system and obtain a closed equation for the first term in the two-scale expansion, the homogenized equation. In order to do this we need an H^1-estimate on the solution $u^\varepsilon(x)$ of the unhomogenized PDE. This enables us to conclude that the two-scale limit is independent of the microscale and consequently a homogenized equation actually exists. Such a scenario does not always occur for multiscale problems. See, for example, the case of linear transport PDE studied in Chapter 21. Other examples can be found in [6, 5]. From this point of view the two-scale system is more fundamental than the homogenized equation, since the former is a well-posed system of equations even when the latter does not exist.

19.6 Exercises

1. Let $a : \mathbb{T}^d \mapsto \mathbb{R}^d$ be a smooth, 1-periodic, mean-zero divergence-free field and consider the problem of homogenization for the steady-state advection–diffusion equation

$$-\Delta u^\varepsilon + \frac{1}{\varepsilon} a\left(\frac{x}{\varepsilon}\right) \cdot \nabla u^\varepsilon = f, \quad \text{for } x \in \Omega, \tag{19.6.1a}$$

$$u^\varepsilon(x) = 0, \quad \text{for } x \in \partial\Omega. \tag{19.6.1b}$$

   Use the method of two-scale convergence to prove the homogenization theorem for this PDE.
2. Carry out the same program as in the previous exercise for the PDE

$$-\Delta u^\varepsilon + \frac{1}{\varepsilon} a\left(\frac{x}{\varepsilon}\right) \cdot \nabla u^\varepsilon + V\left(\frac{x}{\varepsilon}\right) u^\varepsilon = f, \quad \text{for } x \in \Omega,$$

$$u^\varepsilon(x) = 0, \quad \text{for } x \in \partial\Omega,$$

   where $V(y)$ is smooth and 1-periodic with $\int_{y\in\mathbb{T}^d} V(y) > 0$.
3. Carry out the same program as in the previous exercise for the PDE

$$-\Delta u^\varepsilon + \frac{1}{\varepsilon} a\left(\frac{x}{\varepsilon}\right) \cdot \nabla u^\varepsilon + \frac{1}{\varepsilon} V\left(\frac{x}{\varepsilon}\right) u^\varepsilon = f, \quad \text{for } x \in \Omega,$$

$$u^\varepsilon(x) = 0, \quad \text{for } x \in \partial\Omega,$$

   where $V(y)$ is smooth, 1-periodic, and mean zero.
4. Carry out the same program as in Exercise 1 for the PDE (19.6.1) with

$$a(y) = -\nabla_y U(y),$$

   where U is smooth and 1-periodic.
5. State and prove a homogenization theorem for the Neumann problem found by changing Dirichlet boundary conditions to homogeneous Neumann boundary conditions in (19.2.1). Use two-scale convergence to prove the result.
6. Use appropriate energy estimates to prove the homogenization theorem for elliptic PDEs. (See Chapter 20, where this approach is carried out for parabolic PDEs.)

7. (**Homogenization for monotone operators**) Consider the following boundary value problem:

$$-\nabla \cdot \left(a \left(\frac{x}{\varepsilon}, \nabla u^{\varepsilon} \right) \right) = f, \quad \text{for } x \in \Omega, \tag{19.6.2a}$$

$$u^{\varepsilon} = 0, \quad \text{for } x \in \partial\Omega. \tag{19.6.2b}$$

We assume that $f \in L^2(\Omega)$. For the function $a(y, \lambda) : \mathbb{T}^d \times \mathbb{R}^d \mapsto \mathbb{R}$ we make the following assumptions:

i. The map $\lambda \rightarrow a(y, \lambda)$ is measurable and 1-periodic in y for every λ.
ii. The map $y \rightarrow a(y, \lambda)$ is continuous a.e. in $y \in Y$.
iii. There exists a $c > 0$ such that

$$c|\lambda|^2 \leqslant a(y, \lambda) \cdot \lambda, \quad \forall\, y \in Y,\ \forall\, \lambda \in \mathbb{R}^d. \tag{19.6.3}$$

iv. There exists a $c > 0$ such that

$$|a(y, \lambda)| \leqslant c\,(1 + |\lambda|), \quad \forall\, y \in Y,\ \forall\, \lambda \in \mathbb{R}^d. \tag{19.6.4}$$

v. $a(y, \lambda)$ is *strongly monotone*:

$$[a(y, \lambda) - a(y, \mu)] \geqslant c|\lambda - \mu|^2, \quad \forall\, y \in Y,\ \forall\, \lambda, \mu \in \mathbb{R}^d. \tag{19.6.5}$$

a. State and prove an existence and uniqueness theorem for the boundary value problem (19.6.2). (Consult [339, sec. 2.14] if necessary.)
b. State and prove, using two-scale convergence, a homogenization theorem for (19.6.2).
c. State and prove a corrector-type result.

20

Homogenization for Parabolic PDEs: The Convergence Theorem

20.1 Introduction

In this chapter we prove the homogenization theorem for second-order parabolic PDEs of the form studied in Chapter 13. The method of proof is structurally very similar to that used in Chapter 16, where we prove an averaging theorem for Markov chains. To be precise, we obtain an equation for the error in the multiscale expansion and directly estimate the error from this equation. The crucial estimate used in Chapter 16 follows from the fact that Q in that chapter is a stochastic matrix; the analogous estimate in this chapter follows from the maximum principle for parabolic PDEs. Section 20.2 states the basic result and Section 20.3 contains the proof. Extensions of the homogenization theorem presented in this chapter and bibliographical remarks are contained in Section 20.4.

20.2 The Theorem

As in Chapter 13, we set $b^\varepsilon = b\,(x/\varepsilon)$. Consider Equation (13.2.4), namely

$$\frac{\partial u^\varepsilon}{\partial t} = \frac{1}{\varepsilon} b^\varepsilon \cdot \nabla u^\varepsilon + D \Delta u^\varepsilon \quad \text{for } (x,t) \in \mathbb{R}^d \times (0,T),$$

$$u^\varepsilon = f \quad \text{for } (x,t) \in \mathbb{R}^d \times \{0\}.$$

Recall the generator

$$\mathcal{L}_0 = b(y) \cdot \nabla_y + D\Delta_y$$

with periodic boundary conditions and its L^2-adjoint $\mathcal{L}_0^*$, also with periodic boundary conditions. The invariant distribution is in the null space of $\mathcal{L}_0^*$ and the effective diffusivity is given by (13.3.2):

$$\mathcal{K} = DI + 2D \int_{\mathbb{T}^d} \nabla_y \chi(y)^T \rho(y)\, dy + \int_{\mathbb{T}^d} \Big(b(y) \otimes \chi(y)\Big) \rho(y)\, dy.$$

Here ρ is the invariant distribution in the null space of $\mathcal{L}_0^*$, given by (13.2.6). Recall that χ solves the cell problem

$$-\mathcal{L}_0\chi(y) = b(y), \ \ \chi \text{ is 1-periodic}, \ \int_{\mathcal{Y}} \chi(y)\rho(y)dy = 0. \tag{20.2.1}$$

Theorem 20.1. *Let $u^\varepsilon(x,t)$ be the solution of* (13.2.4) *with $b \in C^\infty_{per}(\mathbb{T}^d)$ and $f \in C_b^\infty(\mathbb{R}^d)$. Let $u(x,t)$ be the solution of the homogenized equation*

$$\frac{\partial u}{\partial t} = \mathcal{K} : \nabla\nabla u \quad \textit{for } (x,t) \in \mathbb{R}^d \times (0,T), \tag{20.2.2a}$$

$$u = f(x) \quad \textit{for } (x,t) \in \mathbb{R}^d \times \{0\}. \tag{20.2.2b}$$

Then

$$\|u^\varepsilon - u\|_{L^\infty(\mathbb{R}^d\times(0,T))} \leqslant C\varepsilon. \tag{20.2.3}$$

Thus $u^\varepsilon \to u$ in $L^\infty(\mathbb{R}^d \times (0,T))$.

20.3 The Proof

In Chapter 13 we derived the two-scale expansion

$$u^\varepsilon(x,t) \approx u(x,t) + \varepsilon u_1^\varepsilon(x,t) + \varepsilon^2 u_2^\varepsilon(x,t),$$

where $u_i^\varepsilon(x,t) := u_i(x, x/\varepsilon, t)$, $i = 1, 2$, and

$$u_1(x,y,t) = \chi(y) \cdot \nabla_x u(x,t). \tag{20.3.1}$$

An analysis similar to the one presented in Chapter 12 enables us to obtain the expressions for u_2 (see Exercise 1, Chapter 13) and

$$u_2(x,y,t) = \Theta(y) : \nabla_x\nabla_x u(x,t). \tag{20.3.2}$$

The vector field $\chi(y)$ solves the vector-valued Poisson Equation (20.2.1) and the matrix field $\Theta(y)$ solves the matrix-valued Poisson equation

$$\begin{aligned}-\mathcal{L}_0\Theta &= b(y)\otimes\chi(y) + 2D\nabla_y\chi(y)^T \\ &\quad - \int_{\mathbb{T}^d}\Big(b(y)\otimes\chi(y) + 2D\nabla_y\chi(y)^T\Big)\rho(y)\,dy,\end{aligned} \tag{20.3.3}$$

with periodic boundary conditions. The corrector fields $\chi(y)$, $\Theta(y)$ solve uniformly elliptic PDEs with smooth coefficients and periodic boundary conditions. Consequently, by elliptic regularlity theory, both χ and Θ, together with all their derivatives, are bounded:

$$\|\chi\|_{C^k(\mathbb{R}^d;\mathbb{R}^d)} \leqslant C, \quad \|\Theta\|_{C^k(\mathbb{R}^d;\mathbb{R}^{d\times d})} \leqslant C, \tag{20.3.4}$$

for every integer $k \geqslant 0$. Furthermore, our assumptions on the initial conditions f imply that u, which solves a PDE with constant coefficients, belongs to $L^\infty(\mathbb{R}^d \times (0,T))$, together with all of its derivatives with respect to space and time. This fact, together with estimates (20.3.4), provides us with the bounds

$$\|u_1^\varepsilon\|_{L^\infty((0,T)\times\mathbb{R}^d)} \leqslant C, \quad \|u_2^\varepsilon\|_{L^\infty((0,T)\times\mathbb{R}^d)} \leqslant C, \tag{20.3.5}$$

with the constant C being independent of ε. In writing the preceding, we use the notation $u_i^\varepsilon(x,t) := u_i(x, x/\varepsilon, t)$, $i = 1, 2$.

Let $R^\varepsilon(x,t)$ denote the remainder defined through the equation

$$u^\varepsilon(x,t) = u(x,t) + \varepsilon u_1^\varepsilon(x,t) + \varepsilon^2 u_2^\varepsilon(x,t) + R^\varepsilon(x,t). \tag{20.3.6}$$

Define

$$\mathcal{L}^\varepsilon = \frac{1}{\varepsilon} b^\varepsilon \cdot \nabla_x + D\Delta_x.$$

We want to apply $\mathcal{L}^\varepsilon$ to functions of the form $f(x, x/\varepsilon, t)$. We have

$$\begin{aligned}
\mathcal{L}^\varepsilon &= \frac{1}{\varepsilon^2}\left(b(y)\cdot\nabla_y + D\Delta_y\right) + \frac{1}{\varepsilon}\left(b(y)\cdot\nabla_x + 2D\nabla_x\nabla_y\right) + D\Delta_x \\
&= \frac{1}{\varepsilon^2}\mathcal{L}_0 + \frac{1}{\varepsilon}\mathcal{L}_1 + \mathcal{L}_2,
\end{aligned}$$

with $y = x/\varepsilon$ and

$$\begin{aligned}
\mathcal{L}_0 &= b(y)\cdot\nabla_y + D\Delta_y, \\
\mathcal{L}_1 &= b(y)\cdot\nabla_x + 2D\nabla_x\cdot\nabla_y, \\
\mathcal{L}_2 &= D\Delta_x.
\end{aligned}$$

Recall that u_0, u_1, and u_2 are constructed so that

$$\begin{aligned}
\mathcal{O}(1/\varepsilon^2) \quad & \mathcal{L}_0 u_0 = 0, \\
\mathcal{O}(1/\varepsilon) \quad & \mathcal{L}_0 u_1 = -\mathcal{L}_1 u_0, \\
\mathcal{O}(1) \quad & \mathcal{L}_0 u_2 = -\mathcal{L}_1 u_1 - \mathcal{L}_2 u_0 + \frac{\partial u_0}{\partial t}.
\end{aligned}$$

We apply $\mathcal{L}^\varepsilon$ to the expansion (20.3.6) to obtain, since $u_0 = u(x,t)$,

$$\begin{aligned}
\mathcal{L}^\varepsilon u^\varepsilon &= \mathcal{L}^\varepsilon\left(u + \varepsilon u_1 + \varepsilon^2 u_2\right) + \mathcal{L}^\varepsilon R^\varepsilon \\
&= \frac{1}{\varepsilon^2}\mathcal{L}_0 u + \frac{1}{\varepsilon}\left(\mathcal{L}_0 u_1 + \mathcal{L}_1 u\right) + \left(\mathcal{L}_0 u_2 + \mathcal{L}_1 u_1 + \mathcal{L}_2 u\right) \\
&\quad + \varepsilon\left(\mathcal{L}_1 u_2 + \mathcal{L}_2 u_1\right) + \varepsilon^2 \mathcal{L}_2 u_2 + \mathcal{L}^\varepsilon R^\varepsilon \\
&= \frac{\partial u}{\partial t} + \varepsilon\left(\mathcal{L}_1 u_2 + \mathcal{L}_2 u_1\right) + \varepsilon^2\mathcal{L}_2 u_2 + \mathcal{L}^\varepsilon R^\varepsilon.
\end{aligned}$$

On the other hand,

$$\frac{\partial u^\varepsilon}{\partial t} = \frac{\partial u}{\partial t} + \varepsilon \frac{\partial u_1}{\partial t} + \varepsilon^2 \frac{\partial u_2}{\partial t} + \frac{\partial R^\varepsilon}{\partial t}.$$

We combine these two equations together with the unhomogenized equation to obtain

$$\frac{\partial R^\varepsilon}{\partial t} = \mathcal{L}^\varepsilon R^\varepsilon + \varepsilon F^\varepsilon(x,t).$$

Here

$$F^\varepsilon(x,t) = F(x, x/\varepsilon, t), \tag{20.3.7}$$

$$F(x,y,t) = \mathcal{L}_1 u_2 + \mathcal{L}_2 u_1 - \frac{\partial u_1}{\partial t} + \varepsilon \left(\mathcal{L}_2 u_2 - \frac{\partial u_2}{\partial t} \right). \tag{20.3.8}$$

Furthermore,

$$\begin{aligned} f(x) &= u^\varepsilon(x,0) \\ &= u(x,0) + \varepsilon u_1\left(x, \frac{x}{\varepsilon}, 0\right) + \varepsilon^2 u_2\left(x, \frac{x}{\varepsilon}, 0\right) + R^\varepsilon(x,0), \end{aligned}$$

and consequently, on account of (20.2.2b), we have

$$R^\varepsilon(x,0) = \varepsilon h^\varepsilon(x)$$

with

$$h^\varepsilon(x,0) = -u_1\left(x, \frac{x}{\varepsilon}, 0\right) - \varepsilon u_2\left(x, \frac{x}{\varepsilon}, 0\right). \tag{20.3.9}$$

Putting these calculations together we obtain the following Cauchy problem for the remainder $R^\varepsilon(x,t)$

$$\frac{\partial R^\varepsilon}{\partial t} = \mathcal{L}^\varepsilon R^\varepsilon + \varepsilon F^\varepsilon(x,t) \quad \text{for } (x,t) \in \mathbb{R}^d \times (0,T), \tag{20.3.10a}$$

$$R^\varepsilon = \varepsilon h^\varepsilon(x) \quad \text{for } (x,t) \in \mathbb{R}^d \times \{0\}, \tag{20.3.10b}$$

with $F^\varepsilon(x,t)$ and $h^\varepsilon(x)$ given by (20.3.7) and (20.3.9), respectively.

To prove Theorem 20.1 we will need estimates on F^ε and h^ε.

Lemma 20.2. *Under the assumptions of Theorem 20.1, $F^\varepsilon(x,t)$ and $h^\varepsilon(x,t)$ satisfy*

$$\|F^\varepsilon\|_{L^\infty(\mathbb{R}^d \times (0,T))} \leqslant C \tag{20.3.11}$$

and

$$\|h^\varepsilon\|_{L^\infty(\mathbb{R}^d)} \leqslant C, \tag{20.3.12}$$

respectively, where the constant C is independent of ε.

Proof. We have that

$$\begin{aligned} F(x,y,t) = {} & \mathcal{L}_1 \left(\Theta(y) : \nabla_x \nabla_x u(x,t) \right) + \mathcal{L}_2 \left(\chi(y) \cdot \nabla_x u(x,t) \right) \\ & - \frac{\partial}{\partial t} \left(\chi(y) \cdot \nabla_x u(x,t) \right) + \varepsilon \big(D \Delta_x \Theta(y) : \nabla_x \nabla_x u(x,t) \\ & - \frac{\partial}{\partial t} \left(\Theta(y) : \nabla_x \nabla_x u(x,t) \right) \big). \end{aligned} \tag{20.3.13}$$

Estimate (20.3.4), together with L^∞ bounds on u and its derivatives, imply (20.3.11).

Furthermore,

$$h^\varepsilon(x) = \chi\left(\frac{x}{\varepsilon}\right) \cdot \nabla_x u(x,0) + \varepsilon\Theta\left(\frac{x}{\varepsilon}\right) : \nabla_x \nabla_x u(x,0).$$

The uniform estimates on $\chi(y)$ and $\Theta(y)$, together with our assumptions on the initial conditions of (13.2.4), lead to estimate (20.3.12). □

Proof of Theorem 20.1 The remainder term $R^\varepsilon(x,t)$ satisfies Equation (20.3.10a). We use the maximum principle estimate (7.3.6) from Chapter 7 to obtain

$$\begin{aligned}\|R^\varepsilon\|_{L^\infty(\mathbb{R}^d\times(0,T))} &\leqslant \varepsilon\|h^\varepsilon\|_{L^\infty(\mathbb{R}^d)} + \varepsilon\int_0^T \|F^\varepsilon(\cdot,s)\|_{L^\infty(\mathbb{R}^d)}\,ds \\ &\leqslant \varepsilon C + \varepsilon CT \\ &\leqslant C\varepsilon. \end{aligned} \tag{20.3.14}$$

We combine (20.3.6) with (20.3.14) and (20.3.5) and use the triangle inequality to obtain

$$\begin{aligned}\|u^\varepsilon - u\|_{L^\infty(\mathbb{R}^d\times(0,T))} &= \|\varepsilon u_1 + \varepsilon^2 u_2 + R^\varepsilon\|_{L^\infty(\mathbb{R}^d\times(0,T))} \\ &\leqslant \varepsilon\|u_1\|_{L^\infty(\mathbb{R}^d\times(0,T))} + \varepsilon^2\|u_2\|_{L^\infty(\mathbb{R}^d\times(0,T))} \\ &\quad + \|R^\varepsilon\|_{L^\infty(\mathbb{R}^d\times(0,T))} \leqslant C\varepsilon,\end{aligned}$$

from which (20.2.3) follows. □

20.4 Discussion and Bibliography

Our proof of the homogenization theorem has relied on the maximum principle. This is, perhaps, the simplest approach, provided that one is willing to assume sufficient regularity on the coefficients of the PDE, and it is very well suited for PDEs in unbounded domains. However, other techniques may be used to study homogenization for parabolic equations. These include probabilistic methods [248], energy methods [33, ch. 2], and the method of two-scale convergence [5]. Estimates based on the maximum principle can be used to prove homogenization theorems for nonlinear (even fully nonlinear) elliptic and parabolic PDEs, using the theory of viscosity solutions; see, for example, [203, 95, 96, 97].

The method employed in this chapter, namely the derivation of a PDE for the error term and the use a priori estimates to control the data driving the error equation, can be applied to various problems in the theory of singular perturbations. Some other examples can be found in [240, 246]. The method is often termed *bootstrapping*. For nonlinear parabolic PDEs with random coefficients, a subject of some interest is the existence of traveling fronts and the effect of noise on them. See [333] for references to the literature in this area.

20.5 Exercises

1. Consider the case where $\nabla \cdot b(x) = 0$. Assume that the solution of the Cauchy problem (13.2.4) is smooth and bounded and decays sufficiently fast at infinity.
   a. Consider the inhomogeneous Cauchy problem

$$\frac{\partial R}{\partial t} = \frac{1}{\varepsilon} b^{\varepsilon} \cdot \nabla R + D\Delta R + F(x,t) \quad \text{for } (x,t) \in \mathbb{R}^d \times (0,T),$$
$$R(x,0) = f(x) \quad \text{for } x \in \mathbb{R}^d,$$

   where $f \in L^2(\mathbb{R}^d)$ and $F \in L^2((0,T) \times \mathbb{R}^d)$. Prove the estimate

$$\|R\|^2_{L^2((0,T)\times\mathbb{R}^d)} + C_1\|\nabla R\|^2_{L^2((0,T)\times\mathbb{R}^d)} \leqslant C_2\|f\|^2_{L^2(\mathbb{R}^d)} + C_3\|F\|^2_{L^2((0,T)\times\mathbb{R}^d)}.$$

   b. Use this to prove convergence in $L^2((0,T) \times \mathbb{R}^d)$.
   c. What is the maximum time interval $(0,T)$ over which the homogenization theorem holds?
2. Carry out the same program as in the previous question for the Cauchy problem

$$\frac{\partial u^{\varepsilon}}{\partial t} = \nabla \cdot \left(A\left(\frac{x}{\varepsilon}\right) \nabla u^{\varepsilon} \right) \quad \text{for } (x,t) \in \mathbb{R}^d \times (0,T),$$
$$u^{\varepsilon}(x,0) = u_{in}(x), \quad \text{for } x \in \mathbb{R}^d,$$

   where the matrix $A(y)$ satisfies the standard periodicity, smoothness, and uniform ellipticity assumptions.
3. Consider the initial boundary value problem

$$\frac{\partial u^{\varepsilon}}{\partial t} = \nabla \cdot \left(A\left(\frac{x}{\varepsilon}\right) \nabla u^{\varepsilon} \right) \quad \text{for } (x,t) \in \Omega \times (0,T),$$
$$u^{\varepsilon}(x,0) = u_{in}(x), \quad \text{for } x \in \overline{\Omega}$$
$$u^{\varepsilon}(x,t) = 0, \quad \text{for } (x,t) \in \partial\Omega \times [0,T],$$

   where Ω is a bounded domain in $\mathbb{R}^d$ with smooth boundary and $A(y)$ satisfies the standard assumptions.
   a. Prove the estimate

$$\|u^{\varepsilon}\|^2_{L^{\infty}((0,T)\times\Omega)} + C\|u^{\varepsilon}\|^2_{L^2((0,T);H_0^1(\Omega))} \leqslant \|u_{in}\|^2_{L^2(\Omega)}.$$

   b. Use the preceding estimate to prove the homogenization theorem using the method of two-scale convergence.
   c. Can you apply the bootstrapping method to this problem?
4. Use the maximum principle for elliptic PDEs from Section 7.2.4 and the techniques used in this chapter to prove a homogenization theorem for a second-order uniformly elliptic PDE in nondivergence form, with rapidly oscillating coefficients.

21

Averaging for Linear Transport and Parabolic PDEs: The Convergence Theorem

21.1 Introduction

In this chapter we prove an averaging theorem for linear transport and linear parabolic PDEs with periodic, divergence-free velocity fields. We treat the parabolic case ($D > 0$) and the transport case ($D = 0$) separately. This is because the averaged equations in these two cases are, in general, different.[1]

In Section 21.2 we present our two convergence theorems. We prove the averaging theorem for the case $D > 0$ in Section 21.3. In Section 21.4 we show that, in the case $D = 0$ and when the velocity field does not generate an ergodic flow on the unit torus, the homogenized limit leads to a coupled system of equations, the two-scale system. The proof of both theorems is based on the method of two-scale convergence. We finish the chapter in Section 21.5 with comments and bibliographical remarks.

21.2 The Theorems

As in Chapter 14 we define

$$b^{\varepsilon}(x) = b(x/\varepsilon).$$

As we confine our analysis to divergence-free fields, it is natural to define

$$\overline{b} = \int_{\mathbb{T}^d} b(y)dy.$$

This is the form of $\overline{b}$ given in Result 14.1 when the invariant density ρ is a constant function; this is exactly the case that arises for divergence-free fields.

[1] In the PDE literature, the limiting equations established in this chapter are sometimes referred to as the homogenized equations, rather than the averaged equations. However, as discussed in Chapter 14, we choose to use the consistent terminology introduced in Section 1.3.

Recall the concept of weak solutions for transport equations encapsulated in Definition 7.24 and Remark 7.28. These concepts may also be extended to parabolic equations. Our first theorem concerns the parabolic case.

Theorem 21.1. *Let $u^\varepsilon(x,t)$ be the weak solution of*

$$\frac{\partial u^\varepsilon}{\partial t} - b^\varepsilon \cdot \nabla u^\varepsilon = \varepsilon D \Delta u^\varepsilon \quad \text{for } (x,t) \in \mathbb{R}^d \times \mathbb{R}^+, \tag{21.2.1a}$$

$$u^\varepsilon = g \quad \text{for } x \in \mathbb{R}^d \tag{21.2.1b}$$

and assume that $D > 0$, $g \in C_b^\infty(\mathbb{R}^d)$, and b is smooth, divergence-free, and 1-periodic. Then u^ε two-scale converges to $u_0 \in L^2(\mathbb{R}^+ \times \mathbb{R}^d)$, which is a weak solution of

$$\frac{\partial u_0}{\partial t} - \overline{b} \cdot \nabla u_0 = 0 \quad \text{for } (x,t) \in \mathbb{R}^d \times \mathbb{R}^+, \tag{21.2.2a}$$

$$u_0 = g \quad \text{for } (x,t) \in \mathbb{R}^d \times \{0\}. \tag{21.2.2b}$$

Now we consider the case $D = 0$.

Theorem 21.2. *Let $u^\varepsilon(x,t)$ be the weak solution of*

$$\frac{\partial u^\varepsilon}{\partial t} - b^\varepsilon \cdot \nabla u^\varepsilon = 0 \quad \text{for } (x,t) \in \mathbb{R}^d \times \mathbb{R}^+, \tag{21.2.3a}$$

$$u^\varepsilon = g \quad \text{for } x \in \mathbb{R}^d \tag{21.2.3b}$$

and assume that $g \in C_b^\infty(\mathbb{R}^d)$ and that b is smooth, divergence-free, and 1-periodic. Then u^ε two-scale converges to $u_0 \in L^2(\mathbb{R}^+ \times \mathbb{R}^d; L^2_{per}(\mathbb{T}^d))$, which satisfies the two-scale system

$$\int_{\mathbb{R}^+} \int_{\mathbb{R}^d} \int_{\mathbb{T}^d} b(y) \cdot \nabla_y \phi(x,y,t) u_0(x,y,t)\, dy dx dt = 0, \tag{21.2.4a}$$

$$\phi \in C_0^\infty(\mathbb{R} \times \mathbb{R}^d; C^\infty_{per}(\mathbb{T}^d)), \tag{21.2.4b}$$

$$\begin{aligned} &\int_{\mathbb{R}^+} \int_{\mathbb{R}^d} \int_{\mathbb{T}^d} \left(\frac{\partial \phi(x,y,t)}{\partial t} - b(y) \cdot \nabla_x \phi(x,y,t) \right) u_0(x,y,t)\, dy dx dt \\ + &\int_{\mathbb{R}^d} g(x) \left(\int_{\mathbb{T}^d} \phi(x,y,0)\, dy \right) dx = 0. \end{aligned} \tag{21.2.4c}$$

Furthermore, these equations have a unqiue solution.

Notice that the two-scale system involves an auxiliary function $\phi(x,y,t) \in C_0^\infty(\mathbb{R} \times \mathbb{R}^d; C^\infty_{per}(\mathbb{T}^d))$. As in the elliptic case, to obtain a closed equation for the first term in the multiscale expansion we need to be able to decouple the two-scale system. We can decouple the two-scale system and obtain an averaged transport equation in the case where the function b generates an ergodic flow on the unit torus.

Corollary 21.3. *Assume that b is a smooth, divergence-free vector field, generating an ergodic flow on $\mathbb{T}^d$. Then the two-scale limit is independent of y and is a weak solution of the Cauchy problem*

$$\frac{\partial u_0}{\partial t} - \overline{b} \cdot \nabla u_0 = 0 \quad \textit{for} \ (x,t) \in \mathbb{R}^d \times \mathbb{R}^+, \tag{21.2.5a}$$

$$u_0 = g \quad \textit{for} \ (x,t) \in \mathbb{R}^d \times \{0\}. \tag{21.2.5b}$$

21.3 The Proof: $D > 0$

Notice that, from Theorem 7.22 and since both b^ε and f are smooth and bounded, there exists a unique classical solution to the Cauchy problem (21.2.1). Furthermore, the solution of the problem decays to 0 as $|x| \to \infty$. This fact justifies the integrations by parts that follow.

We first obtain an energy estimate. We multiply Equation (21.2.1a) by u^ε, integrate over $\mathbb{R}^d$, use the fact that b^ε is divergence-free, and integrate by parts to obtain

$$\frac{1}{2}\frac{d}{dt}\int_{\mathbb{R}^d} |u^\varepsilon(\cdot,x)|^2 \, dx + D\varepsilon \int_{\mathbb{R}^d} |\nabla u^\varepsilon(\cdot,x)|^2 \, dx = 0.$$

We now integrate over $(0,T)$ to deduce that

$$\int_{\mathbb{R}^d} |u^\varepsilon(\cdot,t)|^2 \, dx + 2D\varepsilon \int_0^T \int_{\mathbb{R}^d} |\nabla u^\varepsilon|^2 \, dxdt = \int_{\mathbb{R}^d} |g|^2 \, dx.$$

Hence,

$$\|u^\varepsilon\|^2_{L^\infty((0,T);L^2(\mathbb{R}^d))} + C\varepsilon \|\nabla u^\varepsilon\|^2_{L^2((0,T)\times\mathbb{R}^d)} \leqslant C. \tag{21.3.1}$$

Notice that T in this estimate is arbitrary.

In view of Theorem 2.39, Part (iii), estimate (21.3.1) implies that there exists a function $u(x,t) \in L^2((0,T) \times \mathbb{R}^d)$ such that u^ε two-scale converges to $u(x,t)$. Furthermore, the estimate

$$\|\nabla u^\varepsilon\|_{L^2((0,T)\times\mathbb{R}^d)} \leqslant C\varepsilon^{-1/2}$$

implies that $\varepsilon \nabla u^\varepsilon$ converges weakly to 0 in $L^2((0,T) \times \mathbb{R}^d)$.

Now let $\phi^\varepsilon(x,t) \in C_0^\infty(\mathbb{R} \times \mathbb{R}^d)$. The weak formulation of (21.2.1) reads

$$\int_{\mathbb{R}^+}\int_{\mathbb{R}^d} \left(\frac{\partial \phi^\varepsilon}{\partial t} - b^\varepsilon \cdot \nabla \phi^\varepsilon \right) u^\varepsilon \, dxdt + \int_{\mathbb{R}^d} g(x)\phi^\varepsilon(x,0) \, dx$$
$$= \varepsilon D \int_{\mathbb{R}^+ \times \mathbb{R}^d} \nabla u^\varepsilon \cdot \nabla \phi^\varepsilon \, dxdt. \tag{21.3.2}$$

We choose a smooth test function that is independent of the oscillations: $\phi^\varepsilon = \phi(x,t)$. We use this in (21.3.2) and pass to the limit as $\varepsilon \to 0$. We use the fact that the two-scale limit of u^ε is independent of y to obtain

$$\int_{\mathbb{R}^+}\int_{\mathbb{R}^d}\left(\frac{\partial\phi^\varepsilon}{\partial t} - b^\varepsilon\cdot\nabla\phi^\varepsilon\right)u^\varepsilon\,dxdt + \int_{\mathbb{R}^d} g(x)\phi^\varepsilon(x,0)\,dx$$
$$\to \int_{\mathbb{R}^+}\int_{\mathbb{R}^d}\left(\frac{\partial\phi}{\partial t} - \overline{b}\cdot\nabla\phi\right)u\,dxdt + \int_{\mathbb{R}^d} g(x)\phi(x,0)\,dx,$$

where

$$\overline{b} := \int_{\mathbb{T}^d} b(y)\,dy.$$

Furthermore, since $\varepsilon\nabla u^\varepsilon$ converges to 0 weakly in $L^2((0,T)\times\mathbb{R}^d)$, we have that

$$\varepsilon D\int_{\mathbb{R}^+\times\mathbb{R}^d}\nabla u^\varepsilon\cdot\nabla\phi^\varepsilon\,dxdt \to 0.$$

We combine the preceding two limits to obtain

$$\int_{\mathbb{R}^+}\int_{\mathbb{R}^d}\left(\frac{\partial\phi}{\partial t} - \overline{b}\cdot\nabla\phi\right)u\,dxdt + \int_{\mathbb{R}^d} g(x)\phi(x,0)\,dx = 0,$$

which leads to the weak formulation of (21.2.2) (see Remark 7.28). □

21.4 The Proof: $D = 0$

Proof of Theorem 21.2. By Theorem 7.27 there exists a unique weak solution u^ε to (21.2.3) satisfying $u^\varepsilon \in L^2((0,T);H^1(\mathbb{R}^d))$. Furthermore, estimate (7.4.3) gives

$$\|u^\varepsilon\|^2_{L^\infty(\mathbb{R}^+;L^2(\mathbb{R}^d))} \leqslant \|g\|^2_{L^2(\mathbb{R}^d)} \leqslant C. \tag{21.4.1}$$

This estimate implies, by Theorem 2.38, that there exists a subsequence, still denoted by u^ε, that two-scale converges to a function $u_0 \in L^2(\mathbb{R}^+\times\mathbb{R}^d; L^2_{per}(\mathbb{T}^d))$.

Since b is divergence-free, the weak formulation of (21.2.3), Equation (7.4.9), gives

$$\int_{\mathbb{R}^+}\int_{\mathbb{R}^d}\left(\frac{\partial\phi^\varepsilon}{\partial t} - b^\varepsilon\cdot\nabla\phi^\varepsilon\right)u^\varepsilon\,dxdt + \int_{\mathbb{R}^d} g(x)\phi^\varepsilon(x,0)\,dx = 0, \tag{21.4.2}$$

for every $\phi^\varepsilon(x,t)\in C_0^\infty(\mathbb{R}\times\mathbb{R}^d)$. We choose a test function of the form

$$\phi^\varepsilon = \varepsilon\phi\left(x,\frac{x}{\varepsilon},t\right), \quad \phi\in C_0^\infty(\mathbb{R}\times\mathbb{R}^d; C^\infty_{per}(\mathbb{T}^d)).$$

Inserting ϕ^ε in (21.4.2), using the chain rule (ϕ^ε is a function of both x and x/ε), and passing to the limit as ε tends to 0 we obtain:

$$\int_{\mathbb{R}^+}\int_{\mathbb{R}^d}\int_{\mathbb{T}^d} b(y)\cdot\nabla_y\phi(x,y,t)u_0(x,y,t)\,dydxdt = 0.$$

This is Equation (21.2.4a). We now choose a test function

$$\phi^\varepsilon = \phi\left(x, \frac{x}{\varepsilon}, t\right), \quad \phi \in C_0^\infty(\mathbb{R} \times \mathbb{R}^d; C_{per}^\infty(\mathbb{T}^d))$$

such that (21.2.4b) is satisfied. We insert this test function in (21.4.2) and pass to the limit as ε tends to 0 to obtain

$$\int_{\mathbb{R}^+} \int_{\mathbb{R}^d} \int_{\mathbb{T}^d} \left(\frac{\partial \phi(x,y,t)}{\partial t} - b(y) \cdot \nabla_x \phi(x,y,t) \right) u_0(x,y,t)\, dy dx dt$$
$$+ \quad \int_{\mathbb{R}^d} g(x) \left(\int_Y \phi(x,y,0)\, dy \right) dx = 0,$$

which is precisely (21.2.4c). □

Of course, in order for the homogenized system of equations to be of any interest, we need to prove that it has a unique solution. This is the content of the following theorem.

Theorem 21.4. *There exists a unique solution* $u_0(t,x,y) \in L^2(\mathbb{R}^+ \times \mathbb{R}^d \times \mathbb{T}^d)$ *of Equations* (21.2.4).

Proof. The existence of a solution follows from the existence of a two-scale limit for u^ε. Let us proceed with uniqueness. We use precisely the same argument as in the proof of uniqueness of solutions in Theorem 7.27. Let $u_1(x,y,t),\ u_2(x,y,t)$ be two solutions of the homogenized system with the same initial conditions. We form the difference

$$e(x,y,t) = u_1(x,y,t) - u_2(x,y,t).$$

This function satisfies, by linearity, the same system of Equations (21.2.4), with zero initial conditions. We have:

$$\int_0^T \int_{\mathbb{R}^d} \int_{\mathbb{T}^d} \left(\frac{\partial \phi(x,y,t)}{\partial t} - b(y) \cdot \nabla_x \phi(x,y,t) \right) e(x,y,t)\, dy dx dt = 0,$$

where $T > 0$ is arbitrary but fixed. A variant of Theorem 7.27 gives that the difference e satisfies $e \in L^2((0,T); H^1(\mathbb{R}^d); L^2(\mathbb{T}^d))$ with $\frac{\partial e}{\partial t} \in L^2((0,T) \times \mathbb{R}^d \times \mathbb{T}^d))$. Hence, we can use it as a test function in the preceding equation to deduce that

$$\int_0^T \int_{\mathbb{R}^d} \int_{\mathbb{T}^d} \left(\frac{\partial e(x,y,t)}{\partial t} - b(y) \cdot \nabla_x e(x,y,t) \right) e(x,y,t)\, dy dx dt = 0.$$

Assume momentarily that e has compact support. Integration by parts in the second term in the equation gives, using the fact that b is divergence-free,

$$\int_{\mathbb{R}^d} \int_{\mathbb{T}^d} b(y) \cdot \nabla_x e(x,y,t) e(x,y,t)\, dy\, dx$$
$$= - \int_{\mathbb{R}^d} \int_{\mathbb{T}^d} (b(y) e(x,y,t)) \cdot \nabla_x e(x,y,t)\, dy\, dx,$$

from which we deduce that this term vanishes. An approximation argument implies that the same conclusion is valid for $e \in L^2((0,T); H^1(\mathbb{R}^d) \times L^2(\mathbb{T}^d))$. Thus we deduce that

$$\frac{1}{2}\int_0^T \frac{d}{dt}\int_{\mathbb{R}^d}\int_{\mathbb{T}^d} |e(x,y,t)|^2 \, dydxdt = 0,$$

for arbitrary $T > 0$. From this equation, it follows that, since e is zero initially,

$$\int_{\mathbb{R}^d}\int_{\mathbb{T}^d} |e(x,y,t)|^2 \, dy \, dx = 0$$

for every $t > 0$. Consequently, $e(x,y,t) \equiv 0$ and, thus, the solution $u_0(x,y,t)$ is unique. □

Proof of Corollary 21.3. In the case where $b(y)$ is an ergodic, divergence-free vector field, Equation (21.2.4a) is satisfied if and only if u_0 is independent of y. This is because an integration by parts shows that this equation is the weak formulation of the first-order PDE

$$\mathcal{L}_0 u_0 = 0$$

where $\mathcal{L}_0 = -b(y) \cdot \nabla_y$ is equipped with periodic boundary conditions on $\mathbb{T}^d$. Then, we can obtain the (weak formulation of) the averaged Equation (21.2.5) from (21.2.4c) by choosing a test function independent of y. □

21.5 Discussion and Bibliography

Results similar to Theorem 21.1 can be found in, e.g., [5, 203, 253]. See [203, prop. 3.2] for a proof based on the perturbed test function method [96] in the context of viscosity solutions for PDEs. The result holds in the case where the velocity field b depends explicitly on both x and x/ε, i.e.,

$$b^\varepsilon = b\left(x, \frac{x}{\varepsilon}\right),$$

where $b(x,y)$ is 1-periodic in y (see Exercise 5). Now the averaged velocity is a function of x, $\overline{b} = \overline{b}(x)$. It is also possible to obtain the next order correction of (21.2.2) [45, 253]. Indeed, the function $\overline{u}^\varepsilon = u + \varepsilon\langle u_1\rangle$, where $\langle\cdot\rangle$ denotes the average over $\mathbb{T}^d$, satisfies, up to terms of $O(\varepsilon^2)$, the advection–diffusion equation:

$$\frac{\partial \overline{u}^\varepsilon}{\partial t} + \overline{b}(x) \cdot \nabla \overline{u}^\varepsilon = \varepsilon \nabla \cdot \left(\mathcal{K}(x) \nabla \overline{u}^\varepsilon\right).$$

The effective diffusivity $\mathcal{K}(x)$ is obtained through the solution of an appropriate cell problem; notice that it is a function of x.

In the proof of Theorem 21.2 we have followed [80]. A complete (in the L^2 sense) set of test functions was used in [147] to characterize the averaged (sometimes termed homogenized) limit in the general two-dimensional case. It was shown there that the limit was an infinite symmetric set of linear hyperbolic equations. The method of characteristics was used to prove the limit theorem for two-dimensional flows in [312]. The problem of averaging for ODEs (and of the corresponding Liouville equation) of the form

$$\frac{dx}{dt} = -\nabla\left(F(x) + \varepsilon V\left(\frac{x}{\varepsilon}\right)\right)$$

is considered in [226].

21.6 Exercises

1. Consider the following Cauchy problem

$$\frac{\partial u^{\varepsilon}(x,t)}{\partial t} - b\left(x, \frac{x}{\varepsilon}\right) \cdot \nabla u^{\varepsilon}(x,t) = 0 \quad \text{for } (x,t) \in \mathbb{R}^d \times \mathbb{R}^+,$$

$$u^{\varepsilon}(x,0) = g(x) \quad \text{for } x \in \mathbb{R}^d,$$

where $b \in C_b^{\infty}(\mathbb{R}^d, C_{per}^{\infty}(\mathbb{T}^d); \mathbb{R}^d)$ with $\nabla \cdot b(x, x/\varepsilon) = 0$. Use the method of two-scale convergence to prove the averaging theorem.

2. Carry out a similar program as in the previous exercise for the forced transport PDE

$$\frac{\partial u^{\varepsilon}(x,t)}{\partial t} - b(x) \cdot \nabla u^{\varepsilon}(x,t) = g\left(x, \frac{x}{\varepsilon}\right) \quad \text{for } (x,t) \in \mathbb{R}^d \times \mathbb{R}^+,$$

$$u^{\varepsilon}(x,0) = g(x) \quad \text{for } x \in \mathbb{R}^d,$$

with $g(x,y)$ being smooth, periodic in its second argument and $g(x, x/\varepsilon)$ is bounded in $L^2(\mathbb{R}^d)$.

3. Carry out a similar program as in the previous exercise for the transport Equation (21.2.3a) with oscillating initial data

$$u^{\varepsilon}(x,0) = f\left(x, \frac{x}{\varepsilon}\right),$$

where $g(x,y)$ being smooth, periodic in its second argument, and $g(x, x/\varepsilon)$ is bounded in $L^2(\mathbb{R}^d)$.

4. Combine the preceding exercises to prove the averaging theorem for the transport PDE

$$\frac{\partial u^{\varepsilon}(x,t)}{\partial t} - b\left(x, \frac{x}{\varepsilon}\right) \cdot \nabla u^{\varepsilon}(x,t) = g\left(x, \frac{x}{\varepsilon}\right) \quad \text{for } (x,t) \in \mathbb{R}^d \times \mathbb{R}^+,$$

$$u^{\varepsilon}(x,0) = f\left(x, \frac{x}{\varepsilon}\right) \quad \text{for } x \in \mathbb{R}^d,$$

5. Prove the averaging theorem for the advection–diffusion equation

$$\frac{\partial u^{\varepsilon}(x,t)}{\partial t} - b\left(x, \frac{x}{\varepsilon}\right) \cdot \nabla u^{\varepsilon}(x,t) = \varepsilon D \Delta u^{\varepsilon} + g\left(x, \frac{x}{\varepsilon}\right) \quad \text{for } (x,t) \in \mathbb{R}^d \times \mathbb{R}^+,$$

$$u^{\varepsilon}(x,0) = f\left(x, \frac{x}{\varepsilon}\right) \quad \text{for } x \in \mathbb{R}^d,$$

under the assumptions of the previous exercise.

References

1. A. Abdulle. Multiscale methods for advection-diffusion problems. *Discrete Contin. Dyn. Syst.*, suppl:11–21, 2005.
2. A. Abdulle and W. E. Finite difference heterogeneous multi-scale method for homogenization problems. *J. Comput. Phys.*, 191(1):18–39, 2003.
3. A. Abdulle and Ch. Schwab. Heterogeneous multiscale FEM for diffusion problems on rough surfaces. *Multiscale Model. Simul.*, 3(1):195–220 (electronic), 2004/05.
4. R. A. Adams and J. J. F. Fournier. *Sobolev Spaces*. Pure and Applied Mathematics. Elsevier, Oxford, 2003.
5. G. Allaire. Homogeneization et convergence a deux echelles. application a un probleme de convection diffusion. *C.R. Acad. Sci. Paris*, 312:581–6, 1991.
6. G. Allaire. Homogenization and two-scale convergence. *SIAM J. Math. Anal.*, 23(6): 1482–1518, 1992.
7. G. Allaire. *Shape optimization by the homogenization method*, volume 146 of *Applied Mathematical Sciences*. Springer-Verlag, New York, 2002.
8. G. Allaire and M. Briane. Multiscale convergence and reiterated homogenisation. *Proc. Roy. Soc. Edinburgh Sect. A*, 126(2):297–342, 1996.
9. G. Allaire and R. Brizzi. A multiscale finite element method for numerical homogenization. *Multiscale Model. Simul.*, 4(3):790–812 (electronic), 2005.
10. G. Allaire and A. Piatnitski. Homogenization of the Schrödinger equation and effective mass theorems. *Comm. Math. Phys.*, 258:1–22, 2005.
11. R. Aris. On the dispersion of a solute in a fluid flowing through a tube. *Proc. R. Soc. Lond. A*, 235:67–77, 1956.
12. L. Arnold. *Stochastic differential equations: theory and applications*. Wiley-Interscience [John Wiley & Sons], New York, 1974. Translated from the German.
13. V. I. Arnol′d. *Mathematical methods of classical mechanics*, volume 60 of *Graduate Texts in Mathematics*. Springer-Verlag, New York, 1989.
14. V. I. Arnol′d and A. Avez. *Ergodic problems of classical mechanics*. Translated from the French by A. Avez. W. A. Benjamin, Inc., New York-Amsterdam, 1968.
15. V.I. Arnol′d. *Ordinary differential equations*. Springer-Verlag, Berlin, 1992.
16. G. Ben Arous and H. Owhadi. Multiscale homogenization with bounded ratios and anomalous slow diffusion. *Comm. Pure Appl. Math.*, 56(1):80–113, 2003.
17. Z. Artstein. On singularly perturbed ordinary differential equations with measure-valued limits. *Math. Bohem.*, 127(2):139–52, 2002.

18. Z. Artstein and M. Slemrod. On singularly perturbed retarded functional-differential equations. *J. Differential Equations*, 171(1):88–109, 2001.
19. Z. Artstein and A. Vigodner. Singularly perturbed ordinary differential equations with dynamic limits. *Proc. Roy. Soc. Edinburgh Sect. A*, 126(3):541–69, 1996.
20. M. Avellanada and A. J. Majda. Stieltjes integral representation and effective diffusivity bounds for turbulent transport. *Phys. Rev. Lett.*, 62:753–5, 1989.
21. M. Avellanada and A. J. Majda. An integral representation and bounds on the effective diffusivity in passive advection by laminar and turbulent flows. *Comm. in Math. Phys.*, 138:339–91, 1991.
22. M. Avellanada and M. Vergassola. Stieltjes integral representation of effective diffusivities in time-dependent flows. *Phys. Rev. E*, 52:3249–51, 1995.
23. I. Babuška. Homogenization and its application. Mathematical and computational problems. In *Numerical solution of partial differential equations, III (Proc. Third Sympos. (SYNSPADE), Univ. Maryland, College Park, Md., 1975)*, pages 89–116. Academic Press, New York, 1976.
24. I. Babuška. Homogenization approach in engineering. In *Computing methods in applied sciences and engineering (Second Internat. Sympos., Versailles, 1975), Part 1*, pages 137–53. Lecture Notes in Econom. and Math. Systems, Vol. 134. Springer, Berlin, 1976.
25. I. Babuška. Solution of interface problems by homogenization. I. *SIAM J. Math. Anal.*, 7(5):603–34, 1976.
26. I. Babuška. Solution of interface problems by homogenization. II. *SIAM J. Math. Anal.*, 7(5):635–45, 1976.
27. N. Bakhvalov and G. Panasenko. *Homogenisation: averaging processes in periodic media.* Kluwer Academic Publishers Group, Dordrecht, 1989. Mathematical problems in the mechanics of composite materials. Translated from the Russian by D. Leĭtes.
28. R. Balescu. *Statistical dynamics. Matter out of equilibrium.* Imperial College Press, London, 1997.
29. J. M. Ball. A version of the fundamental theorem for Young measures. In *PDEs and continuum models of phase transitions (Nice, 1988)*, volume 344 of *Lecture Notes in Phys.*, pages 207–15. Springer, Berlin, 1989.
30. D. Barkley, I. G. Kevrekidis, and A. M. Stuart. The moment map: nonlinear dynamics of density evolution via a few moments. *SIADS*, 5(3):403–34, 2006.
31. C. Beck. Brownian motion from deterministic dynamics. *Phys. A*, 169(2):324–36, 1990.
32. G. Benettin, L. Galgani, and A. Giorgilli. Realization of holonomic constraints and freezing of high frequency degrees of freedom in the light of classical perturbation theory. I. *Comm. Math. Phys.*, 113(1):87–103, 1987.
33. A. Bensoussan, J.-L. Lions, and G. Papanicolaou. *Asymptotic analysis for periodic structures*, volume 5 of *Studies in Mathematics and Its Applications*. North-Holland Publishing Co., Amsterdam, 1978.
34. N. Berglund and B. Gentz. *Noise-induced phenomena in slow-fast dynamical systems.* Probability and Its Applications (New York). Springer-Verlag, London Ltd., London, 2006. A sample-paths approach.
35. R. N. Bhattacharya, V. K. Gupta, and H. F. Walker. Asymptotics of solute dispersion in periodic porous media. *SIAM J. APPL. MATH*, 49(1):86–98, 1989.
36. P. Billingsley. *Probability and measure*. Wiley Series in Probability and Mathematical Statistics. John Wiley & Sons Inc., New York, third edition, 1995. A Wiley-Interscience Publication.
37. P. Billingsley. *Convergence of probability measures.* Wiley Series in Probability and Statistics: Probability and Statistics. John Wiley & Sons Inc., New York, 1999.

38. D. Blömker. Amplitude equations for locally cubic non-autonomous nonlinearities. *SIAM Journal on Applied Dynamical Systems*, 2(3):464–86, 2003.
39. D. Blömker. Approximation of the stochastic Rayleigh-Bénard problem near the onset of convection and related problems. *Stoch. Dyn.*, 5(3):441–74, 2005.
40. D. Blömker, M. Hairer, and G. A. Pavliotis. Multiscale analysis for stochastic PDEs with quadratic nonlinearities. *Nonlinearity*, 20:1721–44, 2007.
41. D. Blömker, S. Maier-Paape, and G. Schneider. The stochastic Landau equation as an amplitude equation. *Discrete and Continuous Dynamical Systems, Series B*, 1(4):527–41, 2001.
42. J. Bonn and R. M. McLaughlin. Sensitive enhanced diffusivities for flows with fluctuating mean winds: a two-parameter study. *J. Fluid Mech.*, 445:345–75, 2001.
43. F. Bornemann. *Homogenization in time of singularly perturbed mechanical systems*, volume 1687 of *Lecture Notes in Mathematics*. Springer-Verlag, Berlin, 1998.
44. F. A. Bornemann and Ch. Schütte. Homogenization of Hamiltonian systems with a strong constraining potential. *Phys. D*, 102(1-2):57–77, 1997.
45. A. Bourgeat, M. Jurak, and A. L. Piatnitski. Averaging a transport equation with small diffusion and oscillating velocity. *The Mathematics Preprint Server*, 2001.
46. S. Brahim-Otsmane, G. A. Francfort, and F. Murat. Homogenization in thermoelasticity. In *Random media and composites (Leesburg, VA, 1988)*, pages 13–45. SIAM, Philadelphia, 1989.
47. S. Brahim-Otsmane, G. A. Francfort, and F. Murat. Correctors for the homogenization of the wave and heat equations. *J. Math. Pure Appl. (9)*, 71(3):197–231, 1992.
48. A. Braides. *Γ-convergence for beginners*, volume 22 of *Oxford Lecture Series in Mathematics and Its Applications*. Oxford University Press, Oxford, 2002.
49. A. Braides and A. Defranceschi. *Homogenization of multiple integrals*, volume 12 of *Oxford Lecture Series in Mathematics and Its Applications*. The Clarendon Press Oxford University Press, New York, 1998.
50. L. Breiman. *Probability*, volume 7 of *Classics in Applied Mathematics*. Society for Industrial and Applied Mathematics (SIAM), Philadelphia, 1992. Corrected reprint of the 1968 original.
51. Y. Brenier. Remarks on some linear hyperbolic equations with oscillatory coefficients. In *Third International Conference on Hyperbolic Problems, Vol. I, II (Uppsala, 1990)*, pages 119–30. Studentlitteratur, Lund, 1991.
52. M. E. Brewster and G. Beylkin. A multiresolution strategy for numerical homogenization. *Appl. Comput. Harmon. Anal.*, 2(4):327–49, 1995.
53. H. Brezis. *Analyse fonctionnelle*. Collection Mathématiques Appliquées pour la Maîtrise. [Collection of Applied Mathematics for the Master's Degree]. Masson, Paris, 1983.
54. F. Campillo and A. Piatnitski. Effective diffusion in vanishing viscosity. In *Nonlinear partial differential equations and their applications. Collège de France Seminar, Vol. XIV (Paris, 1997/1998)*, volume 31 of *Stud. Math. Appl.*, pages 133–45. North-Holland, Amsterdam, 2002.
55. R. A. Carmona and F. Cerou. Transport by incompressible random velocity fields: simulations and mathematical conjectures. In *Stochastic partial differential equations: six perspectives*, volume 64 of *Math. Surveys Monogr.*, pages 153–81. Amer. Math. Soc., Providence, RI, 1999.
56. R. A. Carmona and L. Xu. Homogenization theory for time-dependent two-dimensional incompressible gaussian flows. *The Annals of Applied Probability*, 7(1):265–79, 1997.
57. J. Carr. *Applications of centre manifold theory*, volume 35 of *Applied Mathematical Sciences*. Springer-Verlag, New York, 1981.

58. S. Cerrai. *Second order PDE's in finite and infinite dimension*, volume 1762 of *Lecture Notes in Mathematics*. Springer-Verlag, Berlin, 2001.
59. X. Chao and A. J. Roberts. On the low-dimensional modelling of Stratonovich stochastic differential equations. *Physica A*, 225:62–80, 1996. http://dx.doi.org/10.1016/0378-4371(95)00387-8.
60. Z. Chen and T. Y. Hou. A mixed multiscale finite element method for elliptic problems with oscillating coefficients. *Math. Comp.*, 72(242):541–76 (electronic), 2003.
61. A. Chertock and D. Levy. On wavelet-based numerical homogenization. *Multiscale Model. Simul.*, 3(1):65–88 (electronic), 2004/05.
62. S. Childress. Alpha effect in flux ropes and sheets. *Phys. Earth and Planet. Int.*, 20:172–80, 1979.
63. S. Childress and A. M. Soward. Scalar transport and alpha-effect for a family of cat's-eye flows. *J. Fluid Mech.*, 205:99–133, 1989.
64. A. J. Chorin and O. H. Hald. *Stochastic tools in mathematics and science*, volume 1 of *Surveys and Tutorials in the Applied Mathematical Sciences*. Springer, New York, 2006.
65. A. J. Chorin, O. H. Hald, and R. Kupferman. Optimal prediction with memory. *Phys. D*, 166(3-4):239–57, 2002.
66. D. Cioranescu and P. Donato. *An Introduction to Homogenization*. Oxford University Press, New York, 1999.
67. D. Cioranescu and J. Saint Jean Pauline. *Homogenization of reticulated structures*, volume 136 of *Applied Mathematical Sciences*. Springer-Verlag, New York, 1999.
68. E. A. Coddington and N. Levinson. *Theory of ordinary differential equations*. McGraw-Hill Book Company, Inc., New York, Toronto, London, 1955.
69. P. Constantin, C. Foias, B. Nicolaenko, and R. Temam. *Integral manifolds and inertial manifolds for dissipative partial differential equations*, volume 70 of *Applied Mathematical Sciences*. Springer-Verlag, New York, 1989.
70. I. P. Cornfeld, S. V. Fomin, and Ya. G. Sinaĭ. *Ergodic theory*, volume 245 of *Grundlehren der Mathematischen Wissenschaften [Fundamental Principles of Mathematical Sciences]*. Springer-Verlag, New York, 1982. Translated from the Russian by A. B. Sosinskiĭ.
71. S. M. Cox and A. J. Roberts. Initial conditions for models of dynamical systems. *Physica D*, 85:126–41, 1995.
72. B. Dacorogna. *Direct methods in the calculus of variations*. Springer-Verlag, Berlin, Heidelberg, 1989.
73. E. De Giorgi and S. Spagnolo. Sulla convergenza degli integrali dell'energia per operatori ellittici del secondo ordine. *Boll. Un. Mat. Ital. (4)*, 8:391–411, 1973.
74. A. De Masi, P. A. Ferrari, S. Goldstein, and W. D. Wick. An invariance principle for reversible Markov processes. Applications to random motions in random environments. *J. Statist. Phys.*, 55(3-4):787–855, 1989.
75. A. Defranceschi. An introduction to homogenization and G-convergence. Lecture notes, http://www.sissa.it/fa/publications/lecture_notes/, 1993.
76. M. Dorobantu and B. Engquist. Wavelet-based numerical homogenization. *SIAM J. Numer. Anal.*, 35(2):540–59 (electronic), 1998.
77. J. Duan, K. Lu, and B. Schmalfuss. Invariant manifolds for stochastic partial differential equations. *Annals of Probability*, 31:2109–35, 2006.
78. J. Duan and B. Nadiga. Stochastic parameterization for large eddy simulation of geophysical flows. *Proc. Ameri. Math. Soc.*, 2006.
79. E. B. Dynkin. *Markov processes. Vols. I, II*, volume 122 of *Translated with the authorization and assistance of the author by J. Fabius, V. Greenberg, A. Maitra, G. Majone.*

Die Grundlehren der Mathematischen Wi ssenschaften, Bände 121. Academic Press Inc., New York, 1965.

80. W. E. Homogenization of linear and nonlinear transport equations. *Communications in Pure and Applied Mathematics*, 45:301–26, 1992.
81. W. E and B. Engquist. The heterogeneous multiscale methods. *Commun. Math. Sci.*, 1(1):87–132, 2003.
82. W. E and B. Engquist. Multiscale modeling and computation. *Notices Amer. Math. Soc.*, 50(9):1062–70, 2003.
83. W. E., B. Engquist, X. Li, W. Ren, and E. Vanden-Eijnden. Heterogeneous multiscale methods: A review. *Communications in Computational Physics*, 2(3):367–450, 2007.
84. W. E., D. Liu, and E. Vanden-Eijnden. Analysis of multiscale methods for stochastic differential equations. *Comm. Pure Appl. Math.*, 58(11):1544–85, 2005.
85. W. E., D. Liu, and E. Vanden-Eijnden. Nested stochastic simulation algorithm for chemical kinetic systems with disparate rates. *J. Chem. Phys.*, 123:194107, 2005.
86. W. E., D. Liu, and E. Vanden-Eijnden. Nested stochastic simulation algorithm for chemical kinetic systems with multiple time scales,. *J. Comp. Phys.*, 221(1):158–180, 2007.
87. R. E. Edwards. *Functional analysis*. Dover Publications Inc., New York, 1995.
88. Y. Efendiev, T. Hou, and V. Ginting. Multiscale finite element methods for nonlinear problems and their applications. *Commun. Math. Sci.*, 2(4):553–89, 2004.
89. Y. Efendiev and A. Pankov. Numerical homogenization of monotone elliptic operators. *Multiscale Model. Simul.*, 2(1):62–79 (electronic), 2003.
90. Y. Efendiev and A. Pankov. Numerical homogenization and correctors for nonlinear elliptic equations. *SIAM J. Appl. Math.*, 65(1):43–68 (electronic), 2004.
91. H. I. Ene and E. Sánchez-Palencia. Équations et phénomènes de surface pour l′écoulement dans un modèle de milieu poreux. *J. Mécanique*, 14:73–108, 1975.
92. B. Engquist. Wavelet based numerical homogenization. In *Proceedings of the International Congress of Mathematicians, Vol. III (Berlin, 1998)*, number Extra Vol. III, pages 503–512 (electronic), 1998.
93. B. Engquist and Y.-H. Tsai. The multiscale hetergeneous methods for stiff ODEs. *Math. Comp.*, 74:1707–42, 2005.
94. S. N. Ethier and T. G. Kurtz. *Markov processes*. Wiley Series in Probability and Mathematical Statistics: Probability and Mathematical Statistics. John Wiley & Sons Inc., New York, 1986.
95. L. C. Evans. *Weak convergence methods for nonlinear partial differential equations*, volume 74 of *CBMS Regional Conference Series in Mathematics*. Published for the Conference Board of the Mathematical Sciences, Washington, DC, 1990.
96. L. C. Evans. The perturbed test function method for viscosity solutions of nonlinear pde. *Proc. Roy. Soc. Edinburgh*, 111A:359–75, 1989.
97. L. C. Evans. Periodic homogenisation of certain fully nonlinear partial differential equations. *Proc. Roy. Soc. Edinburgh*, 120:245–65, 1992.
98. L. C. Evans. *Partial Differential Equations*. AMS, Providence, RI, 1998.
99. G. Falkovich, K. Gawędzki, and M. Vergassola. Particles and fields in fluid turbulence. *Rev. Modern Phys.*, 73(4):913–75, 2001.
100. A. Fannjiang and G. C. Papanicolaou. Convection enhanced diffusion for periodic flows. *SIAM J. APPL. MATH*, 54:333–408, 1994.
101. A. Fannjiang and G. C. Papanicolaou. Convection enhanced diffusion for random flows. *J. Stat. Phys.*, 88:1033–76, 1997.
102. W. Feller. *An introduction to probability theory and its applications. Vol. I*. Third edition. John Wiley & Sons Inc., New York, 1968.

103. W. Feller. *An introduction to probability theory and its applications. Vol. II.* Second edition. John Wiley & Sons Inc., New York, 1971.
104. N. Fenichel. Persistence and smoothness of invariant manifolds for flows. *Indiana Univ. Math. J.*, 21:193–226, 1971/1972.
105. N. Fenichel. Geometric singular perturbation theory for ordinary differential equations. *J. Differential Equations*, 31(1):53–98, 1979.
106. J-P. Fouque, G. C. Papanicolaou, and R. K. Sircar. *Derivatives in financial markets with stochastic volatility.* Cambridge University Press, Cambridge, 2000.
107. G. A. Francfort and F. Murat. Homogenization and optimal bounds in linear elasticity. *Arch. Rational Mech. Anal.*, 94(4):307–34, 1986.
108. G. A. Francfort and P-M. Suquet. Homogenization and mechanical dissipation in thermoviscoelasticity. *Arch. Rational Mech. Anal.*, 96(3):265–93, 1986.
109. M. Freidlin. Some remarks on the Smoluchowski-Kramers approximation. *J. Statist. Phys.*, 117(3-4):617–34, 2004.
110. M. I. Freĭdlin. The Dirichlet problem for an equation with periodic coefficients depending on a small parameter. *Teor. Verojatnost. i Primenen.*, 9:133–9, 1964.
111. M. I. Freidlin and A. D. Wentzell. *Random perturbations of dunamical systems.* Springer-Verlag, New York, 1984.
112. A. Friedman. *Partial differential equations of parabolic type.* Prentice-Hall Inc., Englewood Cliffs, NJ, 1964.
113. A. Friedman. *Partial differential equations.* Holt, Rinehart and Winston, New York, 1969.
114. A. Friedman. *Stochastic differential equations and applications. Vol. 1.* Academic Press [Harcourt Brace Jovanovich Publishers], New York, 1975. Probability and Mathematical Statistics, Vol. 28.
115. F. R. Gantmacher. *The theory of matrices. Vol. 1.* AMS Chelsea Publishing, Providence, RI, 1998. Translated from the Russian by K. A. Hirsch, Reprint of the 1959 translation.
116. B. García-Archilla, J. M. Sanz-Serna, and R. D. Skeel. Long-time-step methods for oscillatory differential equations. *SIAM J. Sci. Comput.*, 20(3):930–63 (electronic), 1999.
117. C. W. Gardiner. *Handbook of stochastic methods.* Springer-Verlag, Berlin, second edition, 1985.
118. J. Garnier. Homogenization in a periodic and time-dependent potential. *SIAM J. APPL. MATH*, 57(1):95–111, 1997.
119. C. W. Gear and I. G. Kevrekidis. Projective methods for stiff differential equations: problems with gaps in their eigenvalue spectrum. *SIAM J. Sci. Comput.*, 24(4):1091–1106 (electronic), 2003.
120. I. I. Gikhman and A. V. Skorokhod. *Introduction to the theory of random processes.* Dover Publications Inc., Mineola, NY, 1996.
121. D. Gilbarg and N. S. Trudinger. *Elliptic partial differential equations of second order.* Classics in Mathematics. Springer-Verlag, Berlin, 2001.
122. D. T. Gillespie. A general method for numerically simulating the stochastic time evolution of coupled chemical reactions. *J. Computational Phys.*, 22(4):403–34, 1976.
123. D. Givon, I. G. Kevrekidis, and R. Kupferman. Strong convergence schemes of projective intregration schemes for singularly perturbed stochastic differential equations. *Comm. Math. Sci.*, 4(4):707–29, 2006.
124. D. Givon and R. Kupferman. White noise limits for discrete dynamical systems driven by fast deterministic dynamics. *Phys. A*, 335(3-4):385–412, 2004.
125. D. Givon, R. Kupferman, and A. M. Stuart. Extracting macroscopic dynamics: model problems and algorithms. *Nonlinearity*, 17(6):R55–R127, 2004.

126. K. Golden, S. Goldstein, and J. L. Lebowitz. Classical transport in modulated structures. *Phys. Rev. Lett.*, 55(24):2629–32, 1985.
127. S. Goldstein. Antisymmetric functionals of reversible Markov processes. *Ann. Inst. H. Poincaré Probab. Statist.*, 31(1):177–90, 1995.
128. O. Gonzalez and A. M. Stuart. *Continuum Mechanics*. Texts in applied mathematics. Cambridge University Press, 2008.
129. R. Graham and A. Schenzle. Stabilization by multiplicative noise. *Phys. Rev. A*, 26(3):1676–85, 1982.
130. G. R. Grimmett and D. R. Stirzaker. *Probability and random processes*. Oxford University Press, New York, 2001.
131. J. Guckenheimer and P. Holmes. *Nonlinear oscillations, dynamical systems, and bifurcations of vector fields*, volume 42 of *Applied Mathematical Sciences*. Springer-Verlag, New York, 1983.
132. R. B. Guenther and J. W. Lee. *Partial differential equations of mathematical physics and integral equations*. Dover Publications Inc., Mineola, NY, 1996.
133. M. Briane, G. W. Milton, and J. R. Willis. On cloaking for elasticity and physical equations with a transformation invariant form. *New J. Phys.*, 8:248–67, 2006.
134. M. Hairer and G. A. Pavliotis. Periodic homogenization for hypoelliptic diffusions. *J. Statist. Phys.*, 117(1-2):261–79, 2004.
135. J. K. Hale. Averaging methods for differential equations with retarded arguments and a small parameter. *J. Differential Equations*, 2:57–73, 1966.
136. J. K Hale. *Asymptotic behavior of dissipative systems*, volume 25 of *Mathematical Surveys and Monographs*. American Mathematical Society, Providence, RI, 1988.
137. P. Hall and C. C. Heyde. *Martingale limit theory and its application*. Academic Press Inc. [Harcourt Brace Jovanovich Publishers], New York, 1980. Probability and Mathematical Statistics.
138. P. Hanggi, P. Talkner, and M. Borkovec. Reaction-rate theory: fifty years after Kramers. *Rev. Modern Phys.*, 62(2):251–341, 1990.
139. R. Z. Has′minskiĭ. *Stochastic stability of differential equations*, volume 7 of *Monographs and Textbooks on Mechanics of Solids and Fluids: Mechanics and Analysis*. Sijthoff & Noordhoff, Alphen aan den Rijn, 1980.
140. S. Heinze. Diffusion-advection in cellular flows with large Peclet numbers. *Arch. Ration. Mech. Anal.*, 168(4):329–42, 2003.
141. V. H. Hoang and Ch. Schwab. High-dimensional finite elements for elliptic problems with multiple scales. *Multiscale Model. Simul.*, 3(1):168–94 (electronic), 2004/05.
142. A. Holmbom. Homogenization of parabolic equations: An alternative approach and some corrector-type results. *Applications of Mathematics*, 42(5):321–43, 1995.
143. M. H. Holmes. *Introduction to perturbation methods*. Springer-Verlag, New York, 1995.
144. W. Horsthemke and R. Lefever. *Noise-induced transitions*, volume 15 of *Springer Series in Synergetics*. Springer-Verlag, Berlin, 1984. Theory and applications in physics, chemistry, and biology.
145. T. Y. Hou and X-H. Wu. A multiscale finite element method for elliptic problems in composite materials and porous media. *J. Comput. Phys.*, 134(1):169–89, 1997.
146. T. Y. Hou, X-H. Wu, and Z. Cai. Convergence of a multiscale finite element method for elliptic problems with rapidly oscillating coefficients. *Math. Comp.*, 68(227):913–43, 1999.
147. T. Y. Hou and X. Xin. Homogenization of linear transport equations with oscillatory vector fields. *SIAM J. APPL. MATH*, 52(1):34–45, 1992.
148. W. Huisinga, Ch. Schütte, and A. M. Stuart. Extracting macroscopic stochastic dynamics: model problems. *Comm. Pure Appl. Math.*, 56(2):234–69, 2003.

149. G. Humer and I. G. Kevrekidis. Coarse molecular dynamics of a peptide fragment: free energy, kinetics and long time dynamics computations. *J. Chem. Phys.*, 118(23):10762–73, 2003.
150. A. R. Humphries and A. M. Stuart. Deterministic and random dynamical systems: theory and numerics. In *Modern methods in scientific computing and applications (Montréal, QC, 2001)*, volume 75 of *NATO Sci. Ser. II Math. Phys. Chem.*, pages 211–54. Kluwer Acad. Publ., Dordrecht, 2002.
151. N. Ikeda and S. Watanabe. *Stochastic differential equations and diffusion processes.* North-Holland Publishing Co., Amsterdam, second edition, 1989.
152. J. Jacod and A. N. Shiryaev. *Limit theorems for stochastic processes*, volume 288 of *Grundlehren der Mathematischen Wissenschaften [Fundamental Principles of Mathematical Sciences].* Springer-Verlag, Berlin, 2003.
153. V. V. Jikov, S. M. Kozlov, and O. A. Oleinik. *Homogenization of differential operators and integral functionals.* Springer, Berlin, 1994.
154. W. Just, K. Gelfert, N. Baba, A. Riegert, and H. Kantz. Elimination of fast chaotic degrees of freedom: on the accuracy of the Born approximation. *J. Statist. Phys.*, 112(1-2):277–92, 2003.
155. W. Just, H. Kantz, Ch. Rödenbeck, and M. Helm. Stochastic modelling: replacing fast degrees of freedom by noise. *J. Phys. A*, 34(15):3199–3213, 2001.
156. I. Karatzas and S. E. Shreve. *Brownian motion and stochastic calculus*, volume 113 of *Graduate Texts in Mathematics.* Springer-Verlag, New York, second edition, 1991.
157. T. Kato. *Perturbation theory for linear operators.* Classics in Mathematics. Springer-Verlag, Berlin, 1995. Reprint of the 1980 edition.
158. J. B. Keller. Darcy's law for flow in porous media and the two-space method. In *Nonlinear partial differential equations in engineering and applied science (Proc. Conf., Univ. Rhode Island, Kingston, R.I., 1979)*, volume 54 of *Lecture Notes in Pure and Appl. Math.*, pages 429–43. Dekker, New York, 1980.
159. J. B. Keller. Effective behavior of heterogeneous media. In *Statistical mechanics and statistical methods in theory and application (Proc. Sympos., Univ. Rochester, Rochester, N.Y., 1976)*, pages 631–44. Plenum, New York, 1977.
160. J. B. Keller and J. Rubinstein. Nonlinear wave motion in a strong potential. *Wave Motion*, 13(3):291–302, 1991.
161. J. Kevorkian and J. D. Cole. *Multiple scale and singular perturbation methods*, volume 114 of *Applied Mathematical Sciences.* Springer-Verlag, New York, 1996.
162. I. G. Kevrekidis, C. W. Gear, J. M. Hyman, P. G. Kevrekidis, O. Runborg, and C. Theodoropoulos. Equation-free, coarse-grained multiscale computation: enabling microscopic simulators to perform system-level analysis. *Commun. Math. Sci.*, 1(4):715–62, 2003.
163. I. G. Kevrekidis, A. Makeev, and D. Maroudas. "coarse" stability and bifurcation analysis using timesteppers. *J. Chem. Phys.*, 116:10083–91, 2002.
164. I. G. Kevrekidis, Y. H. Qian, and K. Theodoropoulos. "Coarse" stability and bifurcation analysis using stochastic simulators: Kinetic Monte Carlo examples. *Proc. Natl. Acad. Sci.*, 97(18):9840–3, 2000.
165. R. Z. Khasminskii. The averaging principle for parabolic and elliptic differential equations and Markov processes with small diffusion. *Teor. Verojatnost. i Primenen.*, 8:3–25, 1963.
166. R. Z. Khasminskii. A limit theorem for solutions of differential equations with a random right hand part. *Teor. Verojatnost. i Primenen*, 11:444–62, 1966.
167. R. Z. Khasminskii and G. Yin. Asymptotic series for singularly perturbed Kolmogorov-Fokker-Planck equations. *SIAM J. Appl. Math.*, 56(6):1766–93, 1996.

168. Y. Kifer. *Random perturbations of dynamical systems*, volume 16 of *Progress in Probability and Statistics*. Birkhäuser Boston Inc., Boston, 1988.
169. Y. Kifer. Averaging in dynamical systems and large deviations. *Invent. Math.*, 110(2):337–70, 1992.
170. Y. Kifer. Limit theorems in averaging for dynamical systems. *Ergodic Theory Dynam. Systems*, 15(6):1143–72, 1995.
171. Y. Kifer. Averaging and climate models. In *Stochastic climate models (Chorin, 1999)*, volume 49 of *Progr. Probab.*, pages 171–88. Birkhäuser, Basel, 2001.
172. Y. Kifer. Stochastic versions of Anosov's and Neistadt's theorems on averaging. *Stoch. Dyn.*, 1(1):1–21, 2001.
173. C. Kipnis and S. R. S. Varadhan. Central limit theorem for additive functionals of reversible Markov processes and applications to simple exclusions. *Comm. Math. Phys.*, 104(1):1–19, 1986.
174. D. L. Koch and J. F. Brady. The symmetry properties of the effective diffusivity tensor in anisotropic porous media. *Phys. Fluids*, 31(5):965–73, 1988.
175. D. L. Koch, R. G. Cox, H. Brenner, and J. F. Brady. The effect of order on dispersion in porous media. *J. Fluid Mech.*, 200:173–88, 1989.
176. B. O. Koopman. Hamiltonian systems of transformations in Hilbert spaces. *Proc. Nat. Acad. Sci. (U.S.A.)*, 17:315–18, 1931.
177. S. M. Kozlov. The averaging of random operators. *Mat. Sb. (N.S.)*, 109(151)(2):188–202, 327, 1979.
178. H-O Kreiss. Problems with different time scales. In *Acta numerica, 1992*, Acta Numer., pages 101–39. Cambridge Univ. Press, Cambridge, 1992.
179. E. Kreyszig. *Introductory functional analysis with applications*. Wiley Classics Library. John Wiley & Sons Inc., New York, 1989.
180. R. Kupferman, G. A. Pavliotis, and A. M. Stuart. Itô versus Stratonovich white-noise limits for systems with inertia and colored multiplicative noise. *Phys. Rev. E (3)*, 70(3): 036120, 2004.
181. T. G. Kurtz. A limit theorem for perturbed operator semigroups with applications to random evolutions. *J. Functional Analysis*, 12:55–67, 1973.
182. T. G. Kurtz. Limit theorems and diffusion approximations for density dependent Markov chains. *Math. Programming Stud.*, 5:67–78, 1976.
183. T. G. Kurtz. Strong approximation theorems for density dependent Markov chains. *Stochastic Processes Appl.*, 6(3):223–40, 1977/78.
184. T. G. Kurtz. The central limit theorems for Markov chains. *The Annals of Probability*, 9:557–60, 1981.
185. C. Landim. Central limit theorem for Markov processes. In *From classical to modern probability*, volume 54 of *Progr. Probab.*, pages 145–205. Birkhäuser, Basel, 2003.
186. C. Landim and S. Olla. *Central limit theorems for Markov processes*. Lecture Notes, 2005.
187. B. Lapeyre, E. Pardoux, and R. Sentis. *Introduction to Monte-Carlo methods for transport and diffusion equations*, volume 6 of *Oxford Texts in Applied and Engineering Mathematics*. Oxford University Press, Oxford, 2003. Translated from the 1998 French original by Alan Craig and Fionn Craig.
188. A. Lasota and M. C. Mackey. *Chaos, fractals, noise*. Springer-Verlag, New York, 1994.
189. P. D. Lax. *Functional analysis*. Wiley, NY, 2002.
190. J. Li, P. G. Kevrekidis, C. W. Gear, and I. G. Kevrekidis. Deciding the nature of the coarse equation through microscopic simulations: the baby-bathwater scheme. *Multiscale Model. Simul.*, 1(3):391–407 (electronic), 2003.

191. S. Lifson and J. L. Jackson. On the self–diffusion of ions in polyelectrolytic solution. *J. Chem. Phys*, 36:2410, 1962.

192. T. Lindvall. *Lectures on the coupling method.* Dover Publications Inc., Mineola, NY, 2002. Corrected reprint of the 1992 original.

193. P. Lochak and C. Meunier. *Multiphase averaging for classical systems*, volume 72 of *Applied Mathematical Sciences.* Springer-Verlag, New York, 1988.

194. M. Loève. *Probability theory. I.* Springer-Verlag, New York, fourth edition, 1977. Graduate Texts in Mathematics, Vol. 45.

195. M. Loève. *Probability theory. II.* Springer-Verlag, New York, fourth edition, 1978. Graduate Texts in Mathematics, Vol. 46.

196. L. Lorenzi and M. Bertoldi. *Analytical methods for Markov semigroups.* CRC Press, New York, 2006.

197. D. Lukkassen, G. Nguetseng, and P. Wall. Two-scale convergence. *Int. J. Pure Appl. Math.*, 2(1):35–86, 2002.

198. T. J. Lyons. Differential equations driven by rough signals. *Rev. Mat. Iberoamericana*, 14(2):215–310, 1998.

199. R. S. MacKay. Slow manifolds. In *Energy Localisation and Transfer*, pages 149–92. World Scientific, 2004.

200. A. J. Majda and P. R. Kramer. Simplified models for turbulent diffusion: Theory, numerical modelling and physical phenomena. *Physics Reports*, 314:237–574, 1999.

201. A. J. Majda and R. M. McLaughlin. The effect of mean flows on enhanced diffusivity in transport by incompressible periodic velocity fields. *Stud. Appl. Math.*, 89(3):245–79, 1993.

202. A. J. Majda, R. V. Abramov, and M. J. Grote. *Information theory and stochastics for multiscale nonlinear systems*, volume 25 of *CRM Mongraph Series.* AMS, Providence, RI, 2005.

203. A. J. Majda and P. Souganidis. The effect of turbulence on mixing in prototype reaction-diffusion systems. *Comm. Pure and Appl. Math.*, 53:1284–1304, 2000.

204. A. J. Majda and I. Timofeyev. Remarkable statistical behavior for truncated Burgers-Hopf dynamics. *Proc. Natl. Acad. Sci. USA*, 97(23):12413–17 (electronic), 2000.

205. A. J. Majda, I. Timofeyev, and E. Vanden Eijnden. A mathematical framework for stochastic climate models. *Comm. Pure Appl. Math.*, 54(8):891–974, 2001.

206. A. J. Majda, I. Timofeyev, and E. Vanden Eijnden. Models for stochastic climate prediction. *Proc. Natl. Acad. Sci. USA*, 96(26):14687–91 (electronic), 1999.

207. A. J. Majda, I. Timofeyev, and E. Vanden-Eijnden. Systematic strategies for stochastic mode reduction in climate. *J. Atmospheric Sci.*, 60(14):1705–22, 2003.

208. A. M. Majda, I. Timofeyev, and E. Vanden-Eijnden. A priori tests of a stochastic mode reduction strategy. *Phys. D*, 170(3-4):206–52, 2002.

209. A. M. Majda, I. Timofeyev, and E. Vanden-Eijnden. Stochastic models for selected slow variables in large deterministic systems. *Nonlinearity*, 19(4):769–94, 2006.

210. X. Mao. *Stochastic differential equations and their applications.* Horwood Publishing Series in Mathematics & Applications. Horwood Publishing Limited, Chichester, 1997.

211. X. Mao and C. Yuan. *Stochastic differential equations with Markovian switching*. Imperial College Press, London, 2006.

212. J. E. Marsden and M. McCracken. *The Hopf bifurcation and its applications.* Springer-Verlag, New York, 1976. With contributions by P. Chernoff, G. Childs, S. Chow, J. R. Dorroh, J. Guckenheimer, L. Howard, N. Kopell, O. Lanford, J. Mallet-Paret, G. Oster, O. Ruiz, S. Schecter, D. Schmidt and S. Smale, Applied Mathematical Sciences, Vol. 19.

213. M.-L. Mascarenhas and A.-M. Toader. Scale convergence in homogenization. *Numer. Func. Anal. Optim.*, 22:127–58, 2001.
214. G. Dal Maso. *An Introduction to Γ–convergence*. Birkhauser, Boston, 1993.
215. A.-M. Matache and Ch. Schwab. Finite dimensional approximations for elliptic problems with rapidly oscillating coefficients. In *Multiscale problems in science and technology (Dubrovnik, 2000)*, pages 203–42. Springer, Berlin, 2002.
216. A.-M. Matache and Ch. Schwab. Two-scale FEM for homogenization problems. *M2AN Math. Model. Numer. Anal.*, 36(4):537–72, 2002.
217. M. R. Maxey and J. J. Riley. Equation of motion for a small rigid sphere in a nonuniform flow. *Phys. Fluids*, 26:883–9, 1983.
218. R. M. Mazo. *Brownian motion*, volume 112 of *International Series of Monographs on Physics*. Oxford University Press, New York, 2002.
219. A. Mazzino. Effective correlation times in turbulent scalar transport. *Phys. Rev. E*, 56:5500–10, 1997.
220. A. Mazzino, S. Musacchio, and A. Vulpiani. Multiple-scale analysis and renormalization for preasymptotic scalar transport. *Phys. Rev. E (3)*, 71(1):011113, 2005.
221. D. W. McLaughlin, G. C. Papanicolaou, and O. R. Pironneau. Convection of microstructure and related problems. *SIAM J. APPL. MATH*, 45:780–97, 1985.
222. R. M. McLaughlin. Numerical averaging and fast homogenization. *J. Statist. Phys.*, 90(3-4):597–626, 1998.
223. R. M. McLaughlin and R. M. Forest. An anelastic, scale-separated model for mixing, with application to atmospheric transport phenomena. *Phys. Fluids*, 11(4):880–92, 1999.
224. I. Melbourne and M. Nicol. Large deviations for nonuniformly hyperbolic systems. Preprint, 2006.
225. I. Melbourne and A. M. Stuart. SDE limits for scale-separated mixing ODEs. Preprint, 2007.
226. G. Menon. Gradient systems with wiggly energies and related averaging problems. *Arch. Ration. Mech. Anal.*, 162(3):193–246, 2002.
227. S. P. Meyn and R. L. Tweedie. *Markov chains and stochastic stability*. Communications and Control Engineering Series. Springer-Verlag London Ltd., London, 1993.
228. I. Mezic, J. F. Brady, and S. Wiggins. Maximal effective diffusivity for time-periodic incompressible fluid flows. *SIAM J. APPL. MATH*, 56(1):40–56, 1996.
229. G. W. Milton. *The theory of composites*, volume 6 of *Cambridge Monographs on Applied and Computational Mathematics*. Cambridge University Press, Cambridge, 2002.
230. G. W. Milton and J. R. Willis. On modifications of newton's second law and linear continuum elastodynamics. *Proc. R Soc. A*, 463:855–80, 2007.
231. P. Ming and X. Yue. Numerical methods for multiscale elliptic problems. *J. Comput. Phys.*, 214(1):421–45, 2006.
232. F. Murat and L. Tartar. H-convergence. In *Topics in the mathematical modelling of composite materials*, volume 31 of *Progr. Nonlinear Differential Equations Appl.*, pages 21–43. Birkhäuser Boston, Boston, 1997.
233. G. Nguetseng. A general convergence result for a functional related to the theory of homogenization. *SIAM J. Math. Anal.*, 20(3):608–23, 1989.
234. G. Nguetseng. Asymptotic analysis for a stiff variational problem arising in mechanics. *SIAM J. Math. Anal.*, 21(6):1394–1414, 1990.
235. J. R. Norris. *Markov chains*. Cambridge Series in Statistical and Probabilistic Mathematics. Cambridge University Press, Cambridge, 1998.
236. B. Øksendal. *Stochastic differential equations*. Universitext. Springer-Verlag, Berlin, 2003.

237. O. A. Oleĭnik, A. S. Shamaev, and G. A. Yosifian. *Mathematical problems in elasticity and homogenization*, volume 26 of *Studies in Mathematics and its Applications*. North-Holland Publishing Co., Amsterdam, 1992.
238. S. Olla. *Homogenization of diffusion processes in random fields*. Lecture Notes, 1994.
239. R. E. O'Malley, Jr. *Singular perturbation methods for ordinary differential equations*, volume 89 of *Applied Mathematical Sciences*. Springer-Verlag, New York, 1991.
240. G. C. Papanicolaou. Some probabilistic problems and methods in singular perturbations. *Rocky Mountain J. Math.*, 6(4):653–74, 1976.
241. G. C. Papanicolaou. Introduction to the asymptotic analysis of stochastic equations. In *Modern modeling of continuum phenomena (Ninth Summer Sem. Appl. Math., Rensselaer Polytech. Inst., Troy, N.Y., 1975)*, pages 109–47. Lectures in Appl. Math., Vol. 16. Amer. Math. Soc., Providence, RI, 1977.
242. G. C. Papanicolaou and W. Kohler. Asymptotic theory of mixing stochastic ordinary differential equations. *Comm. Pure Appl. Math.*, 27:641–68, 1974.
243. G. C. Papanicolaou, D.W. Stroock, and S. R. S. Varadhan. Martingale approach to some limit theorems. In *Papers from the Duke Turbulence Conference (Duke Univ., Durham, N.C., 1976), Paper No. 6*, pages ii+120 pp. Duke Univ. Math. Ser., Vol. III. Duke Univ., Durham, NC, 1977.
244. G. C. Papanicolaou and S. R. S. Varadhan. A limit theorem with strong mixing in Banach space and two applications to stochastic differential equations. *Comm. Pure Appl. Math.*, 26:497–524, 1973.
245. G. C. Papanicolaou. Asymptotic analysis of transport processes. *Bull. Amer. Math. Soc.*, 81:330–92, 1975.
246. G. C. Papanicolaou. Asymptotic analysis of stochastic equations. In *Studies in probability theory*, volume 18 of *MAA Stud. Math.*, pages 111–79. Math. Assoc. America, Washington, DC, 1978.
247. G. C. Papanicolaou and S. R. S. Varadhan. Ornstein-Uhlenbeck process in a random potential. *Comm. Pure Appl. Math.*, 38(6):819–34, 1985.
248. E. Pardoux. Homogenization of linear and semilinear second order parabolic pdes with periodic coefficients: A probabilistic approach. *Journal of Functional Analysis*, 167:498–520, 1999.
249. E. Pardoux and A. Yu. Veretennikov. On the Poisson equation and diffusion approximation. I. *Ann. Probab.*, 29(3):1061–85, 2001.
250. È. Pardoux and A. Yu. Veretennikov. On Poisson equation and diffusion approximation. II. *Ann. Probab.*, 31(3):1166–92, 2003.
251. E. Pardoux and A. Yu. Veretennikov. On the Poisson equation and diffusion approximation. III. *Ann. Probab.*, 33(3):1111–33, 2005.
252. K. R. Parthasarathy. *Probability measures on metric spaces*. AMS Chelsea Publishing, Providence, RI, 2005. Reprint of the 1967 original.
253. G. A. Pavliotis. *Homogenization theory for advection – diffusion equations with mean flow*, Ph.D Thesis. Rensselaer Polytechnic Institute, Troy, NY, 2002.
254. G. A. Pavliotis and A. M. Stuart. White noise limits for inertial particles in a random field. *Multiscale Model. Simul.*, 1(4):527–33 (electronic), 2003.
255. G. A. Pavliotis and A. M. Stuart. Analysis of white noise limits for stochastic systems with two fast relaxation times. *Multiscale Model. Simul.*, 4(1):1–35 (electronic), 2005.
256. G. A. Pavliotis and A. M. Stuart. Periodic homogenization for inertial particles. *Phys. D*, 204(3-4):161–87, 2005.
257. G. A. Pavliotis. A multiscale approach to Brownian motors. *Phys. Lett. A*, 344:331–45, 2005.

258. G. A. Pavliotis and A. M Stuart. Parameter estimation for multiscale diffusions. *J. Stat. Phys.*, 127(4):741–81, 2007.
259. A. Pazy. *Semigroups of linear operators and applications to partial differential equations*. Springer-Verlag, New York, 1983.
260. K. Petersen. *Ergodic theory*, volume 2 of *Cambridge Studies in Advanced Mathematics*. Cambridge University Press, Cambridge, 1983.
261. M. A. Pinsky. *Lectures on random evolution*. World Scientific Publishing Co. Inc., River Edge, NJ, 1991.
262. P. K. Pollett and A. J. Roberts. A description of the long-term behaviour of absorbing continuous time Markov chains using a centre manifold. *Advances Applied Probability*, 22:111–28, 1990.
263. G. Da Prato and J. Zabczyk. *Stochastic equations in infinite dimensions*, volume 44 of *Encyclopedia of Mathematics and Its Applications*. Cambridge University Press, 1992.
264. M. H. Protter and H. F. Weinberger. *Maximum principles in differential equations*. Springer-Verlag, New York, 1984.
265. S. Raimondeau and D. G. Vlachos. Recent developments on multiscale, hierarchical modeling of chemical reactors. *Chem. Eng. J.*, 90:3–23, 2002.
266. M. Reed and B. Simon. *Methods of Modern Mathematical Physics I: Functional Analysis*. Academic Press, New York, 1980.
267. P. Reimann. Brownian motors: noisy transport far from equilibrium. *Phys. Rep.*, 361(2-4):57–265, 2002.
268. P. Reimann, C. Van den Broeck, H. Linke, J. M. Rubi, and A. Perez-Madrid. Giant acceleration of free diffusion by use of tilted periodic potentials. *Phys. Rev. Lett.*, 87(1):010602, 2001.
269. P. Resibois and M. De Leener. *Classical kinetic theory of fluids*. Wiley, New York, 1977.
270. D. Revuz and M. Yor. *Continuous martingales and Brownian motion*, volume 293 of *Grundlehren der Mathematischen Wissenschaften [Fundamental Principles of Mathematical Sciences]*. Springer-Verlag, Berlin, third edition, 1999.
271. H. Risken. *The Fokker-Planck equation*, volume 18 of *Springer Series in Synergetics*. Springer-Verlag, Berlin, 1989.
272. A. J. Roberts. Appropriate initial conditions for asymptotic descriptions of the long term evolution of dynamical systems. *J. Austral. Math. Soc. B*, 31:48–75, 1989.
273. A. J. Roberts. Resolving the multitude of microscale interactions accurately models stochastic partial differential equations. *LMS J. Comput. Math.*, 9:193–221, 2006.
274. J. C. Robinson. *Infinite-dimensional dynamical systems*. Cambridge Texts in Applied Mathematics. Cambridge University Press, Cambridge, 2001.
275. L. C. G. Rogers and David Williams. *Diffusions, Markov processes, and martingales. Vol. 1*. Cambridge Mathematical Library. Cambridge University Press, Cambridge, 2000.
276. L. C. G. Rogers and D. Williams. *Diffusions, Markov processes, and martingales. Vol. 2*. Cambridge Mathematical Library. Cambridge University Press, Cambridge, 2000.
277. H. Rubin and P. Ungar. Motion under a strong constraining force. *Comm. Pure Appl. Math.*, 10:65–87, 1957.
278. G. Samaey, D. Roose, and I. G. Kevrekidis. The gap-tooth scheme for homogenization problems. *Multiscale Model. Simul.*, 4(1):278–306 (electronic), 2005.
279. E. Sánchez-Palencia. *Nonhomogeneous media and vibration theory*, volume 127 of *Lecture Notes in Physics*. Springer-Verlag, Berlin, 1980.
280. J. M. Sancho, M. San Miguel, and D. Dürr. Adiabatic elimination for systems of Brownian particles with nonconstant damping coefficients. *J. Statist. Phys.*, 28(2):291–305, 1982.

281. J. A. Sanders and F. Verhulst. *Averaging methods in nonlinear dynamical systems*, volume 59 of *Applied Mathematical Sciences*. Springer-Verlag, New York, 1985.
282. M Schreier, P. Reimann, P. Hänggi, and E. Pollak. Giant enhancement of diffusion and particle selection in rocked periodic potentials. *Europhys. Let.*, 44(4):416–22, 1998.
283. Z. Schuss. *Theory and applications of stochastic differential equations*. John Wiley & Sons Inc., New York, 1980. Wiley Series in Probability and Statistics.
284. Ch. Schütte and W. Huisinga. Biomolecular conformations can be identified as metastable sets of molecular dynamics. In *Handbook of Numerical Analysis (Computational Chemistry), Vol. X*, 2003.
285. Ch. Schütte, J. Walter, C. Hartmann, and W. Huisinga. An averaging principle for fast degrees of freedom exhibiting long-term correlations. *Multiscale Model. Simul.*, 2(3):501–26 (electronic), 2004.
286. Ch. Schwab. Two scale FEM for homogenization problems. In *Mathematical modeling and numerical simulation in continuum mechanics (Yamaguchi, 2000)*, volume 19 of *Lect. Notes Comput. Sci. Eng.*, pages 91–107. Springer, Berlin, 2002.
287. Ch. Schwab and A.-M. Matache. Generalized FEM for homogenization problems. In *Multiscale and multiresolution methods*, volume 20 of *Lect. Notes Comput. Sci. Eng.*, pages 197–237. Springer, Berlin, 2002.
288. H. Sigurgeirsson and A. M. Stuart. Inertial particles in a random field. *Stoch. Dyn.*, 2(2):295–310, 2002.
289. H. Sigurgeirsson and A. M. Stuart. A model for preferential concentration. *Phys. Fluids*, 14(12):4352–61, 2002.
290. Ya. G. Sinai. *Introduction to ergodic theory*. Princeton University Press, Princeton, NJ, 1976. Translated by V. Scheffer, Mathematical Notes, 18.
291. A. V. Skorokhod. *Asymptotic methods in the theory of stochastic differential equations*, volume 78 of *Translations of Mathematical Monographs*. American Mathematical Society, Providence, RI, 1989.
292. T. H. Solomon and J. P. Gollub. Chaotic particle transport in time-dependent Rayleigh-Benard convection. *Phys. Fluids*, 38(12):6280–86, 1988.
293. T. H. Solomon and J. P. Gollub. Passive transport in steady Rayleigh-Benard convection. *Phys. Fluids*, 31(6):1372–8, 1988.
294. A. M. Soward. Fast dynamo action in a steady flow. *J. Fluid Mech.*, 180:267–95, 1987.
295. A. M. Soward and S. Childress. Large magnetic Reynolds number dynamo action in spatially periodic flow with mean motion. *Proc. Roy. Soc. Lond. A*, 331:649–733, 1990.
296. S. Spagnolo. Sul limite delle soluzioni di problemi di Cauchy relativi all'equazione del calore. *Ann. Scuola Norm. Sup. Pisa (3)*, 21:657–99, 1967.
297. S. Spagnolo. Sulla convergenza di soluzioni di equazioni paraboliche ed ellittiche. *Ann. Scuola Norm. Sup. Pisa (3) 22 (1968), 571-597; errata, ibid. (3)*, 22:673, 1968.
298. C. Sparrow. *The Lorenz equations: bifurcations, chaos, and strange attractors*, volume 41 of *Applied Mathematical Sciences*. Springer-Verlag, New York, 1982.
299. W. J. Stewart. *Introduction to the numerical solution of Markov chains*. Princeton University Press, Princeton, NJ, 1994.
300. R. L. Stratonovich. *Topics in the theory of random noise. Vol. I*. Revised English edition. Translated from the Russian by Richard A. Silverman. Gordon and Breach Science Publishers, New York, 1963.
301. R. L. Stratonovich. *Topics in the theory of random noise. Vol. II*. Revised English edition. Translated from the Russian by Richard A. Silverman. Gordon and Breach Science Publishers, New York, 1967.
302. D. W. Stroock. *Probability theory, an analytic view*. Cambridge University Press, Cambridge, 1993.

303. D. W. Stroock. *An introduction to Markov processes*, volume 230 of *Graduate Texts in Mathematics*. Springer-Verlag, Berlin, 2005.
304. D. W. Stroock and S. R. S. Varadhan. *Multidimensional diffusion processes.* Classics in Mathematics. Springer-Verlag, Berlin, 2006.
305. A. M. Stuart and A. R. Humphries. *Dynamical systems and numerical analysis*, volume 2 of *Cambridge Monographs on Applied and Computational Mathematics*. Cambridge University Press, Cambridge, 1996.
306. F. Takens. Motion under the influence of a strong constraining force. In *Global theory of dynamical systems (Proc. Internat. Conf., Northwestern Univ., Evanston, Ill., 1979)*, volume 819 of *Lecture Notes in Math.*, pages 425–45. Springer, Berlin, 1980.
307. L. Tartar. Compensated compactness and applications to partial differential equations. In *Nonlinear analysis and mechanics: Heriot-Watt Symposium, Vol. IV*, volume 39 of *Res. Notes in Math.*, pages 136–212. Pitman, Boston, 1979.
308. L. Tartar. Cours Peccot au College de France. Unpublished, 1989.
309. L. Tartar. *Nonlocal effects induced by homogenization,* in *PDE and calculus of variations,* F. Culumbini et al., eds. Boston, Birkhauser, 1989.
310. L. Tartar. H-measures, a new approach for studying homogenisation, oscillations and concentration effects in partial differential equations. *Proc. Roy. Soc. Edinburgh Sect. A*, 115(3-4):193–230, 1990.
311. L. Tartar. An introduction to the homogenization method in optimal design. In *Optimal shape design (Tróia, 1998)*, volume 1740 of *Lecture Notes in Math.*, pages 47–156. Springer, Berlin, 2000.
312. T. Tassa. Homogenization of two dimensional linear flows with integral invariance. *SIAM J. Appl. Math.*, 57(5):1390–1405, 1997.
313. G. I. Taylor. Dispersion of soluble matter in solvent flowing slowly through a tube. *Proc. R. Soc. Lond. A*, 219:186–203, 1953.
314. R. Temam. *Infinite-dimensional dynamical systems in mechanics and physics*, volume 68 of *Applied Mathematical Sciences*. Springer-Verlag, New York, 1997.
315. R. Temam. *Navier-Stokes equations.* AMS Chelsea Publishing, Providence, RI, 2001.
316. S. Teufel. *Adiabatic perturbation theory in quantum dynamics*, volume 1821 of *Lecture Notes in Mathematics*. Springer-Verlag, Berlin, 2003.
317. A. N. Tikhonov, A. B. Vasil′eva, and A. G. Sveshnikov. *Differential equations.* Springer Series in Soviet Mathematics. Springer-Verlag, Berlin, 1985.
318. W. Tucker. The Lorenz attractor exists. *C. R. Acad. Sci. Paris Sér. I Math.*, 328(12):1197–1202, 1999.
319. W. Tucker. A rigorous ODE solver and Smale's 14th problem. *Found. Comput. Math.*, 2(1):53–117, 2002.
320. N. G. van Kampen. *Stochastic processes in physics and chemistry.* North-Holland Publishing Co., Amsterdam, 1981. Lecture Notes in Mathematics, 888.
321. N. G. van Kampen. Elimination of fast variables. *Phys. Rep.*, 124(2):69–160, 1985.
322. E. Vanden-Eijnden. Numerical techniques for multi-scale dynamical systems with stochastic effects. *Commun. Math. Sci.*, 1(2):385–91, 2003.
323. M. Vergassola and M. Avellaneda. Scalar transport in compressible flow. *Phys. D*, 106(1-2):148–66, 1997.
324. F. Verhulst. *Methods and applications of singular perturbations*, volume 50 of *Texts in Applied Mathematics*. Springer, New York, 2005. Boundary layers and multiple timescale dynamics.
325. D. G. Vlachos. A review of multiscale analysis: Examples from systems biology, materials engineering, and other fluid-surface interacting systems. *Adv. Chem. Eng.*, 30:1–61, 2005.

326. J. Walters and Ch. Schuette. Conditional averaging for diffusive fast-slow systems: a sketch for derivation. In *Analysis, Modeling, and Simulation of Multiscale Problems*, pages 647–82. Springer, 2006.
327. P. Walters. *An introduction to ergodic theory*, volume 79 of *Graduate Texts in Mathematics*. Springer-Verlag, New York, 1982.
328. W. Wang, D. Cao, and J. Duan. Effective macroscopic dynamics of stochastic partial differential equations in perforated domains. *SIAM J. Math. Anal.*, 38:1508–1527, 2007.
329. W. Wang and J. Duan. Invariant manifold reduction and bifurcation for stochastic partial differential equations. preprint, 2006.
330. J. C. Wells. Invariant manifolds on non-linear operators. *Pacific J. Math.*, 62(1):285–93, 1976.
331. S. Wiggins. *Introduction to applied nonlinear dynamical systems and chaos*, volume 2 of *Texts in Applied Mathematics*. Springer-Verlag, New York, 2003.
332. E. Wong and M. Zakai. On the convergence of ordinary integrals to stochastic integrals. *Ann. Math. Statist.*, 36:1560–4, 1965.
333. J. Xin. Front propagation in heterogeneous media. *SIAM Review*, 42:161–230, 2000.
334. D. Xiu and I.G. Kevrekidis. Equation-free, multiscale computation for unsteady random diffusion. *Multiscale Model. Simul.*, 4(3):915–35 (electronic), 2005.
335. G. G. Yin and Q. Zhang. *Discrete-time Markov chains*, volume 55 of *Applications of Mathematics (New York)*. Springer-Verlag, New York, 2005.
336. G. G Yin and Q. Zhang. *Continuous-time Markov chains and applications*, volume 37 of *Applications of Mathematics (New York)*. Springer-Verlag, New York, 1998.
337. K. Yosida. *Functional Analysis*. Springer-Verlag, Berlin-Heidelberg, 1995.
338. L.-S. Young. What are SRB measures, and which dynamical systems have them? *J. Stat. Phys.*, 108:733–54, 2002.
339. E. Zeidler. *Applied Functional Analysis*, volume 108 of *Applied Mathematical Sciences*. Springer-Verlag, New York, 1995.
340. Ya. B. Zel'dovich. Exact solution of the problem of diffusion in a periodic velocity field, and turbulent diffusion. *Sov. Phys. Dokl.*, 27(10):797–9, 1982.
341. V. V. Zhikov. On two-scale convergence. *Tr. Semin. im. I. G. Petrovskogo*, (23):149–87, 410, 2003.
342. R. Zwanzig. Diffusion in a rough potential. *Proc. Natl. Acad. Sci. USA*, 85:2029–30, 1988.

Index

Texts in Applied Mathematics

(continued after page ii)

31. *Brémaud*: Markov Chains: Gibbs Fields, Monte Carlo Simulation, and Queues.
32. *Durran*: Numerical Methods for Wave Equations in Geophysical Fluids Dynamics.
33. *Thomas*: Numerical Partial Differential Equations: Conservation Laws and Elliptic Equations.
34. *Chicone*: Ordinary Differential Equations with Applications.
35. *Kevorkian*: Partial Differential Equations: Analytical Solution Techniques, 2nd ed.
36. *Dullerud/Paganini*: A Course in Robust Control Theory: A Convex Approach.
37. *Quarteroni/Sacco/Saleri*: Numerical Mathematics.
38. *Gallier*: Geometric Methods and Applications: For Computer Science and Engineering.
39. *Atkinson/Han*: Theoretical Numerical Analysis: A Functional Analysis Framework, 2nd ed.
40. *Brauer/Castillo-Chávez*: Mathematical Models in Population Biology and Epidemiology.
41. *Davies*: Integral Transforms and Their Applications, 3rd ed.
42. *Deuflhard/Bornemann*: Scientific Computing with Ordinary Differential Equations.
43. *Deuflhard/Hohmann*: Numerical Analysis in Modern Scientific Computing: An Introduction, 2nd ed.
44. *Knabner/Angermann*: Numerical Methods for Elliptic and Parabolic Partial Differential Equations.
45. *Larsson/Thomée*: Partial Differential Equations with Numerical Methods.
46. *Pedregal*: Introduction to Optimization.
47. *Ockendon/Ockendon*: Waves and Compressible Flow.
48. *Hinrichsen*: Mathematical Systems Theory I.
49. *Bullo/Lewis*: Geometric Control of Mechanical Systems: Modeling, Analysis, and Design for Simple Mechanical Control Systems.
50. *Verhulst*: Methods and Applications of Singular Perturbations: Boundary Layers and Multiple Timescale Dynamics.
51. *Bondeson/Rylander/Ingelström*: Computational Electromagnetics.
52. *Holmes*: Introduction to Numerical Methods in Differential Equations.
53. *Pavliotis/Stuart*: Multiscale Methods: Averaging and Homogenization.
54. *Hesthaven/Warburton*: Nodal Discontinuous Galerkin Methods.
55. *Allaire/Kaber*: Numerical Linear Algebra.

GPSR Compliance

The European Union's (EU) General Product Safety Regulation (GPSR) is a set of rules that requires consumer products to be safe and our obligations to ensure this.

If you have any concerns about our products, you can contact us on ProductSafety@springernature.com

In case Publisher is established outside the EU, the EU authorized representative is:

Springer Nature Customer Service Center GmbH
Europaplatz 3
69115 Heidelberg, Germany

Batch number: 10682323

Printed by Printforce, the Netherlands